Sustainable Development Goals Series

The **Sustainable Development Goals Series** is Springer Nature's inaugural cross-imprint book series that addresses and supports the United Nations' seventeen Sustainable Development Goals. The series fosters comprehensive research focused on these global targets and endeavours to address some of society's greatest grand challenges. The SDGs are inherently multidisciplinary, and they bring people working across different fields together and working towards a common goal. In this spirit, the Sustainable Development Goals series is the first at Springer Nature to publish books under both the Springer and Palgrave Macmillan imprints, bringing the strengths of our imprints together.

The Sustainable Development Goals Series is organized into eighteen subseries: one subseries based around each of the seventeen respective Sustainable Development Goals, and an eighteenth subseries, "Connecting the Goals," which serves as a home for volumes addressing multiple goals or studying the SDGs as a whole. Each subseries is guided by an expert Subseries Advisor with years or decades of experience studying and addressing core components of their respective Goal.

The SDG Series has a remit as broad as the SDGs themselves, and contributions are welcome from scientists, academics, policymakers, and researchers working in fields related to any of the seventeen goals. If you are interested in contributing a monograph or curated volume to the series, please contact the Publishers: Zachary Romano [Springer; zachary.romano@springer.com] and Rachael Ballard [Palgrave Macmillan; rachael.ballard@palgrave.com].

Peter Neema-Abooki
Editor

The Sustainability of Higher Education in Sub-Saharan Africa

Quality Assurance Perspectives

Editor
Peter Neema-Abooki
Department of Educational Foundations
University of South Africa (UNISA)
Pretoria, South Africa

ISSN 2523-3084 ISSN 2523-3092 (electronic)
Sustainable Development Goals Series
ISBN 978-3-031-46241-2 ISBN 978-3-031-46242-9 (eBook)
https://doi.org/10.1007/978-3-031-46242-9

Color wheel and icons: From https://www.un.org/sustainabledevelopment/, Copyright © 2020 United Nations. Used with the permission of the United Nations.

The content of this publication has not been approved by the United Nations and does not reflect the views of the United Nations or its officials or Member States.

© The Editor(s) (if applicable) and The Author(s), under exclusive licence to Springer Nature Switzerland AG 2024
This work is subject to copyright. All rights are solely and exclusively licensed by the Publisher, whether the whole or part of the material is concerned, specifically the rights of translation, reprinting, reuse of illustrations, recitation, broadcasting, reproduction on microfilms or in any other physical way, and transmission or information storage and retrieval, electronic adaptation, computer software, or by similar or dissimilar methodology now known or hereafter developed.
The use of general descriptive names, registered names, trademarks, service marks, etc. in this publication does not imply, even in the absence of a specific statement, that such names are exempt from the relevant protective laws and regulations and therefore free for general use.
The publisher, the authors, and the editors are safe to assume that the advice and information in this book are believed to be true and accurate at the date of publication. Neither the publisher nor the authors or the editors give a warranty, expressed or implied, with respect to the material contained herein or for any errors or omissions that may have been made. The publisher remains neutral with regard to jurisdictional claims in published maps and institutional affiliations.

This Palgrave Macmillan imprint is published by the registered company Springer Nature Switzerland AG.
The registered company address is: Gewerbestrasse 11, 6330 Cham, Switzerland

Paper in this product is recyclable.

This book is dedicated with profound felicitations to

His Eminence Antoine Cardinal Dr Kambanda
Classmate-n-Friend, Former Bishop of Kibungo
Archbishop of Kigali and Prince of the Church
Umugabo Nyamugabo to whom was bestowed
Ecclesial ranks of Episcopate and Cardinalate

FOREWORD

The Sustainability of Higher Education in Sub-Sahara Africa: Quality Assurance Perspectives is not just a great resource for higher education institutions (HEIs) in the region; it is also for all those who truly want to invest in quality education as a pillar for the sustainability of HEIs in Africa and the globe. Higher education (HE) is an engine of development, thus its quality matters. This book is timely, fascinating, and appealing as it follows a two-year period of disruption of normalcy by the COVID-19 pandemic, imposing new directions for quality assurance (QA) and sustainability of HE. Indeed, it is opportune for Neema-Abooki to call for experiences, research studies, and observations from gurus in the continent to contribute to an invaluable book on QA and sustainability of the HE in the region.

Abooki, with ingenuity and deep experience in leadership and management in HEIs, is adept to be an editor of this book. He is a Professor of Higher Educational Management and Administration with in-depth experiences that weave together pertinent areas of quality and sustainability of HE in Africa. He brings decades of professional experience in these areas having worked in a number of African countries. His ability to work with others has made it possible to attract contributions from authors in different African countries, who, like him, have a good understanding of issues of quality and sustainability in the Sub-Saharan countries. They have contributed valuable experiences, research studies, and practices that make this book a rich resource.

Abooki is an avid author. His other books include *Quality Assessment and Enhancement in Higher Education in Africa ISBN: 9781032308142* Pages: 312 (Copyright Year 2023); and *Quality Assurance in Higher Education in Eastern and Southern Africa: Regional and Continental Perspectives ISBN: 9780367692834; ISBN-10: 036769283X* Pages: 314 (Copyright Year 2022)—both published by Taylor and Francis, Routledge, UK, and USA. I am well aware that several other books are in the offing. Other international publications credited to his name include "Quality Higher Education in Africa: Assessment and Enhancement Perspectives", "Teacher Preparation by Universities and Implications to Quality Education in Uganda", "Management of Research-Related Issues in Higher Education Institutions in Africa: A Case of Makerere and Nairobi Universities in East Africa", "Assessment and Enhancement: Quality Perspectives in Higher Education", "Quality Education in Eastern and Southern Africa: Prospectives and Retrospectives", "Academic Quality in the Public Universities in Eastern and Southern Africa: A Comparative Study of Makerere University and the University of Cape Town", "Internal-Stakeholder-Perception of the Quality of Teacher-Educators at Makerere University", "Administration-Management Distinction in Higher Education: A Quality Management Perspective", "Own-Income Generation: A Pillar of Financial Sustainability in Institutions of Higher Learning", "Cross Border Education and Its Influence on the Quality of Higher Education", "Financing Higher Education: Income Generation in Ugandan Public Universities", "Scope of Quality Assurance in Higher Education Programmes and Projects", "Financial Resource Mobilisation Projects and Its Relationship to Academic Staff Commitment in Uganda Martyrs University in Uganda", "Resource Input Model and Its Effects on Quality of Education in Makerere University", "Gender Issues in Project Planning and Management", "Institutional Autonomy and Governance vis-à-vis the Management of Massification: A Case of Science-Based Faculties at Gulu University in Northern Uganda", Massification Versus Management of Research Publication and Community Engagement in the Science-Based Departments in Uganda: A Case of Gulu University", "Influence of Human Power Planning on Academic Staff and Their Service Delivery in the College of Education and External Studies", "Is Effective Teaching and Learning the Solution to Quality Management of Massification in Science-Based Faculties in Gulu University?", "Academic Staff Competence Development as a Gap in Quality Assurance in

Universities in Uganda", "Pedagogical Knowledge and Effectiveness of Academic Staff in the Teaching-Learning Process at Makerere University", "Mediating Learner-Content Interaction Using Emerging Technologies: A Case of History Education at Makerere University", "Quality Assurance and the Application of Legitimacy Model in Higher Education Institutions", "Mediating Learner-Learner Interaction Using Emerging Technologies: A Case of History Education at Makerere University", "Supervision of Research at Makerere University: Perspective at the College of Education and External Studies", "Student-Research Affairs: Retrospects and Prospects at the College of Education and External Studies at the Premier University in Uganda", "Impact of E-Learning Strategy on Students' Academic Performance at Strathmore University in Kenya", "Managerial Systems as Measures of Quality Management in Universities in Uganda", "Contrapreneurship: A Dilemma in the Management of Higher Education Institutions", "Involvement of the Private Sector in the Financing of Academic Programmes at the Primogenial University in Uganda", "Challenges and Strategies of Improving Staff Development in Higher Education Institutions in Uganda", "Collaboration: A Benchmark in the Management of Universities", "Academic Staff Professional Development at Kyambogo University", "Contrapreneurship in Higher Educational Organisations: Perspectives of Total Quality Management", "Participation as a Prerequisite to Best Practice in the Management of Universities: Perspectives of Total Quality Management", "Students' Personal Characteristics and Completion of Postgraduate Research in the College of Education and External Studies at Makerere University", "E-Learning: A Management-Oriented Fourfold Strategy in Some East African Universities", "Developing a Web Explicit Research Strategy Theory in African Universities: A Cross-Comparison of Specific Regional Efforts Through an Analysis of Research Web Pages", "Characteristics of Management at Selected Universities in Uganda", "Knowledge-Based Economies in Selected Universities in Uganda", "Professional Development: The Case of Academic Staff in a Ugandan Public University", "Policy Initiatives on Science and Technology Education in Uganda: Extent of Implementation at the Post-Basic Level", "Quality Assurance in African Higher Education: Vision and Mission Perspectives", "Quality Assurance in African Higher Education: Environmental Perspectives", "Systemic Design of Instruction on Achieving the Goals of Undergraduate-Level Education in Universities: A

Case Study of Makerere University", "The Multifold Challenge of Quality in Higher Education: Nineteen Ninety African Perspective", "Total Quality Management and the Governance of Educational Institutions in Sub-Saharan Africa: A Case of Universities in Uganda", "Total Quality Management in Organisations: Challenges and Strategies", and "Integration of Total Quality Management in the Management of Universities in Uganda"—just to mention a few. He has now taken another topical focus on exploring the quality and sustainability of HE institutions in Sub-Saharan Africa.

The opening chapter, "Sustainability as an Acme of Quality Assurance in Higher Education", arrays an overview of the prospects of quality assurance (QA) and sustainability of higher education (HE) in the continent and depicts the context, uniqueness, major contribution, and the lessons learned.

In Part I of the book: Chapter 2 advances Digital Learning as a model of QA in HE while Chapter 3 spells out insights on monitoring and evaluation in HE. Chapter 4 castigates the quality of teaching and learning in private institutions and calls for continuous review of the curricula. Chapter 5 probes the relationship between students' virtual learning satisfaction and the virtual teaching service delivery of academic staff. Chapter 6 rules on the role of HE in the implementation of Sustainable Development Goals (SDGs) amidst the triple mission of teaching and learning, research, and community engagements. Chapter 7 counsels on a framework for graduate employability for advancing HE curricula and the SDGs. Chapter 8 focuses on SDG 4 via learners' prior numerical cognition as a predictor of educational performance.

In Part II: Chapter 9 posits that innovation aligns with the scaling features of involving key stakeholders and with indulging in structured pedagogy. Meanwhile, Chapter 10 emphasises the use of sustainable interactive teaching methods in view of enhancing employability skills. Chapter 11 underscores the need for strengthening collaboration between the university, government, and industry. Chapter 12, outlining the role of HE as an engine of societal development, stresses a pluralistic worldview and hybrid model in a bid to decolonise the HE sector. Chapter 13 advocates for participating in and promoting the energy sustainability agenda as one of the emerging critical issues in the twenty-first century. Chapter 14 analyses the change from the neoliberal-driven QA regimes to a QA culture for the sustainability of the quality of African HE. The last chapter (Chapter 15) upholds HE aspects of Africa's readiness for the Fourth Industrial Revolution (4IR) and labour demands for the future of work.

All chapters render a profound *exposé* on HE and the SDGs. Each chapter is self-contained. Readers do not need to read the book in any given sequence. Chapter headings are there as guides to preferred chapters. No matter what your interests may be, there is something for diverse readers.

This book, *The Sustainability of Higher Education in Sub-Saharan Africa: Quality Assurance Perspectives*, tangibly leaves a legacy for all interested in sustainability and quality in HEIs. I, therefore, do recommend it to academia and the entire populace.

PREFACE

Sustainability of education in general, and higher education (HE) in particular, is the basis of development in communities and countries and is expository for navigating a future marked by enormous social changes. Accordingly, this book propagates "sustainability" as one of the key processes that would make HE in Sub-Saharan Africa and the entire continent more relevant to serve as an engine of development.

The context of the book draws from the global debate on the role of HE and its effects on societal development, while the overall scope of the book discusses perspectives as per the variables spelt out in the title and under the two parts, namely "Curriculum and Teaching in Higher Education" and "Higher Education and Innovations". The introductory chapter (Chap. 1), which depicts an overview of the material within the two parts, adds to 15 chapters in total. The chapters address aspects of Sustainable Development Goals (SDGs) and the need to use quality education to achieve these goals. They depict current and useful insights not only on the Sub-Saharan region but on the entire vast continent of Africa. Presented are comparative case studies of quality measures that can enhance the inculcation of employability skills among learners not only in Africa but globally.

Suffice to highlight heretofore that the title of this book mentions "Sub-Saharan Africa" but solely owing to the latitude that almost all, save one of the contributors, hail from this part of the continent. For, the regions represented in the chapters do render an adequately proportionate representation of Sub-Saharan Africa and so do the countries therein. As for Somalia, though the easternmost country in the Horn of Africa, it

xiii

occupies an important geopolitical position adjacent to Sub-Saharan Africa. And, above the Sub-Saharan Africa scenario, the notions advanced reflect and are apt to be generalised on the whole continent. Moreover, slightly more than half of the book focuses directly on continental Africa as substantiated by three chapters in Part I and five chapters in Part II of this compendium.

The book, therefore, explores quality assurance (QA) perspectives from an African context, in particular the models of QA and how sustainable these are to ensure the attainment of graduate attributes as precursors of industrial and/or economic development in the continent. Advanced *inter alia* are issues of pedagogy, entrepreneurship, ethics and culture, internationalisation, and sustainable development, including the "new normal" behest of the COVID-19 phenomenon. The SDGs thesis makes the book up-to-par since quality of HE as the engine of the industrial revolution ought to find its roots and ground itself in a sustainable stance and the stakeholders in HE ought to move with the status quo. This anthology holds that the HE is one of the sectors that calls for these necessary reforms; thus the imperative for HE in the continent to embrace strategies and technological advances to equip the educands for relevancy to the signs of the times.

The main audience encases policymakers in decision-making with regard to education policy formulation, including the faculty and their students on the guide to and in academic analysis. The book is a "must-buy" also to the senior management in higher education institutions (HEIs) in an endeavour to strengthen QA and the production of relevant and quality products. Other related readership and markets herewith are the relevant public and private social sector associations and professional and societal groupings. The most relevant disciplines include and are not limited to Higher Education, African Studies, African Education, Quality Education, Quality Improvement, Sustainable Education, School Leadership, and Interdisciplinary Studies.

This felicitous book is a contribution to global HE studies. Though Africa-context-based, it offers an impetus to think within the dynamics of HE and satisfies the "ought" of bridging the gaps in context, having intro-spected both the African and the global foci as well as the SDGs.

The Lessons ensconced in this treasury are action-oriented and do in turn presuppose recommendations as a way forward.

Pretoria, South Africa   Peter Neema-Abooki

Acknowledgements

The previous two books acknowledged some of the personalities to whom I owe sublime and incessant indebtedness and took cognisance of those who have gratuitously extended a hand in ways sundry ever since my being a resident in Kampala. Needful here to emphasise, this was without prejudice to the phenomenon that many more deserved special mention. In a similar vein, this book, *The Sustainability of Higher Education in Sub-Saharan Africa: Quality Assurance Perspectives*, reiterates sagacious tribute to all explicitly and implicitly enlisted and adds a few other philanthropists that include Mr Gervase Ndyanabo, Prof Charles Niwagaba, Prof Orach of Entebbe, Dr Deus Kamunyu Muhweezi, Rev Sr Dativa Daniela Mukebita—Mother General of the Franciscan Sisters of St Bernadette (FSSB) in Tanzania, Rev Sr Esther Maris Tinu Okoro—Eucharist Heart of Jesus (EHJ) Sisters in Nigeria; serving in the Archdiocese of Johannesburg, Carolin Krammer of Nürnberg in Deutschland, Madam Grace Tusiime of the Ministry of Energy and Mineral Development, and Kenneth Kato Abooki of Germany Embassy in Kampala. The personalities in Boston, namely Fr Richard Kayondo, Fr Emmanuel Rutangusa now in Rwanda, Fr Godfrey Musabe Apuuli, Dr David Nnyanzi, and the families of Andrew and Annet Byaruhanga *both* Amooti, Eugene and Proscovia Nkore, Barnabas and Justen Nkore, Michael and Margret Mukisa, Fred and Mary Ssebugwawo, Madams Agnes Nansubuga, Gladys Mukiibi, Annet Nazziwa, Adela Mary Kyarikunda, and Maria Assumpta Kaweesa, to mention just but a few.

xvi ACKNOWLEDGEMENTS

Singular and profound enunciation is rendered to the happy and perpetual memory of a dignified personality who, as this book was in the final stages of submission for publication, left this life on Tuesday 18 July 2023; just one year shy to turn 80 years of age! He was ordained Priest on 11 July 1971 and coadjutor Bishop on 24 June 1989, succeeded on 23 November 1991, and was appointed Archbishop of Mbarara on 2 January 1999. Acknowledged heretofore is the sublime stance of having treasured him as mentor and friend and as one of the Prelates with whom I am privileged to share successively the Patron Feast Day of Sts Peter and Paul and the Birthday on the following last day of the month of June. This is an orator with a sublime sense of humour, the Archbishop Emeritus of Mbarara Archdiocese, His Grace, The Most Reverend **Paul Kamuza Bakenga**. Remembered and acknowledged in an equidistant vein is yet another *Sacerdos Maximus*, The Right Reverend **Albert Edward Dr Baharagaate Akiiki**, Bishop Emeritus of Hoima Diocese, who lifted off a few months earlier on 5 April 2023; having already celebrated 93 years of age. Highly treasured is his profoundly amicable paternal-cum-fraternal collegiality to my paucity as he resided at Nakulabye Parish in Kampala.

Requiescant in pace.

The book owes superlatively to the contributors, the Publishing team – especially Venkitesan VinodhKumar, and all people of goodwill.

May Divine Benevolence incessantly provide.

Praise for
The Sustainability of Higher Education in Sub-Saharan Africa

"This book project explores quality assurance perspectives from an African context. In particular, it explores the models of quality assurance (QA) and how sustainable these are to ensure the attainment of graduate attributes as precursors of industrial and/or economic development in Africa and elsewhere. The project offers new and useful information as far as specific and therefore contextual African cases are concerned. The case studies that are used for some given countries speak to specific issues which could benefit the reader from a comparative point of view. Sustainable development goals are the central thesis around which the project is based. The project addresses current trends in the area of higher education (HE) and QA through issues of digitalisation for instance which question Africa's readiness for the 4IR. The 4IR is here with Africa and quality HE as the engine of the industrial revolution to find its roots and sustainably ground itself. The policy maker, the researcher, and the learner must live in a comparative space to learn from one another and move with the times. Therefore research from different perspectives on quality HE and its sustainability is imperative for cross-pollination of ideas. The author, who is also the editor, is well placed from his vast academic writing experience. His HE experience, teaching, research supervision, and examination of research projects at the graduate level—among others—all qualify him to produce a high-quality book on this given topic. The writing is of an acceptable quality. My general assessment of the project is that it is viable and timely. This should be a "must-read" to strengthen QA and to produce relevant and quality products."

—Anonymous Reviewer 1, *United Kingdom (UK)*

"In the contemporary world quality education is regarded by many as a catalyst for socioeconomic advancement, particularly in developing countries. The role of quality education in the socio-economic development of African countries is therefore crucial and this makes the book an important project. For, in recent times Higher Education Institutions (HEIs) have increased dramatically in Africa. Apart from public or state-funded universities, there are many private HEIs mushrooming on the continent. Part of the reason is the recognition of higher education (HE) as the vehicle for sustainable development in Africa. With the establishment of many HEIs, quality assurance (QA) has become crucial to ensure that all institutions of higher learning adhere to quality standards so that qualifications (degrees, diplomas, and certificates) are not devalued. Quality education is in line with the UN Sustainable Development Goals (2015) which emphasise the role of education in socio-economic development. This book has an inter- and multidisciplinary nature and it offers useful, valuable, and original contributions to the field of quality education for sustainable development in Africa and beyond. The structure, organisation, and presentation are of good quality. The editor is qualified and has the experience, knowledge, and skills; and he has successfully executed similar projects in the recent past."

—Anonymous Reviewer 2, *United Kingdom (UK)*

"I have shared long background from youth with Prof Peter Neema-Abooki and sat in the same classrooms with him facing Latin Courses, among other subjects. The minimum pass mark for Latin in any examination then in our Institution was 95% and this determined the way forward for other subjects and continuity as a student of that Institute; in other words, failure to achieve the required set marks was tantamount to dismissal; we continued together for more than 4 years. Comparatively, what am referring to for other institutions of learning is Mathematics whose passing mark was not that high like Latin. I don't want to say that Latin is harder than Mathematics since am an educationist. The justification of "harder than another" is another book one can author. For Prof Peter Neema-Abooki, to achieve what he is now, and to author so many books to date in the education arena, is proof that he did well in the Latin subject then; vivid proof will be noticed in the book content as you continue to read it, and for him to request me to endorse this book is a pointer to the dictum: "Birds of the same feathers fly together". I thank him for entrusting me with this noble task. I sincerely appreciate his unwavering trust deep-rooted in amicability. The first Latin word he and I conjugated was *amicus* and he has kept conjugating it practically till today."

"A good book tickles other researchers to see more gaps in research because it opens their analytical minds, which this book does very well as it irrigates yet another appetite even in the author themselves, for, as they conclude one book, they notice another leading area related to what they have written in the current edition. This compendium advances in great proportions the role of HE as an engine for the socio-economic development of Africa and beyond. Without having to pre-empty your inquisitiveness of the contents, I, having read this compendium with keen interest, would rather set the appetite in you to further opening the next page and subsequently next page to next page till the end; as I do recommend this masterpiece as a "must-buy" for academic utilisation and life in general. I am confident that it captures the market in an attempt to depict the well-phrased title and profound content in minute detail."

—Masterjerb Birungi Paul, *BED (Hons)MUK, Adv Dip Distance Education Specialist (UoL), MBA (MKU), Education Capacity Building Expert— Kigali, Rwanda*

"Sustainability of higher education in Africa in general, and in Sub-Saharan Africa in particular, is indeed a contemporary challenge and Dr Prof Peter Akampa Neema-Abooki could not have thought of a better book title to reflect on the issue. This is a trail-blazing book on higher education addressing fundamental issues regarding the sustainability of higher education (HE) on the African continent. With the 2023 UN Report indicating that we are moderately to severely off track in terms of achieving the sustainable development goals (SDGs), the book compels readers to reflect on sustainability through quality systems and contributions of higher education institutions (HEIs) to development, employability, innovativeness, and digitalization, among others. The book is a must-read for policy practitioners, lecturers, students, and all those who are keen on the advancement of quality institutions, be they academic, domestic, or otherwise. It is a fruit of one who is an accomplished potential and an academician of high promise, known to me over thirteen (13) years as a fellow academician grounded in research and publications with a special focus on Quality Assurance in Institutions of Higher Learning, to which field he has widely disseminated findings including the several key-note addresses at academic events in South Africa. His adaptability to cross-culture was depicted not only during my benchmarking visit at Makerere University in Uganda where he hosted me in his capacity as the Dean of the East African School of Higher Education Studies and Development (EASHESD), but also, more evidently, when he returned a visit and lived with me at the University of Venda in the Republic of South Africa."

—Professor Clever Bafana Ndebele, *Senior Director: Learning and Teaching, Office of the DVC Academic Affairs and Research, Walter Sisulu University, South Africa. Formerly held similar Directorates at Northwest University and the University of Venda*

"Sustainability of higher education (HE) is the basis for the development of communities and countries South of the Sahara. The book ably addresses the quality assurance (QA) issues that can enhance the sustainability of HE in Sub-Saharan Africa; doing as it does, taking cognisance of the UN Sustainable Development Goals (2015) as the foundation of socio-economic development. Making an enormous contribution to the sector of Quality HE as a catalyst for socio-economic development, the book offers a very useful, valuable, and original contribution to sustainability in HE. The contributions from the different chapters focus on various aspects of the Sustainable Development Goals (SDGs) and the need to use *Quality Education* to achieve them. This book may appeal to development practitioners, researchers, students, education institutions, and various ministries of education in Africa and the globe due to its inter- and multidisciplinary nature. In view of the relevance of its contribution to the sustainability of HE without which development could be hindered in Sub-Saharan Africa, I fully endorse it."

—Kofi Poku Quan-Baffour, *Professor Extraordinarius, NRF Rated, Adult Education, Department of Community & Continuing Education, University of South Africa (UNISA)*

" This book *The Sustainability of Higher Education in Sub-Saharan Africa: Quality Assurance Perspectives* makes a great contribution to the literature on quality assurance (QA) and higher education (HE). I, with all the confidence that this resource will be much sought after by HE managers, policymakers as well as faculty and students, among others, do take the pleasure and honour to endorse the publication of this authoritative treasury."

—Professor Michael Mawa, *Chief Principal Quality Assurance & Qualifications Framework, Inter-University Council for East Africa, Founding President for the Ugandan Universities Quality Assurance Forum (UUQAF), Founding President of the East African Higher Education Quality Assurance Network (EAQAN)*

CONTENTS

Introduction: Sustainability as an Acme of Quality Assurance
in Higher Education 1
Peter Neema-Abooki

Part I Curriculum and Teaching in Higher Education 13

Models of Quality Assurance: *Towards a Quality Assurance
Framework of Digital Learning in Higher Education in Africa* 15
Erkkie Haipinge and Ngepathimo Kadhila

Monitoring and Evaluation in Higher Education: *Quality
Perspectives in Africa* 43
David Katende

Quality of Teaching and Learning Through Internal Quality
Curriculum Review Mechanisms: *A Case of Private Higher
Education Institutions in Post-conflict Somalia* 73
Abukar Mukhtar Omar and Abdu Kisige

Students Satisfaction and Virtual Learning Service Delivery at
Bugema University 91
Kulthum Nabunya

xxi

xxii CONTENTS

Sustainability and the Triple Mission of the University: *Uganda Martyrs University in Perspective*   115
Christopher Mukidi Acaali

A Framework for Embedding Graduate Employability: *Attributes of Higher Education Curriculum in Africa*   135
Romanus Shivoro, Rakel Kavena Shalyefu, and
Ngepathimo Kadhila

Addressing SDG 4 via Learners' Prior Numerical Cognition: *a Predictor of Educational Performance in a Developing Country in Sub-Saharan Africa*   163
Alexander Michael and Anass Bayaga

Part II Higher Education and Innovations   191

Scaling Education Innovations in Tanzania, Kenya, and Zambia: *Assessing the Design of School In-service Teacher Training*   193
Katherine Fulgence

Mechanisms for Enhancing Employability Skills Among Students Within Vocational Education Training Institutions in Tanzania   223
Mwaka Omar Makame and Katherine Fulgence

Higher Education: *Towards a Model for Successful University-Industry Collaboration in Africa*   251
Ngepathimo Kadhila, Kyashane Stephen Malatji,
and Makwalete Johanna Malatji

Higher Education as an Engine of Development: *Sites of Domination, Contestation, and Struggle in Africa*   281
John Kamwi Nyambe and Ngepathimo Kadhila

CONTENTS xxiii

Energy Sustainability in African Higher Education: *Current Situation and Prospects*   305
Alfred Kirigha Kitawi and Ignatius Waikwa Maranga

Quality Assurance of Higher Education in a Neo-liberal Context: *Towards Transformative Practices in Africa*   327
Joel Jonathan Kayombo, Mjege Kinyota, and Patrick Severine Kavenuke

The Role of Higher Education and the Future of Work in Africa's Fourth Industrial Revolution   349
Kenneth Kamwi Matengu, Ngepathimo Kadhila, and Gilbert Likando

Index   375

NOTES ON CONTRIBUTORS

Christopher Mukidi Acaali is an educationist with an interest in religion and culture. He holds a PhD in Theology from Duquesne University, Pennsylvania, USA, and a Master's Degree in Religious Studies from the University of Portland, Oregon, USA. He has spent most of his time in administration, serving as a registrar at the Mountains of the Moon University in 2005–2013 and Uganda Martyrs University (UMU) in 2014–2021. He was appointed Deputy Vice-Chancellor of UMU from 1 December 2021 to date. Through these assignments, Mukidi has developed a passion for policy and development and has of late shown interest in the development of language (Runyoro-Rutooro language). Besides administration, he has taught and supervised undergraduate and postgraduate students.

Anass Bayaga is currently a cognitive mathematics and STEM cognition professor at Nelson Mandela University. Presently, Bayaga serves as an editorial board member of the *International Journal of Mathematics Teaching and Learning (IJMTL)*. Bayaga is a member of the Membership Committee of the Mixed Methods International Research Association—MMIRA. He was also a member of the International Institute of Informatics and Systemics (IIIS). Bayaga was also a Fulbright researcher at George Washington University, where through predictive modelling, he previously researched cognitive enhancement via mobile computing and applications in STEM, which also presented his research group's niche/focus. His research and teaching interests are mathematics and computational cognition (Neuro-mathematics (STEM), STEM cognitive enhancement via human-computer interaction, and predictive/mathematical modelling.

Katherine Fulgence has specialisation in entrepreneurship education and employability of graduates through skills development programmes. Over the past ten years, she has conducted training, research, and consultancies in the areas of scaling education innovations, digital fluency, teacher education, professional development, and career transitions. She has authored academic writings in peer-reviewed journal articles, book chapters, policy briefs, and consultancy reports.

Erkkie Haipinge is a Deputy Director for eLearning at the Centre for Innovation in Learning and Teaching (CILT), University of Namibia, Namibia. He holds a Master of Arts in Education and Globalisation from the University of Oulu, Finland, and has specialised in educational technologies. He teaches technology integration in learning and teaching and is a proponent of project-based learning. His research interests include social media integration in learning, new digital learning environments, mobile learning, and the application of innovations to learning and teaching. ORCID identifier: https://orcid.org/0000-0003-0445-0124

Ngepathimo Kadhila is Director of Quality Assurance at the University of Namibia, Namibia. He holds a PhD in Higher Education, with a focus on quality assurance from the University of the Free State, South Africa; a Master's Degree in Education and a Bachelor's Degree in Education from the University of Namibia; as well as a Postgraduate Diploma in Higher Education from Rhodes University, South Africa. His research interests include academic development, curriculum development in higher education, teaching and learning in higher education, and quality assurance. ORCID identifier: https://orcid.org/0000-0002-4805-4775

David Katende is a Higher Education Scholar and practitioner, whose specialisation is in education policy, planning, quality assurance, and monitoring and evaluation. He has seventeen years of experience working with the education sector in Uganda, at both secondary and university levels. He holds a Bachelor of Arts degree, a Post-Graduate Diploma in Education, and a Master's Degree in Educational Policy and Planning, with several certificates in doctoral research and higher education management. He is currently a PhD student and a member of Top University Management at Mountains of the Moon University. David has recently graduated with a

First-Class Post-Graduate Diploma in Monitoring and Evaluation and is about to defend a Master's degree dissertation in the same field.

Patrick Severine Kavenuke is a senior lecturer in the Department of Educational Foundations, Management, and Lifelong Learning at Dar es Salaam University College of Education, University of Dar es Salaam, Tanzania. His main areas of research are teacher education, critical pedagogy (education), critical thinking skills in students, international and comparative education, and teacher professional development.

Joel Jonathan Kayombo is a senior lecturer in the Department of Educational Foundations, Management, and Lifelong Learning at the Faculty of Education, Dar es Salaam University College of Education (DUCE), University of Dar es Salaam (UDSM), Tanzania. His research interests include educational policy and reforms, governance and administration in education, educational planning, critical pedagogy (education) and politics of education, sociology of higher education, and globalisation and education.

Mjege Kinyota is Senior Lecturer in STEM Education at Dar es Salaam University College of Education, University of Dar es Salaam, Tanzania. His research interests include teaching and learning in STEM, environmental education, social justice education, teacher education in STEM, and gender in STEM.

Abdu Kisige is a tutor in the Department of Teacher Training at Al-Mustafa Islamic College, Uganda, at the same time occupying the Directorate of Research, helping the College to build and strengthen its research capacity. He is also a member of the College Governing Council. His academic credentials include a Doctor of Philosophy and a Master's degree both in Educational Management, Planning and Administration, and a BA in Education. He has been teaching both undergraduate and postgraduate courses in educational administration and management at the Islamic University, Uganda; Al-Mustafa Islamic College, Uganda; Victoria University; and a visiting lecturer at Cavendish University, Uganda. His current research focuses on cross-border modes of higher education delivery and teacher training in Sub-Saharan Africa.

Alfred Kirigha Kitawi is the director of the Centre for Research in Education at Strathmore University, Kenya. His holds a Doctorate in Higher Education Management from the University of Bath, UK. He has undertaken several monitoring and evaluation projects in the areas of community capacity development, quality assurance, and life-skills development in Kenya and Uganda. He has written articles in the areas of quality assurance in higher education, knowledge management, action research, integration of information and communication technology in higher education, and lifelong learning. He is currently part of the TOTEMK-project team formulating modules to capacitate university lecturers on the delivery of the competency-based curriculum.

Rebecca Nthogo Lekokp is a professor, researcher, and author in the areas of higher education and lifelong learning. She has distinguished herself as a hardworking, committed, and self-driven academic leader, and has contributed significantly to the management and leadership of HE as a member of the university governing council, senate, dean, and departmental head. Her publications are in the areas of higher education leadership, quality assurance, lifelong learning, and inclusive policies—featured in titles such as "HE at the Intersection of Globalisation" and "Technology and Quality Assurance in Higher Education in Sub-Sahara Africa". Her recent work on HE leadership suggests that the capacity to lead in HE includes, among others, finding purpose and direction in what the traditional culture can offer, such as the African culture of interconnectedness.

Gilbert Likando holds a PhD in Adult Education and Lifelong Learning. He is an associate professor and researcher in the Department of Higher Education and Lifelong Learning, School of Education at the University of Namibia, Namibia. His research interests encompass higher education, community studies, teacher education, educational leadership and management, and literacy learning and livelihoods. His experience in teaching at both school and university levels, and interaction with the community, has given him the impetus to understand education as an empowerment tool to leverage social inequality in society. He has authored and co-authored several articles and book chapters in the fields of teacher education and higher education. ORCID identifier: https://orcid.org/0000-0001-7539-2086

Mwaka Omar Makame has a specialisation in educational leadership and policy studies, with her research focusing on the mechanisms for enhancing employability skills among Vocational Education and Training (VET) students. Mwaka demonstrates experience in teaching geography subject at the secondary school level. She also works as a teacher mentor for vulnerable students, especially girls under the Campaign for Female Education (CAMFED) programme in Tanzania.

Kyashane Stephen Malatji is the acting director of the School of Education at Tshwane University of Technology. He holds a PhD in Education, Curriculum Studies from the University of Fort Hare, Master of Education, Bed Hons and BEDSPF, from the University of Limpopo. He completed his Postgraduate Diploma in Higher Education from Rhodes University. He is the author of 66 articles in accredited journals. Malatji has successfully supervised 21 Master and 14 Doctoral students. In 2015, he was awarded a certificate for excellence in research at the University of Venda. In 2018, he was awarded Young Researcher of the Year by Tshwane University of Technology. In 2020, he received an award as Emerging Researcher of the Year awarded by the Education Association of South Africa (EASA). His research interest covers teaching and learning; evaluation of teaching, curriculum development; assessment; quality assurance and teacher development in higher education. He is a coordinator for Buddies Research Empowerment Forum, which is aimed at mentoring upcoming researchers. Malatji is a board member of the journal *African Perspective of Research in Teaching and Learning*. He is also a reviewer for several journals such as the *Journal of Education Studies (JES)* and the *South African Journal of Education (SAJE)*.

Makwalete Johanna Malatji is a lecturer and coordinator of the literacies in the Department of Early Childhood Education at the University of Pretoria. She holds a PhD in Education from the University of Fort Hare. Her focus area is in the teaching of Literacies in the Foundation Phase, parental involvement and teacher education. She is the author of 11 articles in internationally and nationally accredited journals. Malatji has delivered papers at national and international conferences. Malatji is also a co-author of a book titled *Inclusion, Learner Support, and Assistive Technology: Helping Learners Learn – An African Approach*.

Ignatius Waikwa Maranga is an electrical engineer with a specialisation in Electric Power Systems and Renewable Energy Systems. He has a Masters in Electrical Engineering and a Postgraduate Diploma in Project Planning and Management. He is a researcher at Strathmore University, at the Energy Research Center. He has worked on several research projects in solar photovoltaic systems design and integration, electric vehicles and energy management systems. He is passionate about energy and sustainability.

xxx   NOTES ON CONTRIBUTORS

Kenneth Kamwi Matengu holds a PhD in Human Geography from the University of Eastern Finland. He is currently a research professor and vice-chancellor at the University of Namibia. His research interest includes access and equity in education, higher education governance and management. He has published articles, books and book chapters on the coordination of higher education, access with equity in education and higher education governance. He has also published work on tourism, community-based management, decentralisation of rural water supply, local government and community health.

Alexander Michael is a PhD holder from Wits School of Education (University of the Witwatersrand, Johannesburg-South Africa), a lecturer in the Department of Science Education, Taraba State University, Jalingo-Nigeria and a postdoctoral research fellow at Faculty of Education, Nelson Mandela University, Gqeberha-South Africa. His research focuses on the teaching/learning of STEAM and mathematical discourse in classrooms with students who are multilingual and/or learning English.

Kulthum Nabunya PhD, is a lecturer at the Post-Graduate Department of Education at Bugema University, Kampala-Uganda, and Victoria University, Kampala. She holds a masters degree in Educational Policy and Planning and a PhD in Educational Management from Makerere University. Nabunya is an education specialist, researcher and women's rights activist. She is the Officer of Education Services at the Directorate of Education and Social Services in Kampala Capital City Authority (KCCA) where she supervises education operations mainly focusing on quality assurance in both government and private nursery and primary schools. Nabunya is also the proprietor and director of Anwar Baby and Primary School.

Peter Neema-Abooki is Professor of Higher Educational Management and Development Studies, Human Resource Management in Education, Educational Policy and Planning, Educational Foundations, and Curriculum Studies. He has also served as Professor of Business and Management and he is a trained educationist. He is the editor-in-chief of the *International Journal of Progressive and Alternative Education* (based in Nigeria), and editor of *Quality Assurance in Higher Education in Eastern and Southern Africa: Regional and Continental Perspectives* and *Quality Assessment and Enhancement in Higher Education in Africa*. Peter is the co-editor of *Innovating Higher Education*, besides having earlier been a co-editor and Designer of Fields *AFAR* magazine. He is a reviewer at several international fora and a member of several international technical committees. ORCID ID: https://orcid.org/0000-0002-7347-5299

John Kamwi Nyambe is the Associate Dean of the School of Education at the University of Namibia, Namibia. His research interest is in the areas of higher education, covering learning and teaching, curriculum development, assessment, and quality assurance. Further to this he also does research in teacher education and educational reform. He has held various leadership positions in higher education. ORCID identifier: https://orcid.org/0000-0003-0134-270X

Abukar Mukhtar Omar is a lecturer at the Faculty of Education, SIMAD University, Somalia, where he is the also dean of the Faculty. He is also a member of the University's Academic Council. He holds a BSc in Education from Islamic University, Uganda, and an MA in Education Policy and Planning from Makerere University, Uganda. He is currently doing PhD in Education from Makerere University, Uganda. Abukar has taught undergraduate and postgraduate courses in educational administration and management at SIMAD University, Mogadishu, Somalia. His current research focuses on quality assurance in higher education.

Rakel Kavena Shalyefu holds a PhD in Instructional Systems Design, with a specialty in designing and evaluating programmes, and distance education learning materials from Pennsylvania State University, USA. Additional qualifications are an MEd in Adult and Nonformal Education from the University of Massachusetts, a Bachelor's Degree in Pedagogics from the University of Fort Hare, a Bachelor of Education (Honours) from UNISA, and a Postgraduate Diploma in Higher Education for Academic Developers from Rhodes University. She is Associate Professor of Lifelong Learning and Community Development, a Commonwealth Professional Fellow at Cardiff University, and a Country Director for the International Higher Education Teaching and Learning Association (IHETL). ORCID: https://orcid.org/0000-0001-6281-8877

Romanus Shivoro holds a PhD in Education, with a specialty in higher education and graduate employability from the University of Namibia, Namibia, and a Master of Arts Degree in Education from Lucknow University, India. He is the Assistant Director for International Relations at the University of Namibia, managing local and international academic partnerships and cooperation. ORCID: https://orcid.org/0000-0002-3230-0783

ABBREVIATIONS

1IR	First Industrial Revolution
2IR	Second Industrial Revolution
3IR	Third Industrial Revolution
3Ps	Profit, People, Planet
4IR	Fourth Industrial Revolution
AAU	Association of African Universities
AC	Alternating Current
Ads	Advertisements
AfREA	African Evaluation Association
AHEIs	African Higher Education Institutions
AHES	African Higher Education Space
AI	Artificial Intelligence
ANA	Annual National Assessment
e.g. *exempli gratia*	For example
GRN	Government of the Republic of Namibia
SD	Standard Deviations
QA	Quality Assurance
SPSS	Statistical Package for the Social Sciences
IO	Independence of Observation
LR	Homoscedasticity
LCP	Learner-Centered Pedagogy
MIV	Multicollinearity
CD	Significant Outliers
HC	Homogeneity of Covariance
MANOVA	Multivariate Analysis of Variance
SDGs Sr	Sustainable Development Goals Sister
R1	Research Question 1

xxxiii

xxxiv ABBREVIATIONS

R2	Research Question 2
R3	Research Question 3
R4	Research Question 4
AU	African Union
BA	Building Automation Systems
Br.	Brother
CareerEDGE	**Career,** Experience, **D**egree subject knowledge, understanding and skills, **G**eneric skills, **E**motional intelligence
CBC	Competency-Based Curriculum
CBET	Competency-Based Education and Training
CBT	Competence-Based Training
CD-ROM	Compact Disc Read-Only Memory
CE	Circular Economy
CEMS	Centre for Extra-Mural Studies
CEQUAM	Centre for Quality Assurance and Management, University of Namibia
CFL	Compact Fluorescent Lamps
CHET	Centre for Higher Education Transformations
CLOC	Community Lending and Outside Capital
CODESRIA	Council for the Development of Social Science Research in Africa
COL	Commonwealth of Learning
COVID	Corona Virus Disease
COVID-19	Coronavirus Disease 2019
CPD	Continuous Professional Development
CREAM	Clear, Relevant, Economical, Attainable, and Monitorable
CSEEs	Certificate of Secondary Education Examinations
CSOs	Civil Service Organisations
CV	Curriculum Vitae
DAC	Development Assistance Committee
DC	Direct Current
DC	District Council
DMAIC	Define, Measure, Analyse, Improve, and Control
Dr	Doctor
DTVE	Department of Technical Vocational Education
DUCE	Dar es Salaam University College of Education
E-4 Impact	Entrepreneurship for Impact
EADTU	European Association of Distance Teaching Universities
EFA	Education for All
EfS	Education for Sustainability
EGMA	Early Grade Mathematics Assessment
EGRA	Early Grade Reading Assessment
E-learning	Electronic Learning

ENQA	European Association for Quality Assurance in Higher Education
EPC	Energy Performance Contracts
EPRA	Energy Petroleum Regulatory Authority
EQUIP-T	Education Quality Improvement Programme in Tanzania
ESCO	Energy Services Company
ESD	Education for Sustainable Development
ESD & CE	Education for Sustainable Development and Community Engagement
ESDG	Education for Sustainable Development Goal
ESDP	Educational Sector Development Plan
ESG	European Standards and Guidelines for Quality Assurance
ESR	Education for Self-Reliance
ETP	Educational and Training Policy
EV	Electric Vehicle
EVE	Economic Value of Equity
FBOs	Faith-Based Organisations
FDCs	Focal Development Colleges
Fr	Father
GoU	Government of Uganda
GRI	Global Reporting Initiative
GW	Giga Watts
HE	Higher Education
HEI	Higher Education Institution
HEIs	Higher Education Institutions
HIEEP	High Impact Entrepreneurship Education Practices
Https	Hypertext Transfer Protocol Secure
HVAC	Heating, Ventilation, and Air Conditioning
ICT	Information and Communications Technology
ICT	Information and Computer Technology
IDRC	International Development Research Centre
IHEs	Institutions of Higher Education
ILO	International Labour Organisation
IoT	Internet of Things
IP	Intellectual Property
IT	Information and Technology
IUCEA	Inter-University Council for East Africa
JAG	Joint Advisory Group
K	Kindergarten
KNEC	Kenya National Examinations Council
KW	Kilo Watts
KWh	Kilo Watt Hours

LED	Light Emitting Diodes
LMS	Learning Management System
Lt Col	Lieutenant Colonel
MLMS	Moodle Learning Management System
M&E	Monitoring and Evaluation
MAXQDA	Max Weber Qualitative Data Analysis
MC	Municipal Council
MDAs	Ministries, Departments, and Agencies
MDGs	Millennium Development Goals
MEPP	Mathematics Education Primary Programme
MOECHE	Ministry of Education, Culture, and Higher Education
MoEST	Ministry of Education, Science and Technology
MOOCs	Massive Open Online Courses
Mr	Mister
MW	Mega Watts
NACTE	National Accredited Council for Technical Education
NACVET	National Council for Vocational Education and Training
NASMLA	National Assessment System for Monitoring Learner Achievement
NCHE	National Council for Higher Education
NCHE	National Council for Higher Education, Namibia
NCVER	National Centre for Vocational Education Research
NECTA	National Examinations Council of Tanzania
NGO	Non-Government Organisation
NGOs	Non-Governmental Organisations
No.	Number
NPV	Net Present Value
NSDS	National Skills Development Strategies
NSGRP	National Strategy for Growth and Reduction of Poverty
NTA	National Technical Award
NVTA	National Vocational and Training Award
NY	New York
OECD	Organisation for Economic Co-operation and Development
OER	Open Educational Resources
OPM	Office of the Prime Minister
PAR	Participatory Action Research
PC	Personal Computer
Pdf	Portable Document Format
PhD	Doctor of Philosophy
PhD	Degree of Doctor of Philosophy
PME	Participatory Monitoring and Evaluation
PO-RALG	President's Office—Regional Administration and Local Government

pp	Pages
PPAs	Power Purchase Agreements
PRIEDE	Kenya Primary Education Development
Prof	Professor
PSLEs	Primary School Leaving Examinations
PV	Photovoltaic
QA	Quality Assurance
QM	Quality Management
RETs	Renewable Energy Technologies
R&D	Research and Development
Rev	Reverend
SEACMEQ	Southern & Eastern African Consortium for Monitoring Educational Quality
SAPs	Structural Adjustment Plans
SDG	Sustainable Development Goal
SDGs	Sustainable Development Goals
SDSN	Sustainable Development Solutions Network
SEP	Social Enterprise Project
SEQIP	Secondary Education Quality Improvement Project in Kenya
SEQUIP	Secondary Education Quality Improvement Project in Tanzania
SERC	Strathmore Energy Research Center
SILC	Savings and Internal Lending Communities
SIMAD	Somali Institute of Management and Administration Development
SITMS	Strengthening In-service Teacher Mentorship and Support
SITT	School-based In-service Teacher Training
SMART	Specific, Measurable, Achievable, Realistic and Time-bound
SNU	Somali National University
SPHEIR	Strategic Partnerships for Higher Education Innovation and Reform
Sr	Sister
SS	Secondary Schools
Sts	Saints
SSA	Sub-Saharan Africa
STEAM	Science, Technology, Engineering, Arts, and Mathematics
STEM	Science, Technology, Engineering, and Mathematics
TANESCO	Tanzania Electric Supply Company Limited
TCs	Teacher Colleges
TCU	Tanzania Commission for Universities
TESCEA	Transforming Employability for Social Change in East Africa
TET	Technical Education and Training
THE	Times Higher Education
TIE	Tanzania Institute of Education
ToC	Theory of Change

xxxviii   ABBREVIATIONS

TRCs	Teacher Resource Centres
TTU	Tanzania Teachers' Union
TVET	Technical Vocational Education and Training
UMU	Uganda Martyrs University
UN	United Nations
UNAIDS	United Nations Programme on HIV/AIDS
UNAM	University of Namibia
UNESCO	United Nations Educational, Scientific and Cultural Organization
UNICEF	United Nations International Children's Fund
UNSDGs	United Nations Sustainable Development Goals
UOTIA	Universities and Other Tertiary Institutions Act
UPFORD	University Partnership for Research and Development
URT	United Republic of Tanzania
USA	United States of America
USAID	United States Agency for International Development
USEM	Understanding, Skillful Practices, Efficacy Beliefs and Meta-cognition Model
USG	United States Government
UUQAF	Uganda Universities' Quality Assurance Forum
VET	Vocational Education and Training
VETA	Vocational Education and Training Authority
VETCs	Vocational Education and Training Centres
VLE	Virtual Learning Environment
Vol	Volume
VTA	Vocational Training Authority
VTCs	Vocational Training Centres
Web	Website
WEF	World Economic Forum
WIL	Work-Integrated Learning
www	World Wide Web

LIST OF FIGURES

Models of Quality Assurance: *Towards a Quality Assurance Framework of Digital Learning in Higher Education in Africa*

Fig. 1 Process-oriented lifecycle model for quality assurance in digital learning. Adapted from Abdous (2009: 288)   17

Fig. 2 Concepts related to the quality of digital learning (Source: Geoffrey, 2014, p. 92)   28

Fig. 3 Monitoring quality of digital teaching, learning, and assessment on Moodle LMS. Source: University of Namibia (2021)   33

A Framework for Embedding Graduate Employability: *Attributes of Higher Education Curriculum in Africa*

Fig. 1 The essential components of Employability. Source: Pool and Sewell (2007, p. 280)   142

Fig. 2 A proposed framework for embedding graduate employability attributes in the higher education curriculum in Africa   156

Addressing SDG 4 via Learners' Prior Numerical Cognition: *a Predictor of Educational Performance in a Developing Country in Sub-Saharan Africa*

Fig. 1 Revised Bloom's taxonomy theory. Adopted from Anderson et al. (2001)   173

Higher Education: *Towards a Model for Successful University-Industry Collaboration in Africa*

Fig. 1 Triple Helix Model showing the interplay between academia, industry, and government (Triple Helix Model of an innovation). Source: Hermosura (2019: 801) 254

Fig. 2 The Triple Helix Model of university-industry-government linkage for Africa. Source: Ansell and Alison (2007) 271

Quality Assurance of Higher Education in a Neo-liberal Context: *Towards Transformative Practices in Africa*

Fig. 1 Multiple levels analysis of neo-liberalism and quality assurance. Adapted from Kentikelenis and Rochford (2019) 331

The Role of Higher Education and the Future of Work in Africa's Fourth Industrial Revolution

Fig. 1 4IR ecosystem for schools of the future. Source: Adapted from ACET (2018, p. 4) 352

Fig. 2 Jobs of the future. Source: http://quantitative.emory.edu/news/articles/Future-of-Work-Part-I.pdf 359

Fig. 3 Ways that 4IR will influence the three main functions of higher education institutions. Source: Masindei and Roux (2020: 36) 360

LIST OF TABLES

Models of Quality Assurance: *Towards a Quality Assurance Framework of Digital Learning in Higher Education in Africa*

Table 1   Quality assurance framework of digital learning in African higher education   37

Quality of Teaching and Learning Through Internal Quality Curriculum Review Mechanisms: *A Case of Private Higher Education Institutions in Post-conflict Somalia*

Table 1   Internal quality review curriculum reviews   82
Table 2   Internal quality review curriculum reviews   83

Addressing SDG 4 via Learners' Prior Numerical Cognition *a Predictor of Educational Performance in a Developing Country in Sub-Saharan Africa*

Table 1   Summary details of major themes, questions, key gaps   171
Table 2   A sample of the administered questionnaire   176
Table 3   Test for HC   180

Scaling Education Innovations in Tanzania, Kenya, and Zambia: *Assessing the Design of School In-service Teacher Training*

Table 1   Tools and Guides   207

xlii LIST OF TABLES

Mechanisms for Enhancing Employability Skills Among Students Within Vocational Education Training Institutions in Tanzania

Table 1	Institutional mechanisms for enhancing employability skills among VET students	235
Table 2	Students' responses to institutions mechanisms for enhancing employability skills	239
Table 3	Teaching and learning methods used in enhancing the employability skills of students	240

Higher Education: *Towards a Model for Successful University-Industry Collaboration in Africa*

Table 1	A typology of university-industry links, from higher to lower intensity	266

The Role of Higher Education and the Future of Work in Africa's Fourth Industrial Revolution

Table 1	Emergence of industrial revolutions	354
Table 2	Technologies that drive the 4IR	356

Introduction: Sustainability as an Acme of Quality Assurance in Higher Education

Peter Neema-Abooki 🆔

1   PREAMBLE

This chapter spells out an overview of this book whose prospects are on the sustainability of higher education (HE) in Sub-Saharan Africa within the perspectives of quality assurance (QA). Hence, it sums up each chapter in the order of the sub-theme depicted herein as "parts"—namely, curriculum and teaching in higher education (seven chapters) and higher education and innovations (seven chapters). Ultimately, it makes mention of the compendium's uniqueness and major contribution as well as the lessons drawn for and/or from Africa.

P. Neema-Abooki (✉)
Department of Educational Foundations,
University of South Africa, (UNISA), Pretoria, South Africa
e-mail: pneemaster@gmail.com

© The Author(s), under exclusive license to Springer Nature
Switzerland AG 2024
P. Neema-Abooki (ed.), *The Sustainability of Higher Education in Sub-Saharan Africa*, Sustainable Development Goals Series,
https://doi.org/10.1007/978-3-031-46242-9_1

2 Sustainability and Higher Education

The term "sustainability" is broadly used to indicate programmes, initiatives and actions aimed at maintaining and perpetuating an entity, be it physical or juridical. Chankseliani and McCowan (2021) propound that while there are widespread initiatives and an increasing number of universities aligning their activities with the sustainable development goals (SDGs), there is a significant gap in knowledge and evidence. There is still a need to document the wide variety of activities relevant to sustainable development being undertaken by universities, particularly in low- and middle-income countries, and to assess the consonance between activities of teaching, research, community engagement and campus operations. Sustainability, as perceived by Eaton (2021), renders completeness to QA which—as the internal and external examination of the effectiveness and performance of colleges and universities—has always played a crucial role in sustaining and improving the best of what HE has done in the past to build the future. The "best" includes the core values of academic freedom, institutional autonomy and social responsibility.

This chapter subscribes to the age-old contentions of UNESCO (1950) that the spelt-out core values are three indissociable principles for which every university should stand, namely, the right to pursue knowledge for its own sake and to follow wherever the search for truth may lead; the tolerance of divergent opinion and freedom from political interference; the obligation as social institutions to promote, through teaching and research, the principles of freedom and justice, of human dignity and solidarity, and to develop mutually material and moral aid on an international level.

QA in HE is twofold:

- External, which is mandatory to operate as an educational institution and granted by a regulatory body.
- Internal, which refers to the QA system and procedures adopted by the institution for continuous improvement. All QA activities are cyclical and periodic (Azhar, 2022).

The external review of institutional QA, as Azhar (2022) explicates, is a cyclical process repeated every five years; however, the internal processes supporting quality culture are more frequently administered and evaluated. And while the internal QA activities serve as proof of meeting external standards and guidelines, all QA activities require the support and buy-in from the senior management, as well as the understanding and commitment of individual employees, to strengthen the "quality culture" at an institution.

A quality culture is an attitude supporting a set of shared values that guide how continuous improvements are made to everyday working practices, thus leveraging the levels of quality in the outputs; it is a collective responsibility for all employees of the University (Bethlehem University, 2021).

Eaton (2021) advances that QA maintains the key role of sustaining HE even as we adopt new teaching and learning practices, develop new types of institutions, engage new types of education providers and continue our commitment to expand access and equity in HE even amidst the current environment of both the COVID-19 pandemic and the major and painful focus on social change around issues of race and equality, making it clear that QA efforts are more important than ever.

Meanwhile, the key factors influencing the quality of HE are the quality of faculty, curriculum standards, technological infrastructure available, research environment, accreditation regime and the administrative policies and procedures implemented in institutions of higher learning (O'Malley, 2022). For instance, the need for faculty training based on the use of innovative technology, as cited by Abouelenein and Attia (2016), has emerged to enhance their abilities to face the demands of the labour market and employ quality standards of education in general and in their areas of specialisations in particular. As for curriculum standards, much attention is given to content and associated teaching, learning and assessment methodologies. However, an important component of any curriculum is its organisational management, how it is all held together, the way the process is conducted and what mechanisms are applied to ensure quality (Mohamad Kayyal & Trevor Gibbs, 2012). Meantime, technology infrastructure refers to the components that make the operation and management of enterprise IT services and IT environments possible. This infrastructure includes all hardware, software, networks and facilities companies use to create, test, deliver, control and support IT services (Indeed Editorial Team, 2023).

QA in research comprises all the techniques, systems and resources that are deployed to give assurance about the care and control with which research has been conducted. It is typically concerned with the responsibilities of those involved in the research, transparent project planning, the training and competence of research staff, facilities and equipment, documentation of procedures and methods, research records and the handling of samples and materials (University of Reading, n.d.).

HE embraces a central position at the core of mainstream development thinking. With an imperative to promote sustainable development paths, higher education institutions (HEIs) pave the way for sustainable development through knowledge generation and dissemination, research, education and outreach activities (Ramos, 2019). Education for Sustainable Development (ESD), defined by UNESCO (2014), equidistantly allows human beings to acquire the knowledge, skills, attitudes and values necessary to shape a sustainable future. ESD promotes the development of the knowledge, skills, values and actions required to create a sustainable world, which ensures environmental protection and conservation, promotes social equity and encourages economic sustainability. For instance, Sustainable Development Goal 4 propagates for inclusive and equitable quality education and is cognisant that education liberates the intellect, unlocks the imagination and is fundamental for self-respect. It is the key to prosperity and a world of opportunities, making it possible for each of us to contribute to a progressive society (UNESCO, n.d.). For, University social responsibility is about a commitment to maximising the positive societal impact of the entire range of university functions. And, accordingly, SDGs have become a near-universal framework for university social responsibility and civic engagement work (O'Malley, 2022).

Under the consideration of different measures adopted by most United Nations (UN) members, the SDGs included in the 2030 Agenda for Sustainable Development aim to address the main challenges related to social, economic and environmental issues (Jose Manuel Diaz-Sarachaga, Daniel Jato-Espino (2018)). In a similar vein, critical and transformative approaches to education focus on the need to develop new ways of being in the world that are more aligned with the complex nature of current sociological crises. And, the development of virtuous thinking, feeling and acting is key for an education guided by sustainability and the common good (Amparo Merino & Estela Díaz, 2022). Moreover, the approach of "build back better together" must lead our post-COVID sustainability initiatives to act as a catalyst that speeds up the progress towards inclusive and sustainable HE for all (Luis Velazquez, 2022). And, sustainable development is the moral imperative of satisfying needs, ensuring equity and respecting environmental limits including efforts to maximise economic value (Albert Mawonde & Muchaiteyi Togo, 2019).

While a moral imperative is a strongly felt principle that compels that person to act, a kind of categorical imperative, as defined by Immanuel Kant to be a dictate of pure reason in its practical aspect (Stanford Encyclopedia of Philosophy, 2022), the economic value of equity (EVE) is, according to anecdotal evidence, a long-term economic measure/indicator of net cash flow. The EVE is calculated by taking into account the present value of all asset cash flows and subtracting the present value of all liability cash flows. In other words, it is the net present value (NPV) of a bank or a financial institution. NPV, in our case here, is therefore how much an educational investment is worth throughout its lifetime, discounted to today's value.

To help address the challenges of sustainable development, HEIs must transform themselves, bringing together best practice in quality management for tertiary education with best practice in education for sustainable development (Zinaida Fadeeva, Laima Galkute, Clemens Mader, Geoff Scott, 2014).

This chapter nurtures that the notions advanced above ought to be painstakingly borne in mind if Sub-Saharan Africa and, by extension, the entire continent has to merit the stance of recalibrating sustainability in HE.

3 Recapitulation of Content

This edited book has a total of 15 chapters. These include this Introductory chapter that makes a summation of each thus:

In Chapter "Models of quality assurance in higher education: Towards a Quality Assurance Framework of Digital Learning in Higher Education in Africa" Erkkie Haipinge and Ngepathimo Kadhila challenge those students across nations and cultures can study at their times and pace through digital learning. Despite this advantage, the authors underscore the challenge of convincing parents, employers and students that the quality of digital learning is as good as face-to-face learning. To mitigate against this challenge, the duo calls for QA processes for staff and students in HEIs, qualification and professional support of staff, who create and deliver digital courses plus adequate funding and investment in effective technology.

David Katende in Chapter "Monitoring and Evaluation in Higher Education: Quality Perspectives in Africa" describes quality systems and theoretical models of monitoring and evaluation (M&E) for QA in African HE and presents a comprehensive background, approaches, theory, practice and models of QA for M&E in HE. He documents essential steps in central/strategic M&E for QA in HEIs in Africa, depicts the role of M&E from the Organisation for Economic Co-operation and Development/Development Assistance Committee (OECD/DAC) framework perspective; and shares insights on evaluation criteria with emphasis on the M&E data sources, utilisation of such data for QA, data types, data quality, and how to develop M&E indicators for QA in HE.

In Chapter "Assuring Quality of Teaching and Learning through Internal Quality Curriculum Review Mechanisms: A Case of Private Higher Education Institutions in Post-Conflict Somalia" Abukar Mukhtar Omar and Abdu Kisige are concerned with the deteriorating quality of teaching and learning in many private HEIs in post-conflict Somalia, with students not obtaining a good HE and not being competitive on the job, with private universities especially being more concerned with making money than raising educational standards. Their findings revealed that internal quality curriculum review mechanisms determine the quality of teaching and learning and that such initiatives enable graduates to be equipped with subject-specific and transferable skills of the twenty-first century relevant to the demands and needs of global competition. They recommended that managers of HEIs need continuously review their curriculum to weed out outdated content.

Kulthum Nabunya in Chapter "Students' Satisfaction and Virtual Learning Service Delivery in Bugema University" highlights that digitisation in most institutions of higher learning was not widespread, yet many embraced online learning during the lockdowns. In particular, she investigated the relationship between students' virtual learning satisfaction and the virtual teaching service delivery of academic staff at Bugema University and concluded that the online examination procedure, the level of usage of multimedia and the interactivity of the lecturers on the virtual learning environment were generally good, while the internet quality and the level of utilisation of the online resources had a lot of gaps.

Christopher Mukidi Acaali, in Chapter "Sustainability and the Triple Mission of the University: Uganda Martyrs University in Perspective", fronts Uganda Martyrs University (UMU) as a case study on the triple mission of a university, underscores the importance of sustainability (human, social, economic and environmental) especially regarding higher education institutions (HEIs) and the latter's role in the implementation of sustainable development goals (SDGs). He canonises HEIs for having advanced from being ivory towers to the recent developments of redefining their role and becoming critical thinkers to the extent of employing their three-fold mission to address the social and economic challenges in communities and the implementation of the UN SDGs 2030 Agenda.

In Chapter "A Framework for Embedding Graduate Employability: Attributes into Higher Education Curriculum in Africa" Romanus Shivoro, Rakel Kavena Shalyefu and Ngepathimo Kadhila observe that under the context of SDGs, HE policymakers in developing countries have been concerned with the capacity of HEIs producing human capital that is capable of functioning competitively in today's labour market. In addressing this gap, the authors have proposed a framework, based on the USEM Model, for incorporating graduate employability attributes into HE curricula in Africa. They advocate the STEM model as one that contributes to the attainment of the SDGs.

Anass Bayaga and Alexander Michael in Chapter "Addressing SDG 4 via Learners' Prior Numerical Cognition as a Predictor of Educational" examine SDG 4 through learners' prior numerical cognition as a predictor of educational performance. They hold that while SDG 4 seeks to ensure inclusive and equitable education, the challenge is ensuring "that all youth and a substantial proportion of adults achieve literacy and numeracy" by 2030. Thus, the research examines the pre-service teachers' prior mathematical cognition, ease of learning (fluency), mathematics concepts acquisition and inappropriate instruction as predictors of quality educational performance in South Africa. The findings reveal that the underperformance of learners in numeracy at primary schools is associated with inappropriate preparatory practices as well as poor mathematical cognition in their HE.

In Chapter "Scaling Education Innovations in Tanzania, Kenya, and Zambia: Assessing the Design of School In-service Teacher Training" Katherine Fulgence assesses the innovation of School In-service Teacher Training (SITT) currently being scaled up and adopted in selected secondary schools in Tanzania, Kenya and Zambia. Whereas the evaluation of SITT's application in selected primary schools in Tanzania reveals significant progress in students' learning outcomes, the design of the secondary SITT intervention does not systematically apply to scale frameworks. Notwithstanding, the design of the innovation aligns with the scaling features of involving key stakeholders in HEIs and also facilitates further large up-scaling of the innovation and policy uptake. The study informs the design of the scaling strategy as a step towards making the innovation more appealing to governments and potential users.

In Chapter "Mechanisms for Enhancing Employability Skills among Students within Vocational Education" Mwaka Omar Makame and Katherine Fulgence focus on mechanisms for enhancing employability skills within vocational training institutions in Tanzania by identifying eight employability skills' attributes; namely communication skills, teamwork, problem-solving, initiative and enterprise, self-management, planning and organising skills, information communication and technology (ICT) and technical skills. Having concluded that interactive teaching methods, the key to enhancing employability skills, are less applied in vocational education institutions; they recommend such institutions explore more mechanisms for preparing students along the study attributes by using sustainable interactive teaching methods.

In Chapter "Higher Education: Towards a Model for Successful University-Industry Collaboration in Africa" Ngepathimo Kadhila, Kyashane Stephen Malaji Tshwane and Makwalete Johanna Malatji underscore that HE is one of the key engines for achieving a prosperous Africa based on inclusive growth and sustainable development. In their view, this can be achieved by having significant investments in education majorly aimed at developing human and social capital. This will not only spur a skills revolution with emphasis on innovation, science and technology but also will create what the authors called a "helix model" where academia, industries and Government collaborate and partner together for the good of society.

John Kamwi Nyambe and Ngepathimo Kadhila, in Chapter "Higher Education as an Engine of Development: Sites of Domination, Contestation, and Struggle in Africa", advance the role of HE as an engine for the socio-economic development of Africa and beyond. They profess that this role however has been played in the contexts of tension and conflicts between the Eurocentric Western epistemic system with its neoliberal ideology and the African developmental agenda; maintaining as they do that these conflicting forces have pulled HE in opposite directions with dire consequences for the African development agenda. The authors propose a pluralistic worldview and a hybrid model that decolonises HE in Africa.

In Chapter "Energy Sustainability in African Higher Education: Current Situation and Prospects" Alfred Kirigha Kitawi and Ignatius Waikwa Maranga's objective is to document how HEIs within Africa can promote the energy sustainability agenda. The authors dwell on energy programmes as revenue sources for HEIs, and conceive renewable energy as a source of revenue generation for HEIs—not to mention the new research possibilities, consultancies and networks that are likely to be created. They maintain that HEIs can integrate sustainability practices within their processes and, at the same time, their Renewable Energy Technologies (RET) can be used to meet the fourth industrial revolution (4IR) needs in African countries.

Joel Jonathan Kayombo, Mjege Kinyota and Patrick Severine Kavenuke in Chapter "Higher Education Quality Assurance in a Neo-liberal Context: Towards Transformative Practices in Africa" claim that QA in modern HE in Africa is influenced by neoliberal ideologies, with QA philosophies and practices evolving into professional power mechanisms, with an emphasis on compliance, power and responsibility. The triad propagates the change from the neoliberal-driven QA regimes to a QA culture that is aimed at improving quality in HE and posits that the role of QA should shift from a purely technical to a political process to involve more people—including the faculty, administration and other stakeholders—in the design, implementation, and sustenance of QA processes. Thus, they call for a participatory approach rather than an audit one.

In Chapter "The Role of Higher Education and the Future of Work in Africa's Fourth Industrial Revolution" Kenneth Kamwi Matengu, Ngepathimo Kadhila and Gilbert Likando base on the world's strive for a technological revolution which will change how people live, work and interact; and assess Africa's readiness in preparing and producing graduates who are tuned to the changing times of the technological revolution. By analysing programme offerings and labour demands for the future of work under the context of HE, they advocate for the reform of curricula with an emphasis on reskilling and upskilling to prepare a high-level skilled workforce for the fourth industrial revolution (4IR).

4   Uniqueness and Contribution of the Book

This book's uniqueness and major contribution have been manifested by the epistemological and contextual basis that amalgamates theoretical and practical dimensions of the variables, within the specificities of an African context; though borrowing from the international benchmarks yet without being alienated by the latter.

A way forward has been engendered by the requisite lessons that have been drawn bridging the gaps in context, having introspected both the Global and the African foci as well as the sustainable development goals.

REFERENCES

Abouelenein, & Attia, Y. (2016). Training needs for faculty members: Towards achieving quality of University Education in the light of technological innovations. *Academic Journal, 11(13)*, 1180–1193. https://doi.org/10.5897/ERR2015.2377

Azhar, S. (2022). *The importance of quality assurance in higher education.* Retrieved May 3, 2023, from https://www.projectmanagement.com/contentPages/article.cfm?ID=777664&thisPageURL=/articles/777664/the-importance-of-quality-assurance-in-higher-education#_

Bethlehem University. (2021). *Quality assurance policy.* Available: Retrieved May 01, 2023, from https://www.bethlehem.edu/wp-content/uploads/2022/05/BU-Quality-Assurance-Policy.pdf

Chankseliani, M., & McCowan, T. (2021). Higher education and the sustainable development goals. *Higher Education, 81*, 1–8. https://doi.org/10.1007/s10734-020-00652-w

Diaz-Sarachaga, J. M., & Jato-Espino, D. (2018). *Is the sustainable development goals (SDG) index an adequate framework to measure the progress of the 2030 Agenda? Sustainable Development.* Wiley Online Library. Retrieved April 30, 2023, from https://onlinelibrary.wiley.com/doi/abs/10.1002/sd.1735

Eaton, J. S. (2021). The role of quality assurance and the values of higher education. In H. can't Land, A. Corcoran, & D. C. Iancu (Eds.), *The promise of higher education.* Springer. https://doi.org/10.1007/978-3-030-67245-4_28

Fadeeva, F., Galkute, L., Mader, C., & Scott, G. (2014). *Sustainable development and quality assurance in higher education.* Palgrave Studies in Global Higher Education, https://doi.org/10.1057/9781137459145.

Indeed Editorial Team. (2023). *A guide to technology infrastructure: Examples and components.* Retrieved May 01, 2023, from https://ca.indeed.com/career-advice/career-development/technology-infrastructure-examples#:~:text=Technology%20infrastructure%20refers%20to%20

Mawonde, A., & Togo, M. (2019). The role of SDGs in advancing implementation of sustainable development. The case of the University of South Africa, South Africa. In U. Manuel, J. de Miranda Azeiteiro, & P. Davim (Eds.), *Higher education and sustainability: Opportunities and challenges for achieving sustainable development goals* (1st ed.). Taylor & Francis Group. https://doi.org/10.1201/b22452

Merino, A., & Díaz, E. (2022). Transforming ourselves to transform societies: Cultivating virtue in higher education for sustainability. In K. A. A. Gamage & N. Gunawardhana (Eds.), *The Wiley handbook of sustainability in higher education learning and teaching Willey Online Library.* Wiley-Blackwell. https://doi.org/10.1002/9781119852858.ch4

Mohamad Kayyal & Trevor Gibbs. (2012). Applying a quality assurance system model to curriculum transformation: Transferable lessons learned. *Medical Teacher, 34*(10), e690–e697. https://doi.org/10.3109/0142159X.2012.687486

O'Malley, B. (2022). *University social responsibility and the sustainable development goals.* University World News. Available: Retrieved May 02, 2023, from https://www.universityworldnews.com/post.php?story=202210080731555

Ramos, R. P. (2019). *Fourth industrial revolution: Opportunities and challenges on higher education institutions (HEIs) towards 2030 sustainable development goals (SDGS) agenda.* Retrieved Aug 3, 2023, from http://www.brainitiativesph.com/uploads/7/7/6/4/77644974/ramos_rsu_pice2019.pdf

Stanford Encyclopedia of Philosophy. (2022). *Kant's Moral Philosophy.* Retrieved April 30, 2023, from https://en.wikipedia.org/wiki/Moral_imperative#References

UNESCO. (1950). *Academic freedom, university autonomy and social responsibility, International Association of Universities.* Retrieved April 29, 2023, from https://www.iau-aiu.net/IMG/pdf/academic_freedom_policy_statement.pdf

UNESCO. (2014). *Sustainability of higher education.* Retrieved Aug 05, 2023, from https://www.google.com/search?q=SUSTAINABILITY+OF+HIGHER+EDUCATION&ei=tN3sYqy-BoanlwTfgbs4&ved=0ahUKEwis6LXPrK_5AhWG04UKHd_ADgcQ4dUDCA4&uact=5&oq=SUSTAINABILITY+OF+HIGHER+EDUCATION&gs_lcp=Cgdnd3Mtd2l6EAMyBAgAEEcyBAgAEEcyBAgAEEcyBAgAEEcyBAgAEEcyBAgAEEcyBAgAEEcyBA-gAEEc6BwgAEEcQsAN

UNESCO. (n.d.). *Sustainable development goal 4 and its targets.* Retrieved September 23, 2023, from https://www.bing.com/search?q=+Sustainable+Development+Goal+4&form=ANNTH1&refig=808e79fcfbf84b8e8227c205ba4dc180

University of Reading. (n.d.). *Quality assurance in research.* Academic and Governance Services. Retrieved May 01, 2023, from https://www.reading.ac.uk/academic-governance-services/contact-us

Velazquez, L. (2022). COVID-19 disruptions to SDG 4 in higher education institutions. In K. A. A. Gamage & N. Gunawardhana (Eds.), *The Wiley handbook of sustainability in higher education learning and teaching Willey Online Library.* Wiley-Blackwell. https://doi.org/10.1002/9781119852858.ch4

PART I

Curriculum and Teaching in Higher Education

Models of Quality Assurance: *Towards a Quality Assurance Framework of Digital Learning in Higher Education in Africa*

Erkkie Haipinge ⓘ *and Ngepathimo Kadhila* ⓘ

1 INTRODUCTION

Digital connective technologies' prominent positions in the twenty-first century are triggering major changes in many facets of life, signalling that we have entered a new era: the digital age. One of the most important spheres of existence is higher education (HE), which has been re-engineered to adapt to the shifting terrain of what it means to operate in this modern age (Saykili, 2019). Against this backdrop, higher education institutions (HEIs) are confronted with new challenges unique to the twenty-first century, like changing and diverse learner profiles, increased learner mobility, lifelong learning, and increased market-based competition with new tertiary education providers. Distance learning tools, sophisticated learning management systems (Temmerman, 2019), online social

---

E. Haipinge • N. Kadhila (✉)
University of Namibia, Windhoek, Namibia
e-mail: ehaipinge@unam.na; nkadhila@unam.na

© The Author(s), under exclusive license to Springer Nature
Switzerland AG 2024
P. Neema-Abooki (ed.), *The Sustainability of Higher Education in
Sub-Saharan Africa*, Sustainable Development Goals Series,
https://doi.org/10.1007/978-3-031-46242-9_2

networking tools, virtual and augmented reality, OER, and MOOCs are considered affordances brought about by digital connective technologies in the twenty-first century. Digital tools are seen as innovations that contribute to enabling equal educational opportunities for all, accessing quality educational content, and supporting lifelong learning (Abdous, 2009).

On the other hand, the very same innovations offered as solutions potentially prove to be further challenges for HEIs (Saykili, 2019). Particularly in the African context, lack of effective policies and planning, inadequate resource allocation, scarcity of qualified staff for instructional design and technical support, and quick and continual updating needs are the primary reasons for these innovations to function as additional potential challenges (Temmerman, 2019). HEIs are also hesitant to contemplate incorporating these technologies because of concerns about academic dishonesty, plagiarism, and cheating being commonplace examples. However, raising the quality of digital learning has become a necessity and of the utmost importance for all HEIs to lessen the impact the of COVID-19 pandemic and for sustainable digital learning after the post-COVID-19. Therefore, while acknowledging that digital technologies might successfully provide solutions to the challenges that confront HEIs in the twenty-first century, this chapter examines the possible quality challenges brought by the digitalisation of HE and insinuates possible solutions to these challenges. Particularly, the chapter focuses on quality assurance (QA) in digital HE in the African context. The chapter seeks to answer the following question: How is quality understood, and how is QA practised in the digital HE era?

2   Conceptual Framework

This chapter adopted the process-oriented lifecycle model for QA in e-learning as a conceptual framework underpinning the chapter. According to this model, drawing upon the materials described above, a process-oriented model aimed at helping organisations to implement a QA process structured around the foundational process of e-learning development and delivery is proposed. In this regard, it is noteworthy to point out that the design, production, and delivery of e-learning require both a streamlined workflow and the collaboration of several specialists (subject-matter, instructional, and technical) working together in a team environment (Abdous, 2009). Figure 1 (adapted from Abdous, 2009) depicts the process-oriented lifecycle model for QA in online learning.

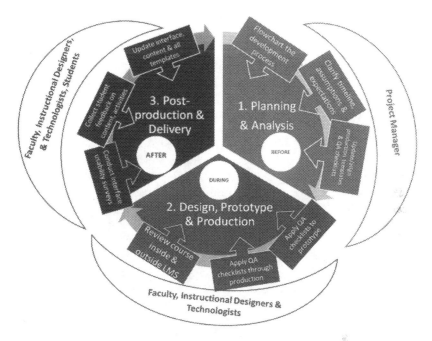

Fig. 1 Process-oriented lifecycle model for quality assurance in digital learning. Adapted from Abdous (2009: 288)

The model in Fig. 1 is structured around three sequential non-linear phases:

1. *Before:* planning and analysis
2. *During:* design, prototype and production; and
3. *After:* post-production and delivery.

In Fig. 1, a project plan and a workflow diagram are used as QA Tools in the planning and analysis phase to flowchart the development process and define the timeframe, assumptions, and expectations. This phase is a must for laying the groundwork for the proposed QA methodology, especially when it comes to refining and upgrading course development templates and checklists. Indeed, the quality standards that support content-gathering templates and production checklists are defined and established during this phase. It defines the constructs that underpin the proposed QA paradigm, in other words. Following that, pre-designed content-gathering templates are utilised during the design/production phase to guarantee that the content is relevant, complete, and consistent. The development of student-centred syllabi, the alignment of objectives with the content matrix, the use of engaging and diverse learning activities, the provision of opportunities for interaction and collaboration, and, of course, the opportunity for meaningful assessment and feedback are all included in these templates.

Team members utilise tailored QA checklists in this step to guarantee that the standards and guidelines identified in the first phase are followed. These checklists, which are used as a screening tool (Abdous, 2009), are built and adapted using evidence from research-based standards and best practices in instructional and web design. Lecturers, instructional designers, and instructional technologists can utilise these templates as self-assessment and improvement methods without sacrificing originality or independence. These templates and checklists are used throughout the process to maintain uniformity and efficiency, as well as to promote a quality culture.

3   CONCEPTUALISING DIGITAL LEARNING

Digital learning is learning facilitated by technology that gives students some element of control over time, place, path, and/or pace (Temmerman, 2019). It refers to a means of disseminating educational content over the Internet in its broadest sense. This can range from downloading content (iTunes university content, digital textbooks, and video or audio materials) to more structured online courses that include examinations and the awarding of a certificate. It is the foundation of many online education programmes and can be a more accessible method of learning for persons seeking a variety of educational options. However, despite popular belief, online and traditional classroom learning are not opposed. Digital learning should be viewed as a distinct teaching and learning method that can be used independently or in conjunction with classroom instruction. Similarly, digital learning does not imply that face-to-face instruction is replicated in an online setting (Abdous, 2009).

Online learning eliminates the time, place, path, and pace constraints that come with face-to-face instruction (Abdous, 2009). In terms of time, learning is no longer limited to the school day or the academic year as students may now learn at any time through the Internet. With regard to the place, learning is no longer limited to the physical classroom as students can now learn from anywhere and on any Internet-connected device. When it comes to the path, learning is no longer limited to the professor's pedagogy. Students can study in their style with interactive and adaptable software, making learning more personal and interesting (Pretorius, 2003). New learning technologies provide professors with real-time data that allows them to tailor lessons to each student's specific needs. In terms of pace, learning is no longer limited to the pace of a whole classroom of students. Students can learn at their speed with interactive and adaptive software, spending more or less time on lessons or subjects to acquire the same level of learning (Temmerman, 2019).

Tavares et al. (2017) espouse that digital learning is more than just providing students with a laptop as it requires a combination of technology, digital content, and instruction. In terms of technology, it serves as how learning materials are delivered, in the process improving the way students receive information. It comprises internet connectivity, software as well as hardware, which can range from a desktop, laptop, iPad, or smartphone. It is a tool but not an instruction. Digital content refers to high-quality academic learning material via technology. It is what students are taught. It is not just a text-only PDF or a PowerPoint presentation but includes everything from innovative interactive, adaptive software to classic literature, video lectures, and games. According to these triple authors, although technology may alter the function of the professor, it will never eliminate the necessity for one. Professors will be able to provide tailored guidance and help to students using digital learning to ensure that they learn and stay on pace to graduate from HEIs as planned. In this regard, a professor may serve as a facilitator for student learning rather than a sage on the platform.

4   Defining Quality in Higher Education

Quality is a disputed notion that means different things to different individuals, with no clear consensus among scholars on what quality is (Netshifhefhe et al., 2016). Even though the term "quality" is frequently employed in everyday life, Harvey and Green (1993) describe it as an elusive idea. Quality in HE, as Harvey and Green (1993) ascertain, is an ambiguous, multidimensional, multilevel, and dynamic notion that can incorporate the contexts of the educational system in which it is applied. Although everyone believes they recognise or understand what quality means to them, there is no singular or objective definition of quality. Consequently, the phrase "*I will know it when I see it*" was coined. And so, the concept of quality is frequently subjectively associated with particular concepts and expectations of what constitutes good performance (Van der Bank & Popoola, 2014).

Quality can be classified into exceptional or excellent, perfection, consistency or zero defect, transformation, value for money, and fitness of purpose (Harvey and Green (1993)):

- *Quality as exceptional or excellence* represents a traditional view of quality and is associated with the concept of providing a product or service that is distinctive, special, meets a minimum set of standards, and confers status on the owner or user.
- *Quality as perfection, consistency, or zero defects* sees quality in terms of a consistent or flawless outcome. This approach has its origin in the issue of quality control in the manufacturing industry where quality is referred to as "zero errors or defects".
- *Quality as fitness for purpose* relates quality to the purpose of a service or product and leads to the fulfilment of the needs and expectation of the customers. Quality is, thus, judged in terms of the extent to which a product or service meets its intended purpose with the customer often specifying the requirements.
- *Quality as fitness of purpose refers* to the ability of an HEI or programme of study to respond to the country's national imperatives like transformation, equity, access, social justice, emancipation, etc.
- *Quality as transformation* perceives quality as a "qualitative change" or a fundamental change of form. Thus, in educational terms, transformation refers to the enhancement and empowerment of students or the development of knowledge and skills with education being about doing something to the student as opposed to doing something for the consumer.
- *Quality as value for money* in the HE context is related to the accountability of HEIs for public expenditure. Thus, this notion perceives quality in terms of a return on investment through efficiency and effectiveness (Harvey & Green, 1993).

As a relative, multidimensional concept, quality, therefore, means different things to different people. Determining the criteria for assessing quality requires that the context and needs of various stakeholders are considered (Harvey & Green, 1993). Also required is the careful interrogation of national QA agencies and HEIs around the purposes of HE. Accordingly, D'Andrea and Gosling (2005) are of the view that it is futile to try to formulate a definitive definition of quality. Thus, quality and its improvement in HE need to be approached in dynamic rather than in static or absolute terms.

5 Defining Quality in the Context of Digital Learning

Tavares et al. (2017) say that the concept of digital learning quality is as complicated as digital learning itself. There is a large body of literature on HE quality that uses a plethora of words and concepts in its reference. It frequently identifies a conflict between two aspects of QA: accountability and quality improvement. Another hot topic is the role of the student in determining quality. Some claim that defining digital learning quality in HE should start with the assumption that digital training is a process of co-production between the online learning environment and the student, with the student perspective serving as the starting point for quality development in all areas of online learning. These tensions become more pressing when new modes of delivery increasingly become a part of traditional campus-based HE, and institutions attempt to cope with these wholly new types of courses using the same procedures (Simunich et al., 2021).

The effectiveness of QA projects in online education is contingent on the successful implementation of the processes, resources, and tools designed to ensure quality (Simunich et al., 2021). However, there are limited studies on the elements that best facilitate the adoption of digital learning in HEIs. So, what defines quality in online learning? Several distinct quality standards or benchmarks have been developed and tested in various situations around the world. Despite the differences in vocabulary and emphasis, a good online learning experience may be identified to consider a set of common features. According to Simunich et al. (2021), these features, among others, include:

> Institutional support (vision, planning, & infrastructure), Course development, Teaching and learning (instruction), Course structure, Student support, Faculty support, Technology, Evaluation, Student assessment, and Examination security.

It is indisputable that HEIs have primary responsibility for the quality of their provision of education and its assurance. Therefore, these quality dimensions may provide HEIs offering digital education with an overarching framework for the assurance and improvement of the quality of digital learning. For this chapter, quality in the context of digital learning is defined as the use of digital applications guided by digital pedagogical methods to support students' development of knowledge and skills while enhancing competencies they will require in the digital age.

6   QUALITY ASSURANCE OF DIGITAL LEARNING

QA in HE entails an "all-embracing term referring to an ongoing, continuous process of evaluating (assessing, monitoring, guaranteeing, maintaining, and improving) the quality of a HE system, institutions, or programmes" (Vlăsceanu et al., 2007, p. 20). It is the systematic internal and external management procedures and mechanisms by which an HEI assures its stakeholders of both its quality and its ability to manage the maintenance and enhancement of such quality (Luckett, 2007). HE has become more accessible to students because of digital learning opportunities, which have gone a long way towards achieving the objective of "education for all". Students can learn from (nearly) anywhere, at their own pace, and at times that suit them. It also offers a fantastic, yet often missed, opportunity to connect with fellow students from other countries and cultures (Temmerman, 2019). Digital learning necessitates student autonomy, self-direction, and good time management, among other things (Cole, 2019). Therefore, digital learning may not suit everyone whereby some students, especially those lacking self-direction, are better suited to face-to-face instruction.

Temmerman (2019) elucidates that some students require social and physical connection with their peers as well as with their professors. The professor and student are physically separated in a digital learning environment, and how this is handled determines whether online learning succeeds or fails. In the digital learning environment, there are numerous stakeholders. The institutions that provide digital education, the professors who teach the courses, the students enrolled in digital education, the parents who pay their children's tuition, potential employers of online course graduates, the ministry or government, and the larger society are all included.

The Commonwealth of Learning (2009) perceive that the digital teaching-learning environment has benefited greatly from technological advancements. It has altered the way we teach and learn, and it has opened up the world of learning and opportunities to individuals who would not otherwise have had it. Hence, governments, institutions, scholars, and students must all commit to and support digital education for it to be successful. The provision of high-quality education is a must and needs well-funded institutions, qualified and motivated staff, effective and ongoing QA processes, and supportive leadership. Therefore, Temmerman (2019) advises that institutions that deliver online courses should have well-defined QA mechanisms in place for both staff and students and ensure that they are followed. The staff who create and offer digital courses must be suitably qualified and professionally supported. Adequate funding and investment in relevant technology must also be provided (Tømte et al., 2019).

7 Benefits of Digital Learning

New technologies and methods are used in digital learning to help students learn in new ways. Powered by digital technologies, digital learning is quickly overtaking the more traditional method of "face-to-face" learning. The usage of tablets or iPads in the classroom instead of traditional notebooks and paper can be a simple way to incorporate digital learning. It can also entail the use of sophisticated digital solutions to enable students to access educational content that was previously unavailable to them. The possibilities for using digital products and services for educational purposes are endless, and it has some tremendous advantages for students.

Digital learning may offer the following benefits (Temmerman, 2019):

- *Personalised learning:* One of the most significant advantages of digital learning is that it allows teachers or course providers to personalise their learning programmes or curriculum for each student. They can consider the student's abilities and how they are progressing, then adjust learning accounting for obstacles or rapid growth.
- *More engaging lessons:* Learning can be delivered in a variety of ways with digital learning. Imagery, audio, and video can all be seamlessly incorporated into a lesson. At the same time, learning activities can be embedded in online courses.

- *Learning flexibility:* Digital learning allows for flexible learning by enabling supplementing of synchronous classroom teaching, which strictly requires students and educators to be at the same venue at the same time with asynchronous delivery that enables self-paced learning. Blended pedagogical approaches—hybrid learning, and Flipped Classrooms—also enable flexibility for the student. The ability for students to access learning through a variety of digital devices, including mobile devices, adds to the flexibility of digital learning.

Opined by Bertrand (2010), students who use digital learning tools are more engaged in the process and more interested in expanding their knowledge base. They may not even realise they are learning because they are using engaging methods like peer education, teamwork, problem-solving, reverse teaching, concept maps, gamification, staging, role-playing, and storytelling. As for Temmerman (2019), digital learning provides better learning context, a larger sense of perspective, and more interesting activities than traditional education techniques since it is significantly more interactive and memorable than huge textbooks or one-dimensional lectures. Digital learning enables students to connect better with learning material, thereby providing them with a more engaging way to absorb information. Considering other factors commensurate with the increase in student learning retention rates and improvement in their motivation and accountability when they can track their development. Rationalised is digital learning benefits students in HE.

8 DIGITAL LEARNING FOR SUSTAINABLE DEVELOPMENT

Sustainable development is defined as "development that meets the needs of the present without compromising the ability of future generations to meet their own needs" (Verma, 2019, p. 1). United Nations (2023) asserts that sustainable development requires three key pillars that support development to operate in harmony, namely economic growth, social inclusion, and environmental protection. Although all seventeen sustainable development goals (SDGs) of the 2030 Agenda for Sustainable Development are imperative to the achievement of sustainable development, three goals are particularly relevant to digital learning and QA in education. These are Goal 4, Quality Education; Goal 5, Gender Equality; and Goal 9, Industry, Innovation, and Infrastructure, and they are discussed further next.

8.1 SDG 4: Quality Education

Sustainable Development Goal 4 is directly linked to the education sector whose call on education is to ensure inclusive and equitable quality education and promote lifelong learning opportunities for all people (SDSN Australia/Pacific, 2017, p. 5). Digital learning directly impacts SDG 4 by enhancing access to quality education through the breaking down of barriers like removing the need for students to reside in localities where HEIs are found, and by enabling students to learn through online and distance modes that are flexible and inclusive. Secondly, digital learning has a positive impact on students of multi-literacies such as digital literacy and media competencies that they need to become successful self-directed students and, later on, effective professionals. As such, digital learning is also an enabler for the implementation of SDGs, in that, as future implementers, students have a better chance of developing "cross-cutting skills whose examples include systems thinking, critical thinking, self-awareness, integrated problem-solving, and anticipatory, normative, strategic and collaboration competencies" (SDSN Australia/Pacific, 2017, p. 12).

8.2 Goal 5: Gender Equality

Goal 5 focuses on achieving gender equality and empowering all women and girls (SDSN Australia/Pacific, 2017, p. 5). Gender inequality is significant to the achievement of sustainable development. Around the world, "women are more likely to be poor, unemployed, and doing unpaid work" (Clark, 2017). Most of these conditions are the result of unequal access to quality education, and digital learning provides enhanced opportunities for this access. Educational programmes that are offered through flexible delivery modes such as distance or online tend to benefit those that suffer from societal exclusions, including exclusion from education. Due to the flexibility and far-reaching capabilities of digital-enabled learning, women who may be juggling other roles as mothers or who may have missed educational opportunities earlier in their lives benefit from digital learning.

8.3 Goal 9: Industry, Innovation, and Infrastructure

Sustainable Development Goal 9 refers to building resilient infrastructure, promoting inclusive and sustainable industrialisation and fostering innovation (SDSN Australia/Pacific, 2017, p. 5). Many economies in the world, the Namibian economy and those of other SADC countries included, are transitioning towards knowledge-based and industrial economies. These economies require skills that are different from those in demand for economies based on the primary sector or service industries. The fourth industrial revolution which is characteristic of knowledge-based economies requires skills: creativity and innovation, critical thinking, collaboration, and digital skills. Digital learning, apart from serving as the vehicle through which learning is delivered, also helps to cultivate students' soft and hard skills, including digital competencies, collaboration, and media literacies that are required for participation in new industries and in contributing to innovation.

Sustainable development seeks to drive and achieve balanced development that does not result in the progress of one sector (e.g., economy) at the expense of another including social exclusion or environmental damage. Light of the three SDGs discussed above can contribute to the framework and inform perspectives when the QA of digital learning is being considered and implemented. Quality digital learning should be that which enhances the quality of learning and improves educational inclusion of those excluded in particular women, marginalised communities, and displaced people, and it should cultivate twenty-first-century skills required in knowledge-based economies and the fourth industrial revolution. Therefore, sustainable development can be addressed by digital learning by serving as one of the criteria for assuring the quality of education that is delivered digitally.

9   Challenges for Quality Assurance in Digital Learning

Apart from the benefits discussed earlier, digital learning also brings with it challenges that may be linked to the student or their learning environment. Therefore, HEIs must ensure that digital learning solutions adhere to best practices in terms of both development methodologies and delivery techniques. In the perception of Geoffrey (2014), if institutions must ensure quality learning, the following are some of the important questions that HEIs should think about when adopting digital learning:

- How learning materials are currently designed and developed
- What quality assessment processes are in place

- Presence of formal quality procedures
- Guidelines and methodologies in place for designing and developing learning
- Policies and guidelines for accessibility and usability
- Maintenance and updating of learning materials

Assessment is the ongoing process of establishing clear, measurable expected student learning outcomes; ensuring that students have enough opportunities to achieve those outcomes; systematically gathering, analysing, and interpreting evidence to determine how well students' learning matches the expectations; and using the resulting information to understand and improve student learning (Suskie, 2009). However, the quality of digital learning experiences is difficult to assess since technology often supports just a portion of the learning activities that students engage in (Bertrand, 2010). As a result, evaluating the role of technology in blended and online learning experiences necessitates well-researched and constructed approaches that are sufficiently sensitive to recognise and acknowledge the technology's related nature to learning quality. There are various difficulties, according to Geoffrey (2014), institutions need to focus on as far as QA is concerned. These are presented in Fig. 1.

To adequately assess and ensure the quality of learning in digital learning, institutions of higher learning must focus on numerous components of the learning process, as shown in Fig. 2. While the necessity for evaluation is obvious, as Suskie (2009) points out, there is a gap in implementation between the expected outcomes and how institutions should get there. This gap makes it difficult for African HEIs to design an effective evaluation plan that can produce useful data about the students, course, programme, and institution. The gap's length and width may differ from one institution to the next.

Other challenges may include a *lack of digital skills:* by default, digital learning requires both learners and educators to have a certain degree of digital skills for them to navigate comfortably and effectively using digital platforms. If digital skills are lacking, this leads to the inability to take advantage of opportunities offered by digital learning. However, digital skills alone are not the only issue because digital learning requires specialised skills. *Digital divide:* access to the Internet is another challenge due to inadequacy or unavailability of quality and affordable Internet connection. The cost of the Internet and data is particularly high in developing countries and in Sub-Saharan Africa in particular. These costs are

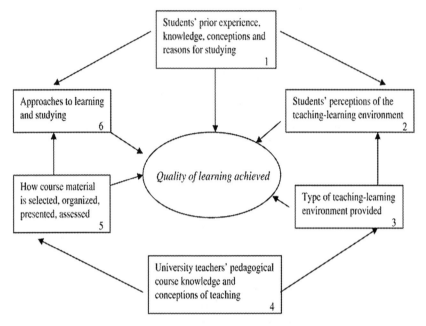

Fig. 2 Concepts related to the quality of digital learning (Source: Geoffrey, 2014, p. 92)

prohibitive towards the use of Internet services in general, reducing the adoption of digital learning in the process. Gaps in educational opportunities between the haves and have-nots widened during the COVID-19 pandemic. Students with Internet access, computers, and supportive families fared better than those who do not (Simunich et al., 2021).

10　A Case from the University of Namibia

Imperative to first and foremost state that the University of Namibia (UNAM) is the oldest in Namibia. It was established in 1992 through an Act of Parliament (Act 18 of 1992). The core business of UNAM is teaching, research, innovation, and community service. In addition, UNAM serves multiple citizens worldwide through varied activities by building links with industry, employers, schools, and government agencies. UNAM comprises (12) campuses across the country, with an enrolment of more than 30,000 students and a staff complement of about 2200, making it the biggest HEI in the country. UNAM offers an array of internationally benchmarked programmes that aim to meet the contemporary and future needs of society and industry. It aims to create industry leaders and entrepreneurs.

The University offers its academic programmes through four faculties, namely Faculty of Agriculture, Engineering and Natural Sciences; Faculty of Commerce, Management and Law; Faculty of Education and Human Science; and Faculty of Health Sciences and Veterinary Medicine.

To gauge the climate of QA at the UNAM, the authors engaged with fellow academics. One was an academic leader in the faculty and two members of one Department. The academic leader, who is a member of the faculty's QA committee, showed advanced knowledge about QA processes and practises at the macro level. She also established experience in conducting programme reviews at peer institutions under the auspices of the National Council for Higher Education (NCHE). The lecturers' views on QA were more specific to teaching and learning, naturally because that is their level of experience and practice.

Gathered from the academic leader was that national external QA agencies like the NCHE are robust and objective in their accreditation processes. However, she expressed dissatisfaction with the culture of QA at the University itself, which has sometimes resulted in undesirable consequences thus a few programmes that missed accreditation and even stopped running. She mentioned that the areas that need improvement include ensuring that programmes seek accreditation for all delivery modes apart from the traditional face-to-face mode (e.g., distance and online).

The other issue the academic leader pointed out was the attitude of institutional leaders towards the accreditation processes is viewed with suspicion and almost with an expectation that each review process should result in positive results—that is, accreditation of programmes. Institutions tend to over-sell themselves and exaggerate the reality to paint a favourable image for reviewers to enhance their chances of passing accreditation assessments. Another challenge she mentioned was that due to the small size of the country's HEIs landscape, there is a small pool of experts from which accreditation authorities can draw. This results in institutions being reviewed by institutional peers, resulting in feelings of distrust, especially when programmes being reviewed are in direct competition with those at the home institutions of reviewers. In some cases, it breeds hostilities.

Lecturer X who is based at one of the UNAM Campuses sees QA in teaching and learning as measures put in place to ensure that teaching activities by lecturers are being carried out as expected. Therefore, the lecturer is comfortable with the current monitoring mechanisms on the quality of teaching and learning; through student evaluation of lecturers. This, he says, would help to check on lecturers' preparedness, ability to engage with students, the quality of learning materials, and alignment of teaching and assessment to the learning outcomes. The sentiments of lecturer X reflect the constructive alignment idea of John Biggs (2003, p. 1) whereby teaching should involve setting up a "learning environment that supports learning activities appropriate to achieving desired learning outcomes" so that the "teaching methods used, and the assessment tasks are aligned to learning activities".

However, lecturer X was concerned about the current practice of lecturer evaluation in that it may not serve the purpose of improving teaching and learning due to the time it is administered, the methods used and the validity of the evaluation instruments. For example, the fact that these evaluations are carried out towards the end of the semester makes it impossible for lecturers to use the feedback to improve their teaching in the current semester. This timing may also put the reliability of the data in question as students may be emotionally driven in their feedback relative to the status of the grades they have attained in the course in question. He recommends that the question items on the evaluation instruments should be moderated by experts in fields of industrial, behavioural, and educational psychology, while such lecturer evaluations should be carried out several times a semester and results or reports be shared with lecturers so that it can fulfil the quality enhancement role. This is to say that, instead of lecturer evaluation assuming the retrospective QA approach that "looks back to what has already been done and makes a summative judgement against external standards" (Biggs, 2001, p. 222), lecturer evaluation should rather apply prospective QA, thereby being "concerned with assuring that teaching and learning does now, and will continue to fit the purpose of the institution".

The second lecturer, Lecturer Y, teaches at another UNAM campus. Lecturer Y expressed concern about the state of quality in teaching and learning at his campus, indicating that

> Everybody complains about this at our campus. Student involvement in class discussion is virtually none, assignments are getting more and more

plagiarised, and cheating during tests and exams. Attendance of lessons not taken seriously, and then demand (by students) for block teaching or remedial lessons.

Lecturer Y voiced that he is not very content with the QA Department in regard to certaining that quality is assured. He claims that the QA representative is not visible on campus, and he expects such a person to engage with lecturers and even observe lectures so that recommendations for improvement can be devised. It is apparent from Lecturer Y that he sees QA as the responsibility of the QA unit, the Centre for Quality Assurance and Management (CEQUAM). The concern here is that, although the University's QA Policy clearly states that QA is everyone's responsibility at various levels, this appears not to be the case on the ground. This "lack of policy ownership by academic staff has serious implications in terms of policy implementation" (Mhlanga, 2008, p. 314).

10.1 Monitoring of Quality in Digital Learning Using Technology

With the continued rollout of e-learning at UNAM, particularly when all learning moved online due to the COVID-19 pandemic, concerns were raised about how the quality of learning and teaching will be assured online. There were also concerns from both University leadership, national QA and accreditation bodies, parents, and other stakeholders who wanted to be assured that the quality of teaching and learning taking place online was comparable to that of the familiar traditional face-to-face teaching. To address these concerns, the authors led initiatives and interventions that sought to enhance and assure quality in online and blended learning. These were the development of:

1. Open, Distance and eLearning (ODeL) Policy
2. Quality Standards for Online and Blended Course Design, and
3. Procedure for Monitoring of Quality of Learning, Teaching and Assessment

The ODeL Policy of UNAM's main goals is, amongst others, to:

- *Ensure that programmes developed and implemented by Faculties to be delivered via ODL and Online modes comply with ODeL requirements.*

- *Provide enhanced and equitable access to academic programmes through ODeL best practices and consider the use of MOOCs where appropriate.*
- *Strive for quality, relevance, and responsive programmes that are supported through ODeL practices.*
- *Ensure that all social groups, including the marginalised and people with disabilities, are provided with the same opportunity to learn.*
- *Provide for the development of quality distance and online study materials.*
- *Provide guidelines for the preparation and conversion of programmes and courses from face-to-face learning to online, blended, or distance modes*

As per the statements above, the quality aspects addressed by this ODeL Policy focus on compliance with set standards, and the enhancement of equitable access to quality learning and teaching using technology. Recognising the challenges faced by academics in moving from traditional to online teaching, the Policy also provides guidelines on how this transition can be done informed by established best practices.

The Quality Standards for Online and Blended Course Design in essence seek to ensure that the design of courses online matches the expected standards of effective digital teaching and learning practices. The idea is to not only provide a framework for assessing the quality of courses but also to guide lecturers and students on how to continually enhance the quality of their courses and to evaluate the quality of courses respectively. Without standards set, it would be difficult to attain consistency in quality across courses and programmes, as each lecturer may end up doing their own thing.

Finally, the *Procedure for Monitoring of Quality of Learning, Teaching and Assessment* was born out of necessity to involve the QA structures in the process of the assurance and enhancement of quality in online courses. Figure 3 outlines the roles played by all stakeholders involved in monitoring the quality of learning, teaching, and assessment using the Moodle learning management system. These include the e-learning and learning design service providers who reside in my office, the lecturers, heads of departments, deans, and the QA Centre itself. The procedure provides a structure that in the long run hopefully results in a quality culture.

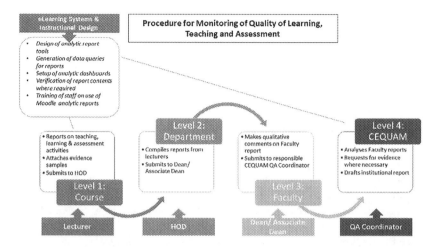

Fig. 3 Monitoring quality of digital teaching, learning, and assessment on Moodle LMS. Source: University of Namibia (2021)

10.2 Monitoring of Quality in Digital Assessment Using Technology

Assessment is key in enhancing the overall quality of teaching and learning in HE. The security of academic standards is central to HE and the key to securing standards is the integrity of the assessment. Therefore, secure assessment practices provide assured stakeholders, including employers and professional bodies, who have achieved the expected academic standards. Types of academic dishonesty include:

- *Contract cheating* can involve ghost-writers that are family, friends, or other students.
- *Essay mills* involve commercial entities that use sophisticated marketing techniques to target students to perform academic work on their behalf at a cost:

Many of these companies have sought to capitalise on the concerns and uncertainties faced by students as a consequence of COVID-19—often using social media to advertise discounts to students for essay writing services, suggesting they could fill a gap caused by students receiving a lack of supervision.

Technology can support the detection of academic dishonesty in digital assessment. Therefore, HEIs in Africa are encouraged to use artificial intelligence to deter cheating in digital assessments. For example, text matching/plagiarism software. Turnitin, for instance, can be used to identify copied text. Learning analytics can be used to analyse student and lecturer learning and assessment activities in the Learning Management System. Remote invigilation applications (e.g., webcams or facial recognition software) can also be employed, and some HEIs go as far as installing this software on student computers when students are taking remote assessments.

UNAM has developed guidelines and regulations to support the effort of QA and enhancement in online assessment—assessment with the use of networked and Internet-based digital technologies. Two guidelines, namely the *Guide on Enhancing Reliability and Diversity of Online Assessment* and the *Measures to Academic Dishonesty in Online Assessments*, were developed. The *Guide on Enhancing Reliability and Diversity of Online Assessment* sought to particularly address the concern of the University leadership and other stakeholders involved regarding the reliability of assessment when using technology. To equip lecturers with a set of skills on how to enhance reliability, the focus was on methods to limit chances of cheating, the fact being that students took their exams remotely with minimal to no supervision.

UNAM uses various ways and measures to curb academic dishonesty in digital assessments. These include the types of assessment questions, the creation of question banks, the randomisation of test questions, etc. It was revealed that technology itself, even the use of other solutions like proctoring tools that act as surveillance machines to monitor students' online activities, does not promote honesty in the long run. Students comply out of fear of getting caught, but not necessarily because they believe it to be an ethically wrong thing to do (cheating that is). There is therefore a need to equally pay attention to changing behaviour, apart from using technology to make dishonest behaviour hard to do. This chapter propounds how HEIs in Africa can prevent student academic dishonesty as follows:

- Ensuring students are prepared and that they have developed the skills necessary to succeed in their assessments like academic writing and referencing skills
- Discussing academic misconduct with students so they are clear about what is prohibited and why

- Ensuring students understand the long-term and ethical benefits of completing their work;
- Discussing the risks of engaging with contract cheating services; and
- Ensuring students are aware of the institutional processes and potential consequences if they are caught cheating.

When students are well-prepared; have skills necessary to succeed in their assessments, serving as academic writing and referencing skills; and are aware of the risks of committing academic dishonesty, they will have no reason to engage in those activities.

11   Lessons for Africa

Per process-oriented lifecycle model for QA in digital learning (see Fig. 1), African HEIs wishing to improve the quality of their digital learning environment could learn some valuable lessons from this model as follows:

1. Clarifying quality expectations during the *planning phase* aids in determining the QA implementation approach. It's important to present a complete picture of the overall quality criteria, expectations, and procedure. Three production issues must be kept in mind in this regard:
   - Gain academic buy-in by describing the entire content-collecting process and emphasising the value of each content-gathering template. Academic staff resistance is frequently caused by a lack of comprehension of the process' objective and a refusal to accept new approaches to designing courses.
   - Ascertain that the academic staff is aware of the meanings of the various quality checklist items.
   - Assist the production team members in developing a shared understanding of the checklist items so that they can be applied consistently. Quality is determined by how academics use and experience it in practice.
2. Adding more quality checklists to the production team throughout the *design/production process* is likely to be unproductive unless roles and responsibilities are well-defined and understood. In addition, an information system that tracks and facilitates clerical chores must be used to support QA implementation. The effectiveness of the QA model implementation depends on having a flexible and efficient information system.
3. A double consideration is necessary during the *delivery phase*:

- Academic staff preparation and online teaching talents have a substantial impact on how e-learning content is delivered. As a result, for a good e-learning experience, both development opportunities and continuing technical support are essential.
- On the student side, their level of interaction with the content is influenced by their preparation, technical literacy, study plans, and tactics. The author avows that the outcome of any learning process is dependent on student enablement and empowerment, which may be achieved by offering online orientations, ongoing support, and regular and systematic feedback gathering.

With these guidelines in mind, it is pivotal to emphasise that the model proposed by Abdous (2009) is a sort of blueprint that aids HEIs in implementing effective and systemic QA methods in the digital learning environment. However, essential enabling variables like the clarity of quality criteria, common knowledge of QA checklists, and support from both professors and students are prerequisites to its success. Particularly, raising the quality of digital learning has become a necessary and top priority for all universities to mitigate the impact of the COVID-19 pandemic and ensure long-term digital learning post-COVID-19. Therefore, this chapter proposes a Quality Assurance Framework of Digital Learning comprising QA indicators for digital learning, for possible use in African HEIs as follows:

Table 1 depicts nine dimensions of QA, namely, educational philosophy which requires HEIs to educational philosophy to their vision and mission; programme learning outcomes that must be aligned to the needs of stakeholder and graduate employability attributes; sound and effective digital learning methods; conducive digital learning environment; wholistic and authentic digital assessment; appropriate, adequate, and accessible technology; appropriate and adequate infrastructure; and support for staff and students alike.

This chapter advances that lecturers need to be well-prepared and trained for being able to cater for students with diverse backgrounds, expectations, and needs. Therefore, all staff teaching in HEIs should receive training in relevant digital technologies and pedagogies as part of initial training and continuous professional development. To be effective, digital professors need to develop skills and knowledge specific to digital educational systems. Particularly, professors need to be able to prove proficiency in (Saykili, 2019):

Table 1 Quality assurance framework of digital learning in African higher education

QA indicators for digital learning

Educational philosophy	• Alignment of educational philosophy to context, university's vision and mission
	• Translation of educational philosophy to curriculum
Programme and learning outcomes	• Achievement of programme and learning outcomes
	• Employment and employability
	• Future-ready skills
	• Professional mobility
	• Contribution to society and community
	• Constructive alignment of curriculum
	• QA and assessment of online education and study programmes
Learning design/learning methods	• Effectiveness of individuals, groups, and communities learning
	• Quality of learning and teaching
	• Constructive alignment of learning to learn outcomes
Learning environment	• Positive student experience
	• Student engagement and motivation
	• Device time
	• Global experience
Learning assessment	• Holistic and authentic assessment
	• Academic integrity
	• Marking scheme and rubrics
	• Constructive alignment of assessment to learning outcomes
Technology	• Adoption and penetration of technology
	• Blended synchronous and asynchronous learning
	• Uses of technology
	• User technology experience
Infrastructure	• Accessibility, connectivity and reliability of networks, digital resources, and tools
	• Quality and availability of digital resources and tools
	• Privacy and security of networks and data
Support services for lecturers	• Pedagogy and technology training and support
	• Quality of support services
	• Percentage of courses online

(*continued*)

Table 1 (continued)

QA indicators for digital learning

Support services for students	• Study skills assistance • Online educational counselling • Ongoing programme advising • Access for students with disabilities • Quality of feedback and guidance • Quality of support services • Course completion rates

Source: Adapted from Bin (2020)

- Accessibility standards for technology in the classroom and digital teaching are completely understood.
- Possess the ability to assess Internet resources.
- Understanding the many features of copyright privileges and infringements.
- Proper etiquette and commonly acceptable online use standards are modelled and monitored.
- Making appropriate lesson plans for online students and putting them into action.
- Ability to troubleshoot minor technical issues and, when necessary, resort to technical support professionals.

12   Conclusion

The concept of quality in teaching, learning, and assessment in HE is a slippery and multifaceted term. Different stakeholders perceive the quality differently, and the concept has also changed its essence over the years. Conceptualising and ensuring quality is more challenging in digital learning than in a conventional face-to-face learning environment. Digital education has experienced tremendous growth in the last decade. In response to rising demand for HEIs and severe shortages of academic staff and infrastructure in an increasing number of HEIs, some campus-based face-to-face programmes are being delivered using blended learning approaches rather than traditional learning methods. HEIs around the world are expanding their digital learning programme offerings, accompanied by increased scrutiny of quality and accountability. QA systems for assessing the quality of traditional face-to-face programmes and digital learning programmes are seen as two different entities.

A recurrent debate in HE circles is whether digital learning requires a different QA than face-to-face learning. It is doubtful that the philosophy, principles, and standards customarily applied in evaluating and accrediting face-to-face programmes can be used without major adjustments for assessing the quality and effectiveness of digital learning programmes (The World Bank, 2002). As a result, appropriate and reliable QA systems are required to convince the public that digitally delivered programmes satisfy acceptable academic and professional standards comparable to those delivered face-to-face. This will help in the creation of stakeholder trust and confidence in the quality of programmes received through digital education, which currently exists. This study proposed a QA Framework of Digital Learning comprising QA indicators for digital learning, for possible use in African HEIs to improve the current practice.

REFERENCES

Abdous, M. (2009). E-learning quality assurance: A process-oriented lifecycle model. *Quality Assurance in Education, 17*(3), 281–295.

Bertrand, W. E. (2010). Higher education and technology transfer: The effects of techno sclerosis on development. *Journal of International Affairs, 64*(1), 101–119.

Biggs, J. (2001). The reflective institution: Assuring and enhancing the quality of teaching and learning. *Higher Education, 41*(3), 221–238.

Biggs, J. (2003). Aligning teaching and assessing to course objectives. *Teaching and Learning in Higher Education: New Trends and Innovations, 2,* 13–17.

Bin, J. O. C. (2020). *Quality assurance of digital learning in the disruptive era, education quality international.* Accessed Mar 14, 2023, from https://johnsonongcheebin.blogspot.com/2020/12/quality-assurance-of-digital-learning.html?m=1

Clark, H. (2017). What will it take to achieve the sustainable development goals? *Journal of International Affairs, 53*–59.

Cole, N. L. (2019) *How sociologists define human agency.* Accessed May 29, 2022, from https://www.thoughtco.com/agency-definition-3026036

Commonwealth of Learning (COL). (2009). *The quality assurance toolkit for distance higher institutions and programs.* COL.

D'Andrea, V., & Gosling, D. (2005). *Improving teaching and learning in higher education: A whole institutional approach.* Society for Research into Higher Education/Open University Press/Bell and Brain.

Geoffrey, N. M. (2014). Challenges of implementing quality assurance systems in blended learning in Uganda: The need for an assessment framework. *Huria Journal, 18*, 87–99.

Harvey, L., & Green, D. (1993). Defining quality. *Assessment & Evaluation in Higher Education, 18*(1), 9–34.

Luckett, K. (2007). The introduction of external QA in South African HEIs: An analysis of stakeholder response. *Quality in Higher Education, 13*(2), 97–116. https://doi.org/10.1080/13538320701629129

Mhlanga, E. (2008) *Quality assurance in higher education in Southern Africa: The case of the Universities of the Witwatersrand, Zimbabwe and Botswana*, University of the Witwatersrand. Accessed Mar 29, 2022, from http://wiredspace.wits.ac.za/bitstream/handle/10539/7599/PhD%20Thesis%20Quality%20Assurance%20in%20Higher%20Education.pdf?sequence=1

Netshifhefhe, L., Nobongoza, V., & Mamposa, C. (2016). Quality assurance in teaching and learning processes in higher education: A critical appraisal. *Journal of Communication, 7*(1), 65–78.

Pretorius, R. (2003). Quality enhancement in higher education in South Africa: Why a paradigm shift is necessary: Perspectives on higher education. *South African Journal of Higher Education, 17*(3), 129–136.

Saykili, A. (2019). Higher education in the digital age: The impact of digital connective technologies. *Journal of Educational Technology and Online Learning, 2*(1), 1–15.

SDSN Australia/Pacific. (2017) *Getting started with the SDGs in Universities: A Guide for Universities, higher education institutions, and the academic sector. Australia, New Zealand and Pacific Edition*. Sustainable Development Solutions Network.

Simunich, B., McMahon, E. A., Hopf, L., Altman, B. W., & Zimmerman, W. A. (2021). Creating a culture of online quality: The people, policies, and processes that facilitate institutional change for online course quality assurance. *American Journal of Distance Education, 36*, 36. https://doi.org/10.1080/08923647.2021.2010021

Suskie, L. (2009). *Assessing student learning: A common sense guide*. Jossey-Bass.

Tavares, O., Videira, C. S. P., & Amaral, A. (2017). Academics' perceptions of the impact of internal quality assurance on teaching and learning. *Assessment & Evaluation in Higher Education, 42*(8), 1293–1305.

Temmerman, N. (2019) *The quality of online higher education must be assured*. World University News. Accessed Feb 28, 2022, from https://www.university-worldnews.com/post.php?story=20190917120217325

Tømte, C. E., Fossland, T., Aamodt, P. O., & Degn, L. (2019). Digitalisation in higher education: Mapping institutional approaches for teaching and learning. *Quality in Higher Education, 25*(1), 98–114.

Van Der Bank, C. M., & Popoola, B. A. (2014). A theoretical framework of total quality assurance in a university of technology. *Academic Journal of Interdisciplinary Studies, 3*(4), 401–408.

Verma, A. K. (2019). Sustainable development and environmental ethics. *International Journal on Environmental Sciences, 10*(1), 1–5.

Vlăsceanu, L., Grünberg, L., & Pârlea, D. (2007). *Quality assurance and accreditation: A glossary of basic terms and definitions.* UNESCO-CEPES.

United Nations. (2023). *The sustainable development Agenda.* Accessed Feb 26, 2023, from https://www.un.org/sustainabledevelopment/development-agenda/

World Bank. (2002). *Constructing knowledge societies: New challenges for tertiary education.* World Bank.

Monitoring and Evaluation in Higher Education: *Quality Perspectives in Africa*

David Katende 🆔

1 INTRODUCTION

Quality in higher education (HE) has been promoted for several reasons (Brown, 2017), from accountability to providing information to students in their decision-making process for application and admission to HEIs. This is owing to the phenomenon that quality assurance (QA) has become a generic term for external quality monitoring and accreditation (Harvey, 2020). The perceived need for external QA reflects a demand for accountability globally by stakeholders (Kinser, 2014). Elken and Stensaker (2018) add that quality work takes a practice-oriented approach and focuses on the formal and informal processes of different actors that continuously shape the daily practice to enhance quality in HE.

Quality of teaching and learning has become a strategic issue in HEIs across the globe over the past decades (Harvey & James, 2010; Enders & Westerheijden, 2014). Under the new public management paradigm, comparison of educational outcomes, rankings, and a higher degree of

---

D. Katende (✉)
Mountains of the Moon University, Fort Portal, Uganda
e-mail: davkatende@gmail.com

© The Author(s), under exclusive license to Springer Nature Switzerland AG 2024
P. Neema-Abooki (ed.), *The Sustainability of Higher Education in Sub-Saharan Africa*, Sustainable Development Goals Series,
https://doi.org/10.1007/978-3-031-46242-9_3

43

university autonomy and accountability have become an integral part of university managers' day-to-day work (Broucker & Kurt de, 2015; van Vught & de Boer, 2015). Comparability of individual universities' provisions have become a core part of the reforms carried out as part of the Bologna process (Bollaert, 2014). This has resulted in the establishment of formalised external and internal QA mechanisms.

Despite the existence of widely acceptable external and internal QA mechanisms, many universities suffer from a bureaucratic burden and illegitimate interference from the central management, which holds managerial power in regulating and disciplining academics (Lucas, 2014). These mechanisms are supposed to draw on certain sets of quality standards, most importantly the Standards for QA in the European Higher Education Area (ESG) (ENQA, 2015). QA officials are in the awkward position of having to justify their approaches and methods (Markus & Philipp, 2018). They (QA practitioners) try to make their research instruments more sophisticated to keep pace with the methodological debate, thereby challenging the methodological and managerial approaches of quality managers' work.

Suffice also to note that there is still an insufficient capacity of stakeholders to participate effectively during monitoring and evaluation processes. There are also no clear incentives for participating in the monitoring and evaluation processes in the HE sector (Onyango, 2018). The challenges highlighted here require cautious consideration to fix the missing links by managers of HEIs in Africa.

Any efforts to evaluate the quality of teaching and learning in HEIs have to take into account the students' input (Markus & Philipp, 2018), since students' achievements vary not only with the quality of the teaching but also with other sources of variance including student's aptitude or the time budget they are spending on extra-curricular activities or jobs.

Quality as a concept means different things to different stakeholders (Henriette & Cecilia, 2020). And, research on the impact of QA and Quality Management (QM) in HE and particularly on the quality managers' perceptions of QA and QM effectiveness in HEIs is still rather rare. Although Markus and Philipp (2018) present results on the perceived effectiveness of QM in teaching and learning in German HEIs, their data are based on a nationwide survey among quality managers conducted in 2015. This presents a need to document the matter of M&E for QA in HE, focusing on approaches, theories, models, and practices,

2 Approaches, Theory, Practice, and Models

In the field of M&E, the term approach refers to those even broader conceptual collections, often representing groupings of models sharing similar principles. It can also mean ways of thinking about and conducting evaluation studies (Gullickson & King, 2019). Evaluation theories are not like the theories of science, which provide empirically testable predictions; they are conceptual positions or arguments posing a particular resolution to some underlying fundamental question about evaluation practice.

Models are conceptual frameworks that incorporate positions on various underlying theories about fundamental issues. Sometimes these positions are rendered explicit or implicit (Gullickson & King, 2019). For example, responsive evaluation takes a particular position on theoretical issues as; legitimate purposes of an evaluation study and benefits that will accrue. Stufflebeam and Coryn (2014) convince that evaluation theories, models, and approaches can each be evaluated on conceptual and moral grounds, but collecting empirical evidence of evaluation feasibility and effectiveness requires the implementation and assessment of discrete identifiable actions.

M&E emerged as a profession in the second half of the twentieth century. Initially, its approaches were quite rudimentary and focused on the assessment of inputs and outputs, but later the analysis of outcomes and impacts soon followed (Ministry of General Education, 2018). Monitoring is a continuous function that uses the systematic collection of data on specified indicators to provide management and main stakeholders of an ongoing development intervention with the extent of progress and achievement of objectives and progress in the use of resources (Ministry of General Education, 2018). Monitoring is the routine process of data collection and measurement of progress towards set objectives at institutional, policy, programme, or project levels.

Monitoring and evaluation of the HE sub-sector should ultimately inform policymakers and management at different levels. This framework outlines information products to be generated out of the data and points to how the information products will be reported (Council on Higher Education, 2013). Data for this M&E system will herewith be drawn from various sources,

3 Data Sources and Utilisation of Quality Monitoring and Evaluation

To define and implement an effective monitoring system, HEIs first need to answer the following questions: (1) Why are we collecting data? (2) What monitoring data do we collect? (3) How do we collect this monitoring data? (4) Who collects monitoring data? (5) What are some of the challenges in collecting the data, and how do we overcome them?

Having responded to the above questions, the institution moves on to understand the relationship between institutional data collected, data processing, and information (Mayanja, 2020). Managers of HEIs should note that data are recorded facts and figures about events and transactions of an institution's activities (Ministry of General Education, 2018). These can be dates or other numbers (quantitative data) or names and events (qualitative data). They are the input raw materials from which information is produced and hence they are often referred to as raw or basic data. Information is data that have been processed in a way to be useful to the recipient.

3.1 Types of Data

There are two types of data that HEIs should consider: (1) quantitative and (2) qualitative (UNESCO, 2013):

Quantitative data is "numerical in nature" and helps the QA office of an institution among others to answer questions such as when? How much? In addition, how many? How often? It is useful in testing statistical relationships between a problem and likely causes as observed by the Inter-University Council for East Africa (2010). However, numerical data may not explain the underlying causes of problems (e.g., why there is low performance in Science related courses in Ugandan Universities at the undergraduate level?) (UNESCO, 2013). Despite the limitations of this kind of data, the QA office should consider it vital since it delivers a statistical relationship between the variables in the spotlight as affirmed by the Inter-University Council for East Africa (2010). For example, quantitative data in conducting an institutional internal self-assessment of the QA unit can be in the form of the amount of money allocated to QA, the number of staff in the QA unit, and the number of items and equipment in the QA office among others.

Qualitative data is "descriptive in nature" and helps the QA office of an institution to answer the "why and how?" questions. It can be much more than just words or text (e.g., photographs, videos, sound recordings, etc.). It provides insights into institutional stakeholders' beliefs, attitudes, and practices (UNESCO, 2013). The argument raised by UNESCO (2013) is right since QA deals with evidence, and qualitative data in the form of photographs, videos, and sound recordings provide tangible justification. Qualitative data is data that deals with the description and can be observed or self-reported, but not always precisely measured, less structured. The advantage of qualitative data is that it is easier to develop and can provide "rich data" which is detailed and widely applicable (Inter-University Council for East Africa, 2010). However, data have limitations: being challenging to analyse, being labour-intensive to collect, and usually generating longer reports. Despite the limitations noted above, qualitative data is very essential in providing the institution with narrative or in-depth information. HEIs must consider triangulating both types of data as a strategy to provide any possible missing link during and after the data collection exercise. Triangulation of both qualitative and quantitative data greatly checks oversights and provides a comprehensive picture of the matter in the spotlight for inquiry.

3.2 Data Quality

In HEIs, the quality of data is an essential aspect of M&E. This implies that a monitoring system in an HEI is only as good as the quality of the data it generates. Poor quality data (data that is missing or incorrect or invalid) undermines the entire monitoring system. It is therefore vital that efforts are made by managers of HEIs to safeguard the quality of data, as guided by USAID (2007). Data quality is determined by its level of relevance, validity, and reliability in line with the data needs to purposefully address the matter in the spotlight. The salient features of quality data can be elaborated by the characteristics of good measure of data quality. Characteristics of good measures of data quality include relevance of data, credibility of data, data validity, and reliability.

Regarding relevance, the key question asked here is whether the measure captures what matters (Wandiembe, 2010). HEIs should never consider measuring what is easy instead of what is needed. This is because any tendency to measure what is easy rather than what is needed either for convenience or lack of human resource capacity breeds serious oversights.

The same may cause risks like failure to properly track the needed changes which in the long run leads to wastage of resources on unintended activities and hence poor accountability.

Credibility as a unit measure of data quality concentrates on whether the measure considered is believable (Yin, 2012). This requires further probing to ascertain if the measure considered will be viewed as a reasonable and appropriate way to capture the information sought. Therefore, HEIs in Africa should seriously ensure that data credibility is properly accounted for, in ensuring data quality for M&E and QA.

Another aspect that is key in ensuring data quality for M&E and QA in HEIs in Africa is "internal validity". This looks at how well the measure captures what it is supposed to be captured. In this case, internal validity pays attention to the instruments used. For example, are waiting lists a valid measure of demand, is a campus a valid measure of direction, or are class attendance lists a valid measure of quality learning outcomes (OECD, 2015)? This aspect is very important in guiding the process of formulating valid data collection instruments for M&E and QA in HEIs.

The concept of reliability, as Oniye (2017) professes, measures precision, and stability, that is to say, the extent to which the same result would be obtained with repeated trials, and how reliably students acquired competencies in measuring learning outcomes. Therefore, HEIs have to check for consistency of results obtained by a relevant tool or method in assessing an attribute.

In South Africa, participatory M&E is embraced (Ogochukwu, 2012) as is the case with Uganda, where even the National Policy on Public Sector M&E of Uganda (OPM, 2011) reiterates that Ministries, Departments, and Agencies (MDAs), including the Ministry of Education and Sports, have different roles and responsibilities when it comes to monitoring and evaluation. Participatory Monitoring and Evaluation (PME) is the most emphasised approach in the Ugandan context of HE. This process postulates that stakeholders at various levels engage in monitoring and evaluating, as stressed by Onyango (2018). Key among the activities for M&E where stakeholders play a role is indicator development, where institutional capacity indicators are collectively developed and defined in line with the institutional values. Various indicators can be developed, and these include input, output, activity, outcome, and impact indicators. Suffice to elucidate that the indicator development process is not a one-off but a rigorous process as elaborated below.

4 Developing Monitoring and Evaluation Indicators

An indicator is a quantitative or qualitative variable that measures change. It can also be referred to as a variable that shows a change from a baseline level at the start of the intervention to another level after the intervention has had time to make its impact (USAID, 2007). Indicators provide relevant information on performance, achievement, and accountability.

4.1 Characteristics of Good Indicators

1. *SMART*—Specific, Measurable, Achievable, Realistic & Time-bond

By specific, reference is made to how the institutionally established target reflects a specific area for improvement (Chalmers, 2007). For example, in the case of HEIs, this can be categorically broken down into areas of research, teaching, and community engagement, respectively.

"Measurable" is characterised by concrete criteria for measuring progress towards the attainment of the intended institutional key outputs as supported by Mayanja (2020). This should be categorised to inform a breakdown of activities geared to attaining specific core functions in line with the overarching goal of the institution.

"Achievable" relates to the institutional resource levels and the institutional external environment (Mayanja, 2020). This may further influence the capacity of an HEI, in its ability to perform and achieve its set objectives.

"Realistic" is about how the established indicator (s) is a real representation of the intended results. Chalmers (2007) explains that this should be as defined by the existing basket of indicators present. For HEIs, the key indicator benchmarks can be derived from documented literature from UNESCO and also from the National and International QA Frameworks as documented by the national and international regulatory bodies for HE; for instance, the NCHE for Uganda in particular, the Inter-University Council for East Africa (IUCEA), and the African Higher Education Space.

"Time-bound" focuses on a time frame linked to a target date for achieving a particular objective. This is very important in HEIs, whose functions keep facing reforms in terms of the nature of programmes, activities, and quality requirements for deliverables of HE. This is very true and important to note because the HE market is driven by the invisible hands of the market, hence an appeal to all institutions of HE to comply with the same.

5 *CREAM*: CLEAR, RELEVANT, ECONOMICAL, ATTAINABLE, AND MONITORABLE (YARBROUGH ET AL., 2010)

"Clear" emphasises the need for clarity and accuracy of designed indicators as variables that show a change from a baseline level at the start of the intervention to another level after the intervention has had time to make its impact. Therefore, in HEIs, indicators should be clear and concise. For, relevance is linked with the desired change to be promoted by a proximate intervention. This could be in research or teaching or even community engagement. At the end of the day, relevancy will also reflect on the community needs: local, national, and international needs. Their further contention is that economics is focused on the cost implication of the indicator, to contextually pay attention to the institutional economic capacity to perform the core functions of research, teaching, and community engagement (Yarbrough et al., 2010). Needful to note hereby is that "Attainable" is as shown above in the SMART framework, under achievable.

By "monitorable", attention is put on the fact that indicators provide necessary information on performance, achievement, and accountability (Yarbrough et al., 2010). Therefore, such indicators selected have to reflect performance (activity-wise monitorable), achievement (output-wise and outcome-wise monitorable), and finally accountability. Below are the steps followed in developing HE capacity indicators.

5.1 Steps Followed in Developing Capacity Indicators

Step 1 HEIs should clarify the result statements: Identify what needs to be measured; this should be in line with the strategic plan (Onyango, 2018). This is important for HEIs because it will guide them in building strategic institutional decisions for the institutional strategic direction and for reviewing the strategic plan of the institution as justified.

Step 2 HEIs should develop a list of possible indicators for results through brainstorming and research (Yarbrough et al., 2010). For HEIs, indicators need to focus on a single issue providing relevant information on a situation, particularly information that offers strategic insight required for effective planning and sound decision-making.

Step 3 HEIs should assess each possible indicator. Onyango (2018) pleads that HEIs should make this assessment to build proper judgement on whether the selected indicators correctly bring out the exact measure of the attribute(s) intended.

Step 4 HEIs should select the best indicators and develop a basket of indicators (UNESCO, 2013). For HEIs, guidance is that selecting good indicators can be done from benchmarks comparatively with the UNESCO basket of indicators.

Step 5 HEIs should draft indicator protocols (World Bank, 2013), which should properly profile where to find which indicators, how to use the indicators, and the indicator definitions in the narrative summary.

Step 6 HEIs should collect baseline data for the basket of indicators (UNAIDS, 2015). This is because indicators are also defined by the feasibility of collecting meaningful and credible data for them.

Step 7 Managers and key stakeholders of HEIs should fully participate in refining the indicators and protocols. UNESCO (2013) advances that finalising the whole selection process must consider this step. In this regard, HEIs should involve students in the process of refining indicators for teaching and learning; the community in refining the indicators for community engagement; and employers, politicians, and regulatory bodies in refining curriculum and research protocols.

The process of indicator formulation flows through a rigorous process characterised by internal brainstorming, identifying and building on the experience of other similar organisations, consulting with beneficiaries and experts, and identifying existing secondary data sources (data collected by someone else but that you can use to measure a result). Referring to already existing indicators developed by government, UNAIDS, USG, Global Fund, World Bank, among others, becomes inclusive and allows sufficient opportunity for a free flow of ideas and creativity (World Bank, 2011 and 2013). This process is based on data needs, which determine the data sources as elaborated in the next section.

Smith (2010) holds that monitoring is conducted for three core reasons: to ensure day-to-day management decision-making for performance improvement, accountability for resources being used, and to inform the evaluation process. Monitoring can take place at the input level (looking at institutional grants), activity level (looking at teaching, learning, and faculty/school/college inspection), output level (looking at constructed lecture room space, procurement of learning materials, and equipment), and outcome level (looking at the quality of teaching and learning, focusing on learning outcomes or the teaching environment). Therefore, monitoring is typically undertaken regularly and covers the majority of programme/project components.

HEIs should conduct monitoring and evaluation to effectively, efficiently, and sustainably establish institutional assessment mechanisms for determining the intended institutional goal. As observed by Borg (2018), evaluation is the systematic and objective assessment of an ongoing or completed project, programme, or policy, including its design, implementation, and results. The aim is to determine the relevance and fulfilment of objectives, development efficiency, effectiveness, impact, and sustainability. An evaluation should provide information that is credible and useful, enabling the incorporation of lessons learned into the decision-making process of both recipients and donors.

Despite the need to document both monitoring and evaluation plans separately, the two are inseparable. The impression here is that whereas monitoring provides routine, descriptive information on where a project is at any time relative to respective targets and outcomes, evaluation gives evidence of why targets and outcomes are or are not being achieved and involves making judgements and reaching conclusions (Ministry of General Education, 2018). Evaluations might be used to inform problem definition and programme design, whereas monitoring is used to set and track implementation progress. HEIs should consider two forms of evaluation: formative evaluation which sheds further light on strengths and weaknesses in the implementation processes, and summative evaluation which can be used to determine the overall impact of the programme.

The M&E framework for any institution should be guided by the theory of change which defines the steps necessary to bring about a long-term goal of the institution. In M&E, a theory of change is designed to guide how problems identified in the conceptual framework could be addressed to reach the desired goal. If an HEI wishes to develop and implement an M&E plan, certain steps are recommended as explained below.

6 RECOMMENDED KEY STEPS FOR AFRICA

1. Stakeholder consultation and participation, which involves advocating for the need for M&E among stakeholders. This ensures an understanding of programme goals and objectives and identifying user needs and perspectives. This leads to learning about existing data collection systems and their quality, understanding indicators that are being collected, and determining the capacity for collecting and using data. Stakeholder participation entails involving stakeholders in developing the M&E framework, selecting indicators, setting targets, and reviewing results (Daniel et al., 2005). This whole process requires building consensus and commitment and maintaining effective relationships with intended users in institutions of higher learning in Africa.

2. Translating the problem statement, programme goals, and objectives into M&E frameworks. This helps to establish the scope of the M&E plan (Daniel et al., 2005). The importance of Daniel and others' argument is that it helps managers of QA in HE not to rush for solutions but to first understand what the problem is. Here approaches acting as the problem tree analysis can be used to generate answers to questions: how do we know that there is a problem, why is it a problem, and how can it be solved?

3. Developing M&E Framework (basic structure underlying a system). This helps in determining elements to be monitored and evaluated (Daniel et al., 2005). The M&E structure, therefore, provides clarity on positioning the M&E unit and its functions, roles of personnel, reporting lines, and mandate of M&E in the institution. This, therefore, guides HE managers on matters of institutional strategy for M&E, system, staffing, skills, and shared values.

4. Defining indicators and identifying data sources. This is important in determining M&E methods for data and information collection, developing a data collection plan, and determining M&E responsibilities (Yarbrough et al., 2010). HE managers ought to follow the capacity indicator protocols in evaluating the strategic issues in line with the strategic objectives set to ensure the realisation of the institutional goal(s).

5. Setting institutional targets following the institutional vision and mission is essential (Yarbrough et al., 2010). Therefore, HE manag-

54   D. KATENDE

ers should follow the institutional vision and mission in generating costed activity plans and budgets, geared to achieving the vision and mission of the institution. The implementation of the activity plan should be monitored from the start to the end, following the vision and mission.

6. Defining the reporting system, utilisation, and dissemination of results (Marelize & Jody, 2010). HE managers should develop clear reporting systems to remedy clashing mandates and above all streamline communication and linkages among M&E staff. In addition, managers ought to clarify who reports what, where, and how? Including what to report and when. Various ways of reporting results can be adopted, and these include meetings, conferences, media, and documented reports.

7. Planning for mid-course adjustments. This can be done through departmental review meetings, which may come up with proposals to make reviews (Yarbrough et al., 2010). Reviews are vital in addressing emerging trends that may negatively affect activity implementation processes and planned resources for M&E.

Monitoring and evaluation in HEIs can only be a success story if it is institutionalised. It is imperative to note that the institutionalisation of M&E in HE is not a one-off exercise but a process that starts with developing an institutional system of M&E as illustrated below through the suggested checklist of the twelve components of an M&E system.

7   Suggested Checklist for a Monitoring and Evaluation System

Although Marelize and Jody (2010) propose twelve components of a monitoring and evaluation system for any organisation, the first ten are key for HEIs in Africa:

1. The structure and organisation alignment for monitoring and evaluation systems. This focuses the institution on ensuring that their staff understand the organisation's overall goals and strategies and understand the role of monitoring and evaluation and the ability of the staff to execute their responsibilities (Marelize & Jody, 2010). In the medium term, clarity and relevancy of job descriptions for

staff, and an adequate number of skilled staff. It further ensures effective leadership and commitment to ensure performance. It also looks at incentives for individuals to be involved in ensuring the M&E system performance and defined career paths for its professionals. Therefore, regulatory bodies of HEIs should direct that; right away from the design of the organ gram (organisation structure), the M&E microstructure should be reflected.

2. The human capacity for M&E systems in the institution. This focuses the institution to ensure adequate and skilled M&E staff that can effectively and efficiently complete all activities defined in the M&E work plan. It further ensures the establishment of a defined skill set for individuals responsible for M&E functions, assessment, including career paths for M&E, and the ability to develop plans. It looks at standard curricula for M&E capacity building, supervision, in-service training, and mentoring as mechanisms for continuous capacity building (Marelize & Jody, 2010). Managers in HEIs ought to recruit staff with M&E skills, plan to build capacity for M&E, and engage the majority of their staff, especially the heads of units, in M&E activities and training.

3. The significance of M&E partnerships. The long-term focus here is to ensure internal and external partnerships to strengthen the M&E system are established and maintained. In the medium-term, results should be given to ensure an inventory of all M&E stakeholders, a mechanism to coordinate and communicate with all M&E stakeholders, participation in a National M&E Technical Working Group, local leadership, and capacity for stakeholder coordination. The duo advise that institutions ought to strategically develop M&E plans with a long-term intention to ensure that the plan will be updated periodically. The plan should address data needs, national standardised indicators, data collection tools and procedures, and roles and responsibilities to implement a functional M&E system. The medium-term results should seek to ensure the participation of all relevant stakeholders in developing the M&E plan. The M&E plan and its revisions should be based on the findings of periodic M&E system assessments, should be sector-specific, and should be linked to the national M&E plan (Marelize & Jody, 2010). Managers of HEIs ought to generate costed-work plans to aid in generating proper M&E activities for the institution and a realistic budget.

56   D. KATENDE

4. M&E work plans designed ought to have a multi-partner and multi-year M&E work plan used as the basis for planning, prioritising, and costing; mobilising resources; and funding all M&E activities. In the medium term, the M&E work plan should contain activities, responsible implementers, time frames, activity costs calculated using a unit cost table, and identified funding. The plan should also ensure that resources are committed to implementing the M&E work plan. The M&E work plan should be developed/updated annually based on performance monitoring. Here, managers of HEIs should note that this aids in building proper mechanisms for evaluating the performance of the institutional strategic plans, making choices, and determining institutional priorities with existing institutional resources.

5. Advocacy is also very important in this matter since it emphasises communications and culture for the institutional M&E systems. The long-term intention here is to ensure that knowledge of, and commitment to, M&E and the M&E system among policymakers, programme managers, programme staff, and other stakeholders. An M&E communication and advocacy plan is part of the national communication strategy (if it exists). The plan should identify the objectives of the advocacy and communications efforts, the target audiences, key messages to be communicated, and the communication and advocacy channels. M&E should be reflected in the planning and policies of all programmes being monitored.

6. The Institution should ensure that there are M&E champions among high-level officials who endorse M&E actions. An M&E champion is usually a senior decision-maker or person of influence who is positive about and fully values and understands the benefits of using information for decision-making. An M&E champion would speak, act, and write on behalf of those responsible for managing the programme's M&E, to promote, protect, and defend the creation of a functional M&E system. Targeted, structured, and planned M&E advocacy activities should be put in place with M&E materials for different audiences to communicate key M&E messages as defined in the M&E advocacy and communications plan. In this regard, managers of HEIs should strategically determine the means through which their results should be communicated. This should take into account the communication policy of the institution and the institutional values.

7. The need for routine monitoring which, in the perception of Marelize and Jody (2010), in the long term should focus on timely and high-quality routine data. In the medium term, routine monitoring forms, data flow, and manuals should be established, entailing defined data management processes for routine data as well as routine procedures for data transfer from the institutions at the local level to those at the national levels. What is important for institutions of HE to note here is the need to build a strong institutional database for M&E with clear guidelines on how to access, what to access, and proper data protocols for access and security. This is essential because institutional data is the source of its strength and thus needs appropriate protection mechanisms.

8. Periodic Surveys that answer relevant questions and are unbiased and accurate, generalised, ethical, and economical are undertaken as required by the programme data needs. Inventory of relevant surveys that have already been conducted. Formulation of Specified schedules for future surveys (either to be conducted by the institution or from where the institution should draw its data). Protocols for all surveys based on international or national standards (if in existence) should be incorporated. Managers of HEIs should strongly note the importance of determining which ethical standards for M&E the institution should adopt, bearing in mind the interests of institutional project grants partners, and the philosophy of the institution.

9. The importance of databases for Monitoring and Evaluation (M&E) systems. De Silva and Ratnadiwakara (2018) emphasise the role of databases in improving the effectiveness of M&E systems. Databases should be developed and maintained to enable stakeholders to access relevant data for policy formulation, programme management, and improvement. They should ably respond to the decision-making and reporting needs of stakeholders in HE, with well-defined and managed database(s) for collating, verifying, cleaning, analysing, and presenting programme monitoring data from all levels and sectors. With emerging technology, Management Information Systems as forms of integrated information management systems are recommended for all HEIs to handle information flow and effecting decision-making processes.

10. Very important also is supportive supervision and data auditing. Here, data quality (valid, reliable, comprehensive, and timely) is emphasised, and data is externally verified periodically. Guidelines for supportive supervision are developed, data auditing protocols are followed, and supportive supervision visits, including data assessments, and feedback should take place. Data audit visits should take place periodically to ensure that supervision reports and data audit reports are produced. Englebert and Gueye (2015) introspect that a sort of chain of command for the M&E function ought to be generated in the institution to enhance a clear task relationship and task performance mechanism in line with the set objectives of the institution.

The last two ought to also be noted by HEIs:

11. Evaluation and research should be emphasised to ensure that results are used to inform HE policy. In this regard, an inventory of completed and ongoing program evaluation and research studies and an inventory of local programme evaluation and research capacity should be established. Programme evaluation and research agenda, ethical approval procedures, and standards should be put in place with clear guidelines, standards, and methods. Programme research and evaluation findings should be disseminated and discussed evidence the of use of evaluation/research findings referenced in planning documents.

12. Using information to improve results by involving HE stakeholders in the process, whose involvement in the programme enhances learning from the data presented and gaining knowledge about the programme, therefore, enabling them to make better decisions about how to achieve the HE programme results as per the M&E Plan.

Conclusively, planning is a process of thinking about and organising the activities required for achieving a desired goal (Ministry of General Education, 2018). It is one of the most important programme/project management and time management techniques. HEIs should note that M&E for QA requires a well-developed plan which makes implementation simple. Planning for M&E, therefore, is planning to succeed, and vice versa, and planning is comprised of key components as explained in the section below.

8 Components of a Monitoring and Evaluation Plan Document in Higher Education

The following components are recommended for HEIs:

The Introduction section, as guided by the Ministry of General Education (2018), should provide background information about the HEI and programme in the spotlight, vision, mission, values, philosophy, target group(s), and geographical area of operation. In this section, the institution should consider adding information about the motivations of the internal and external stakeholders and the extent of their interest, commitment, and nature of participation in the institutional programme/project in question.

Kusek and Rist (2018) propose the programme description and framework section. This section will detail the specifics of the programme in the spotlight for evaluation in HE. For example, it may be a teaching program, research programme, or community outreach programme. The details here include the stakeholders, program inputs, program activities, outputs, outcomes, and foreseen impact of the intervention. It is within this very section where the programme log frame matrix can be availed.

Third is a detailed description of the plan indicators, which in the mind of Adebowale (2018) is a section that presents key institutional indicators that need to be tracked for programme achievements to be measured. The indicators are at Impact, Outcome, and Output levels and are linked to the strategic objectives of the institution and those in the programme/project in the spotlight. The M&E plan will depend on data sources comprising both routine and periodic data education sources to calculate these indicators.

Fourth is the institutional data collection plan section which Akinyemi (2021) refers to as a section that provides additional information about the processes of data collection for the institution. A diagrammatic view of the movement of data from one level to another, including feedback, is useful in describing the tools to be used for data collection and a clear institutional system for data management. The responsibilities for carrying out the tasks should be specified within this plan, and the necessary institutional capacity to perform the function needs to be developed through the training of programme staff or procured from outside of the organisation.

The fifth is a section of the plan for monitoring, which explains the types of monitoring, who will monitor, when to monitor, and how to monitor. Arhin (2019) stresses that the section describes the monitoring activities that will be carried out throughout the life cycle of the institution. This section is very important, as it profiles all the monitoring details that make it clear for both new and old M&E technical human resources for the institution to refer to. In the same way, it simplifies review processes for the monitoring plan of the institution.

The sixth component is that of the evaluation plan. It is documented as a second sub-section under the section plan for monitoring and evaluation. Nevertheless, this chapter strongly recommends that it should be treated as a separate independent section, to avoid any possible oversight (OECD/DAC, 2006). The section describes the evaluation activities, which include how baseline data is going to be collected, follow-up evaluations (mid-term, end-term, etc.), and any other special studies, including operational research.

The seventh section is that of the data analysis plan. This section should provide details of the set of standardised or customised data collection tools and protocols for the institution (OECD/DAC, 2006). It is under this section where the institution's data collection tools as questionnaires and interview guides, among others, are detailed. This can be included in an annexe to assist with actual M&E plan implementation.

The eighth component is the plan for the utilisation of the information gained by the institution after monitoring and evaluation have been conducted and a report documented with lessons learned. Marelize and Jody (2010) stress that this section describes how the M&E data will be presented and used by the institution for its day-to-day management decision-making and reporting to various stakeholders.

This chapter proposes the following way of presenting and disseminating M&E data in HEIs: departmental meetings, seminars and conferences, public dissemination, radio and television broadcasts, public education, brochures, flyers, posters, websites, etc. Irrespective of the method used, when communicating M&E results, findings ought to be reported concisely and to the point. The report should follow a logical sequence of presentation avoiding unsupported statements and recommendations. The lessons learned should be pragmatic and constructive to make the final product attractive. Where need be, graphics can be used.

The section on the mechanism for updating the plan is also very essential. This is a common oversight by many M&E institutional plans. However, this MUST be included, indicating strategies for plan review in line with changing trends in HEIs as advocated by OECD/DAC (2006).

Therefore, all HEIs ought to treat the institution's M&E plan as a living document, which should be reviewed periodically to catch up with changing trends in HE.

Final section is that of the budget that outlines the required budget to implement the M&E plan. This section includes the cost drivers like survey and census design and administration, data storage costs (including software and hardware requirements in the institution), costs associated with carrying out evaluations (whether outsourced or internal to the organisation), M&E dissemination costs as well as the training and development needs for participating staff to perform M&E duties, as supported by Manda et al. (2022). A costed-work plan in this regard is imperative in facilitating purposeful monitoring and evaluation.

There are great lessons to learn from the African continent. The most important issue to consider in building an M&E plan for any HEIs in Africa is the capacity and resources for M&E. Marelize and Jody (2010) insist that at least 10 per cent of the institution's resources should be devoted to M&E, that is, for costs related to data collection systems and information dissemination.

9 Monitoring and Evaluation in Higher Education and the Sustainable Development Goals

M&E is an essential aspect of any programme, including HE ones. It is necessary to measure and evaluate the performance of the programmes against their objectives, assess their impact, and make necessary changes to ensure they are meeting the needs of the stakeholders. The SDGs set by the United Nations play a significant role in shaping the M&E practices in HE. This chapter provides a review of the current literature on M&E in HE and the SDGs, focusing on the African perspective and covering the period from 2015 to 2023.

M&E in HE programmes is essential to ensure that they are meeting their objectives and are relevant to the changing needs of the students, employers, and society. Othman et al. (2015) observe that M&E practices in HE have evolved from a focus on inputs and outputs to outcomes and impact. They advance that outcome-based M&E is more effective in measuring the success of HE programmes as it assesses the actual results achieved by the students rather than just their inputs and outputs. Othman and colleagues' argument is in line with the fact that a purposeful measure of results helps to tell the level of success from failure, hence justifying the relevance of M&E in HE.

Similarly, Kusek and Rist (2018) construe that HEIs need to move beyond traditional M&E practices and adopt innovative methods that are more responsive to the needs of the stakeholders. They recommend using technology-based tools akin to online surveys, mobile data collection, and real-time monitoring to collect data on student performance and programme outcomes. The above observations are key especially now with the advancement in technology, which has seen the emergence of new M&E software, both in the project-related operations and in HE.

In addition to traditional M&E practices, HE institutions also need to incorporate stakeholder engagement in their M&E processes. According to Hardison and Gibbons (2019), involving stakeholders serving as students, faculty, employers, and community members in the M&E processes can provide valuable insights into the effectiveness of HE programmes and their impact on the community. The above argument is a strong reflection on the relevance of community engagement, as the third mission of HEIs in Africa and the world at large.

9.1 Sustainable Development Goals, Monitoring, and Evaluation in Higher Education

The SDGs have significant implications for M&E practices in HE. HEIs in Africa and the world at large play an irreplaceable role in achieving the SDGs by producing skilled graduates who can contribute to sustainable development (Etzkowitz & Zhou, 2017). According to UNESCO (2017), M&E practices in HE should be aligned with the SDGs and focus on measuring the impact of HE programmes on sustainable development. This argument is in support of the fact that the measure of the third mission of universities is purposed for assessing how much HEIs contribute towards social economic development in Africa and globally.

Several studies have explored the link between the SDGs and M&E in HE. For example, Rahman and Khaled (2019) hold that M&E practices in HE need to be reoriented towards the SDGs to ensure that they are contributing to sustainable development. In their view, HEIs should align their curricula, research, and community engagement activities with the SDGs to achieve this goal. Similarly, Moktan and Mahat (2019) believe that the SDGs provide a framework for assessing the relevance and impact of HE programmes. Therefore, HEIs in Africa and the world at large should use the SDGs as a basis for designing their M&E systems to ensure that they are aligned with the global agenda of sustainable development.

The specific SDGs referred to in this chapter are the Sustainable Development Goals set by the United Nations, focused on institutions of HE. These goals include:

No Poverty (SDG1): This goal aims to eradicate extreme poverty and reduce inequality by ensuring equal access to resources, services, and opportunities for all. In Africa, poverty remains a significant challenge, with approximately 416 million people living in extreme poverty (UNESCO, 2017). To address this issue, various programmes and initiatives have been implemented to provide access to relevant HE, healthcare, and social protection for the poorest and most vulnerable populations. Therefore, the design of HE in Africa and the world at large ought to focus on developing relevant skills to address the aspects highlighted above.

Zero Hunger (SDG2): This goal propagates the need for everyone to have access to sufficient, safe, and nutritious food. In Africa, hunger and malnutrition remain undisputable, with approximately 256 million people suffering from hunger (UNESCO, 2017). To address this issue, various HE academic programmes have been implemented to provide skills in improving agricultural productivity, increasing food security, and access to nutrition education and support.

Good Health and Well-being (SDG3): Moktan and Mahat (2019) advance that this goal aims to ensure that everyone has access to affordable and quality healthcare services, including maternal and child health services, disease prevention, and treatment. In Africa, healthcare systems face numerous challenges, including inadequate infrastructure, limited resources, and a shortage of healthcare professionals. Various initiatives have been implemented to avert these challenges, including the provision of appropriate skills in areas in the same manner as running mobile health clinics, providing telemedicine services, and management of community-based healthcare programmes.

Quality Education (SDG4): This goal advocates for access to quality education and lifelong learning opportunities for all. In Africa, access to quality education remains a significant challenge, with approximately thirty million children out of school (Moktan & Mahat, 2019). The situation is worse when it comes to access to HE. However, various initiatives have been implemented to improve access to HE, including the provision of Universal Secondary Education, the establishment of community schools, and the use of technology-based education programmes to boost access to HE. It is important to note that M&E is needed to measure access to quality HE, to win public support.

Gender Equality (SDG5): This goal calls for equal opportunities and rights regardless of gender. In Africa, gender inequality remains a significant challenge, with women and girls facing discrimination and limited access to education, healthcare, and employment opportunities (UNESCO, 2017). To address gender inequality, academic programmes in HE have been developed in such a way to include relevant training on women's empowerment, gender-sensitive education, and healthcare services, and the promotion of women's participation in decision-making processes has been implemented.

Clean Water and Sanitation (SGD6): This goal regards access to safe and affordable water and sanitation services for the entire populace. In Africa, access to clean water and sanitation remains a significant challenge, with approximately 663 million people lacking access to clean water and 2.4 billion lacking access to basic sanitation (Moktan & Mahat, 2019). Academic programmes in HEIs have been developed in such a way to provide skills for development and preservation of safe water sources, the construction of sanitation facilities, and the promotion of hygiene education, as a strategy to avert the situation. Therefore, a measure of achievement of the intended results here calls for M&E.

Affordable and Clean Energy (SDG7): This goal propagates everyone's access to affordable and clean energy sources. In Africa, access to reliable and affordable energy remains a significant challenge, with approximately 600 million people lacking access to electricity (UNESCO, 2017). In this regard, HEIs in Africa and the world at large have come up with relevant academic programmes geared at providing appropriate skills in developing renewable energy sources, developing off-grid energy systems, and promoting energy-efficient technologies.

Decent Work and Economic Growth (SDG8): This goal promotes sustained and inclusive economic growth, employment, and decent work for all (Moktan & Mahat, 2019). In Africa, unemployment and underemployment remain acute, particularly among young people. Various initiatives have been implemented to counter this challenge, including the use of HE to develop entrepreneurship and the development of vocational and technical education programmes. It should be noted that M&E ought to be used to measure progress in the areas highlighted above.

Industry, Innovation, and Infrastructure (SDG9): This goal is in amity with sustainable industrialisation, innovation, and infrastructure development. In Africa, limited access to infrastructure and technology remains rampant, particularly in rural areas. Interventions to respond to this challenge include the development of rural infrastructure, the promotion of innovation and technology transfer, and the establishment of public-private partnerships for infrastructure development.

Reduced Inequalities (SDG10): This goal challenges inequalities within and among countries, particularly by promoting social, economic, and political inclusion. This has been seen through the promotion of affirmative action on women and girl child education through the provision of a 1.5 point to boost access by girls to HE in Uganda. However, all this ought to be measured through effective M&E in HE, to recognise success, consolidate it, and perhaps reward success.

Adopting the appropriate use of M&E by external and internal regulatory organs in HE is therefore essential to ensure that HE academic programmes are meeting their objectives and are relevant to the changing needs of the stakeholders. HE institutions need to adopt innovative and outcome-based M&E practices that are responsive to the needs of the stakeholders. Therefore, Rahman and Khaled's (2019) argument that a focus on the SDGs should be considered when developing M&E strategies does provide a framework for assessing the relevance and impact of HE academic programmes, which should be integrated into the M&E processes in HE. Aligning M&E practices in HE with the SDGs can ensure that HE institutions are contributing to sustainable development and addressing the global challenges of our time.

10 Lessons from Africa for the Rest of the World

The present wave of globalisation is rapidly developing into a complex system of exchange, interactive dynamics, and structures that collectively interact to effect rapid changes in human life. The consequent changes in HE relative to evaluation approaches and quality control have accordingly become an important issue for debate (Mayanja, 2020). UNESCO (2013) observes that globalisation has brought with it an increased level of academic fraud. This poses a danger to employers since they end up recruiting substandard candidates for superior assignments hence dwindling productivity (UNESCO, 2013). Recruiting lukewarm staff members also leads to no achievement of the organisation's set goals (Ndi, 2021). Therefore, the following recommendations should be considered:

1. Participatory M&E (PME) is an approach that institutions of HE should use to ensure close follow-up on targets, goals, and objectives so that these institutions maintain quality standards in their academic programmes, research, consultancies, outreach services, and administrative functions (World Bank, 2013). PME is a process through which stakeholders at various levels engage in M&E (Onyango, 2018). The process involves monitoring and evaluating particular project interventions; sharing control over the content, the process, and the results of the M&E activity; and engaging in taking or identifying corrective action

2. To ensure proper standardisation and ensuring uniformity (mimetic), regulated institutional value (normative reflection), and shared regulation (coerciveness), all HEIs should consider centralised regulation (UOTIA, 2001). Borrowing from the experience of Uganda's HE sector, all HEIs in Uganda are regulated by the National Council for Higher Education (NCHE) which makes emphasis on ensuring the involvement of key stakeholders during educational service delivery. The Council also makes emphasis on ensuring that all these institutions are closely monitored for compliance with the set standards.

3. HEIs throughout the world should consider adopting a programme-based approach to M&E. Borrowing from the Ugandan experience where the National Policy on Public Sector M&E of Uganda (OPM, 2011) reiterates that ministries, departments, and agencies, including the Ministry of Education and Sports, have different roles and responsibilities when it comes to M&E. Such roles include programme performance. The policy advocates for the utilisation of participatory approaches during M&E undertakings of the respective MDAs and local governments. HEIs should ensure that there is sufficient involvement of stakeholders during M&E if quality programmes are to be ensured (UNAIDS, 2015). Stakeholder engagement is important at all stages including curriculum development and review, accreditation, implementation, and assessment.

4. Accordingly, the triple helix model should be successfully operationalised by bringing on board government, through the ministries or departments of education as a means of ensuring national isomorphism in HE M&E of performance. Borrowing from Uganda still, NCHE (2018) promotes that an HEI must involve the Ministry of

Education and Sports, which is the parent ministry, with different roles and responsibilities when it comes to M&E.

11 Conclusions

Literature has shown that the HE sector globally has over the years rapidly developed into a complex system of exchange, interactive dynamics, and structures that collectively interact to effect rapid changes in all aspects of human life. This chapter has rallied how quality systems and theoretical models in HE face a lot of challenges to implement at a time when the HE space across the globe is competitively growing uncontrollably. However, M&E has not been well disseminated among key players in QA in HE. Thus, the quality systems and theoretical models have been kept on shelves of QA players in HE instead of referring to them.

Emphasis on quality systems and theoretical models in the HE sector throughout the world would come with many merits. There should be enhanced ownership of the sector's interventions among stakeholders, while still constraints should be appreciated faster, and this should easily guide decision-making towards objectives achievement in HE. Although M&E is being promoted in most HEIs, a participatory M&E approach should be adopted in all HEIs. In this way, key stakeholders in HE and QA players should fully participate and own the whole process and approaches used.

Policy enforcement, boosting the capacity of participants, and incentivising M&E should be adopted by HEIs (Bello & Olusegun, 2020). Governments need to ensure that their National Policies on Public Sector M&E are widely disseminated among key players in the public sector, including the HE sector. The policies should be reviewed to compel all accounting officers to implement the M&E recommended systems as guided by the established frameworks for M&E.

12 General Recommendations for Best Practice

1. Participation of stakeholders during M&E should be enhanced through an incentivised system (OPM, 2011). Such a system ought to entail timely feedback on recommendations from the M&E activities, making the undertakings as flexible as possible, and building participant capacity in QA and M&E. Accounting officers who do

not comply with implementing the M&E function in HE should be reprimanded by respective supervisors. Incentives like the reference to performance as presented by the M&E function in the institution should be used during consideration for promotions in the institution, as a means of popularising the M&E function for QA in HE rewards, and recognition schemes may also consider the performance of QA and M&E officials.

2. All government departments, education standards agencies, ministries of education, and local, national, and international regulatory bodies, namely Uganda Universities Quality Assurance Forum (UUQAF), the NCHE in Uganda, the Inter-University Council of East Africa (IUCEA), the African Higher Education Space (AHES), European Quality Assurance Standards agencies, and individual HEIs, should reflect on, adapt, and institutionalise M&E, templates, principles, and standards documented in this chapter as important benchmarks.

References

Adebowale, B. A. (2018). Monitoring and evaluation in higher education institutions in Nigeria. *Journal of Educational and Social Research*, 8(1), 95–102.

Akinyemi, A. O. (2021). Monitoring and evaluation of academic programmes in Nigerian universities. *Journal of Applied Research in Higher Education*. ahead-of-print (ahead-of-print).

Arhin, A. A. (2019). Monitoring and evaluation of academic programmes in Ghanaian universities. *Journal of Education and Practice*, 10(6), 37–45.

Bello, S. A., & Olusegun, S. (2020). Monitoring and evaluation of academic programmes in Nigerian universities: Challenges and prospects. *International Journal of Research in Education and Science (IJRES)*, 6(1), 197–204.

Bollaert, L. (2014). *A manual for internal quality Assurance in Higher Education–With a special focus on professional higher education*. EURASHE.

Borg, S. (2018). Evaluating the impact of professional development. *RELC Journal*, 22, 003368821878437. https://doi.org/10.1177/0033688218784371

Broucker, B., & Kurt de, W. (2015). New public Management in Higher Education. In J. Huisman, H. de Boer, D. D. Dill, & M. Souto-Otero (Eds.), *The Palgrave international handbook of higher education policy and governance* (pp. 57–75). Palgrave Macmillan.

Brown, J. T. (2017). The seven silos of accountability in higher education: Systematising multiple logics and fields. *Research & Practice in Assessment*, 11, 41–58.

Chalmers, D., (2007). *A review of Australian and international quality systems and indicators of learning and teaching*, August, V1.2. (Chippendale, NSW, Carrick Institute for Learning and Teaching in Higher Education).

Council on Higher Education. (2013). *Higher education monitoring and evaluation framework.*

Daniel, A. W., Bob, D., Tina, J., Robert, B. K., Miller, J., & Unwin, T. (2005). *Monitoring and evaluation of ICT in education projects.* A Handbook for Developing Countries.

De Silva, S., & Ratnadiwakara, D. (2018). Development of a database system for monitoring and evaluation of development projects. In *2018 IEEE 9th Annual Information Technology, Electronics and Mobile Communication Conference (IEMCON)* (pp. 1115–1121). IEEE.

Elken, M., & Stensaker, B. (2018). Conceptualising 'quality work' in higher education. *Quality in Higher Education, 24*(3), 189–202.

Enders, J., & Westerheijden, F. D. (2014). Quality assurance in the European policy arena. *Policy and Society, 33*(3), 167–176.

Englebert, P., & Gueye, B. (2015). Developing and implementing a monitoring and evaluation system for the Africa RISING program in West Africa. *African Journal of Agricultural and Resource Economics, 10*(1), 11–25.

ENQA (European Association for Quality Assurance in Higher Education). (2015). *Standards and guidelines for quality Assurance in the European Higher Education Area.* ENQA.

Etzkowitz, H., & Zhou, C. (2017). *The triple helix: University-industry-government innovation and entrepreneurship.* Routledge.

Gullickson, A., & King, J. A. (2019). The current state of evaluator education: A situation analysis and call to action. *Evaluation and Programme Planning, 75,* 20. https://doi.org/10.1016/j.evalprogplan.2019.02.012

Hardison, J., & Gibbons, M. (2019). Enhancing the rigour and relevance of program evaluation in higher education: Integrating stakeholder engagement. *New Directions for Evaluation, 164,* 71–83.

Harvey, L., & James, W. (2010). Fifteen years of quality in higher education (part two). *Quality in Higher Education, 16*(2), 81–113.

Harvey, L., (2004–2020). *Analytic quality glossary, Quality research international.* Accessed Dec 18, 2019, from http://www.qualityresearchinternational.com/glossary/

Henriette, L., & Cecilia, C. (2020). Engagement for quality development in higher education: A process for quality assurance of assessment. *Quality in Higher Education, 26*(2), 135–155. https://doi.org/10.1080/1353832 2.2020.1761008

Inter-University Council for East Africa. (2010). *A road map to quality. Handbook for quality assurance in higher education*, Volume 4.

Kinser, K. (2014). Questioning quality assurance. *New Directions for Higher Education, 168*, 55–67.

Kusek, J. Z., & Rist, R. C. (2018). *Ten steps to a results-based monitoring and evaluation system: A handbook for development practitioners.* World Bank Publications.

Lucas, L. (2014). Academic resistance to quality assurance processes in higher education in the UK. *Policy and Society, 33*(3), 215–224.

Manda, J., Chivuno-Kuria, S., & Mwakapenda, W. (2022). Monitoring and evaluation of academic programmes in Tanzanian universities. *International Journal of Education and Development using Information and Communication Technology (IJEDICT), 18*(1), 19–34.

Marelize, G., & Jody, Z. K. (2010). *Making monitoring and evaluation systems work: A capacity development toolkit.* World Bank Publication.

Markus, S., & Philipp, P. (2018). Assessing quality assurance in higher education: Quality managers' perceptions of effectiveness. *European Journal of Higher Education, 8*(3), 258–271. https://doi.org/10.1080/2156823 5.2018.1474777

Mayanja, C. S. (2020). Participatory monitoring and evaluation for quality programmes in higher education: What is the way for Uganda? *International Journal of Educational Administration and Policy Studies, 12*(1), 52–59. https://doi.org/10.5897/IJEAPS2020.0637

Ministry of General Education. (2018). *Monitoring & evaluation (M&E) in the education sector.* A Course Reader produced as part of a joint MoGE-UNZA capacity-building programme.

Moktan, S., & Mahat, S. R. (2019). Monitoring and evaluation of higher education for sustainable development in Nepal: An exploration of opportunities and challenges. *Journal of Education for Sustainable Development, 13*(1), 53–68.

National Council for Higher Education [NCHE]. (2018). *Annual performance report and financial statement of financial year 2017/18.*

Ndi, M. E. (2021). Monitoring and evaluation of academic programmes in Cameroon: Challenges and prospects. *Journal of Education and Learning, 10*(4), 299–306.

OECD, (2015). *Education at a glance, interim report: Update of employment and educational attainment indicators.* Accessed Jan 20, 2015, from https://www.oecd.org/edu/EAG-Interim-report.pdf

OECD/DAC. (2006). Sourcebook on emerging good practices in managing for development results.

Ogochukwu, N. (2012). Participatory monitoring and evaluation of governance in Africa: Lessons from the civil society African peer review mechanism monitoring project (AMP) in South Africa. *Africa Insight, 41*(4).

Oniye, O. A. (2017). Basic steps in conducting educational research. In A. Y. Abdulkareen (Ed.), *Introduction in research method in education*. Ibadan AgboAreo Publisher.

Onyango, R. O. (2018). Participatory monitoring and evaluation: An overview of guiding pedagogical principles and implications on development. *International Journal of Novel Research in Humanity and Social Sciences, 5*(4), 428–433. Accessed July-August, 2018, from www.noveltyjournals.com

OPM. (2011). *National policy on public sector monitoring and evaluation*. Accessed Mar 20, 2011, from https://usaidlearninglab.org/sites/default/files/resource/files/Attachment_J.15_M%26E_Policy_Final_Draft.pdf

Othman, J., Rashid, A. S. A., Sulong, N. B., & Bakar, N. A. (2015). Outcome-based education and the importance of monitoring and evaluation in higher education institutions. *Procedia-Social and Behavioural Sciences, 204*, 153–160.

Rahman, M. H., & Khaled, M. A. (2019). Monitoring and evaluation of higher education in Bangladesh: A sustainable development goals perspective. *Journal of Education and Practice, 10*(5), 106–114.

Smith, N. L. (2010). Characterising the Evaluand in evaluating theory. *American Journal of Evaluation, 31*(3), 383–389. https://doi.org/10.1177/1098214010371820

Stufflebeam, D. L., & Coryn, C. L. S. (2014). *Evaluation theory, models, and applications* (Vol. 2nd). Jossey-Bass.

U.S. Agency for International Development (USAID). (2007). *Data quality assurance tool for programme-level indicators*. USAID. Accessed May 30, 2009, from http://www.pepfar.gov/documents/organisation/79628.pdf

UNAIDS. (2015). Terminology guidelines. In *Joint United Nations Programme on HIV/AIDS 20 avenue Appia 1211*.

UNESCO. (2017). *Education for sustainable development goals: Learning objectives*. UNESCO.

United Nations Educational Scientific and Cultural Organisation [UNESCO]. (2013). *Review of the International Institute for Educational Planning (IIEP). Internal Oversight Service Evaluation Section Final Report*. Accessed Aug 1, 2013, from http://www.iiep.unesco.org/sites/default/files/iiepunesco_review.pdf

Universities and Other Tertiary Institutions Act. (2001). (as amended in, 2003 and 2006), enacted by the parliament of the Republic of Uganda as Act 7. Statutory Instrument 2007, No.1, pp. 1–4.

van Vught, F., & de Boer, H. (2015). Governance models and policy instruments. In J. Huisman, H. de Boer, D. D. Dill, & M. Souto-Otero (Eds.), *The Palgrave international handbook of higher education policy and governance* (pp. 38–56). Palgrave Macmillan.

Wandiembe, P. (2010). *Sample survey theory introduction* (2nd ed.). Kampala.

72 D. KATENDE

World Bank. (2011). *Impact evaluation in practice*. Available online at: http://siteresources.worldbank.org/EXTHDOFFICE/Resources/5485726-1295455628620/Impact_Evaluaion_in_Pracice.pdf

World Bank. (2013). *Designing a results framework for achieving results: A how-to guide*. Independent Evaluation Group.

Yarbrough, D. B., Caruthers, F. A., Shulha, L. M., & Hopson, R. K. (2010). *The Programme evaluation standards: A guide for evaluators and evaluation users* (3rd ed.). Sage.

Yin, R. K. (2012). *Applications of case study research* (3rd ed.). Thousand Oaks, CA: Sage.

Quality of Teaching and Learning Through Internal Quality Curriculum Review Mechanisms: *A Case of Private Higher Education Institutions in Post-conflict Somalia*

Abukar Mukhtar Omar ⓘ *and Abdu Kisige* ⓘ

1   INTRODUCTION

At the time of independence, Somalia had only one University College, later what became the Somali National University (SNU), established in 1954 in the trust territory of Somalia. It obtained official university status in 1969 ([SNU], 2014). This was the only university until the collapse of Siad Barre's regime in 1991, leading the Somali National University, ceased its operations, thus creating a gap in the higher education (HE) sector in the country. This happened due to certain forms of conflict,

---

A. M. Omar
Faculty of Education, SIMAD University, Mogadishu, Somalia
e-mail: abukar@simad.edu.so

A. Kisige (✉)
Al-Mustafa Islamic College, Kampala, Uganda
e-mail: kisiabdu@gmail.com

© The Author(s), under exclusive license to Springer Nature Switzerland AG 2024
P. Neema-Abooki (ed.), *The Sustainability of Higher Education in Sub-Saharan Africa*, Sustainable Development Goals Series,
https://doi.org/10.1007/978-3-031-46242-9_4

73

educational institutions become part of the battleground of the conflict, and often, the purposes of education become dislocated by war as students and teachers were recruited in conflict, interrupting their studies and significantly altering their lives (Reimers & Chung, 2010).

Conversely, the civil war and the subsequent collapse of the Somali state destroyed the country's education infrastructure (Aynte, 2013). To this effect, HE in conflict-affected contexts faced numerous challenges directly or indirectly conflict, including physical destruction, population displacement, war-related conditions, and low resilience of the sector (Milton & Barakat, 2016). This is corroborated by Babury and Hayward (2013), who affirm that the conflict also affected the physical infrastructure of the education system where most buildings got damaged or in disrepair, equipment was ruined or missing, most laboratories were inoperative, and libraries were stripped.

IIEP (2010) advances that the effects of the instability caused by the civil war were seen in Somali HE as the majority of the faculty members fled the country during the early years of the civil war. Violence against academics and a high level of displacement remain in the country's post-conflict phase, escalating sectarian violence in 2006 exacerbated what was already a large-scale academic "brain drain", displacing an estimated 5000 academics. At the end of the civil war, the HE system was left in destruction and decay, in which many of the best faculty members had fled, been imprisoned, or were killed (Babury & Hayward, 2013). Though private HEIs continue to provide HE to the masses, they fail to provide quality education to society. This can be evident, for example, by the type of graduates produced by private universities in Somalia whose educational needs to gain legitimate employment are not met. Their jobs may be available, but the quality of skills offered by the institution may not match the fourth industrialized labour market (Ainebyona, 2016), an aspect that might be linked to the low quality of teaching and learning. Further, noted fears that many stakeholders are concerned that students are not obtaining a good HE and are not competitive on the job, with private universities being more concerned with making money than raising educational standards. It is, therefore, suitable to isolate the reasons that are associated with the low quality of teaching and learning in private HEIs in Somalia.

Several scholars castigate the low quality of teaching and learning (Ndirangu & Udoto, 2011; Fabiyi & Uzoka, 2009; Ainebyona, 2016; Aynte, 2013; Eno et al., 2015). This study anticipated that internal quality curricular review mechanisms might explain the low quality of teaching and learning. Hence, this study on the role of assuring the quality of teaching and learning through internal quality curricular review mechanisms in higher learning institutions in post-conflict Somalia.

2 Objectives of the Study

The sole objective of this study was:

> To establish the effect of internal quality curricular review mechanisms on the Quality of teaching and learning in private HEIs in post-conflict Somalia.

Hypothesis
The study hypothesis is that:

> Internal quality curricular review mechanisms have a significant effect on the quality of teaching and learning in HEIs in post-conflict Somalia.

3 Review of Related Literature

Responsiveness in tertiary teaching and learning has dimensions functioning as curricula and pedagogy (Endut, 2014). This results in today's global competition for knowledge required by many economies, thus forcing higher education (HE) managers to update their curriculum for both the institution and students, hence a significant permanent undertaking (Saint, Hartnett & Strassner, 2003). However, Clark (2001) avouches that the former can be done at the university department level where they need to view the curriculum requires to be done every two to three years to ensure that the content of teaching reflects the rapid advance in scientific knowledge. This would enable university administrators and academic staff to develop pedagogy innovations that could accommodate the increasingly diverse student population resulting from the massification of HE due to the policy of accessibility of HE (El-Khawas, 2001). This supports the earlier submission by Kasozi (2006) that higher education institutions (HEIs) need to introduce programmes that enhance students' practical skills with an internship relevant to the world of work. Quality curriculum for HE is imperative for the re-sharpening of the minds of citizens to enable them to take constructive criticism and meaningful contribution towards national development. Through an appropriate curriculum review, citizens of any country can acquire relevant cognitive, social, communication, and life skills to improve a nation's economic growth, productivity, and global competitiveness (Asiyai, 2020). Indeed, the quality of the curriculum is an increasingly significant issue in HE (Grainger et al., 2019).

According to Boitshwarelo and Vemuri (2017), improved clarity concerning the curriculum in HE is needed to address issues concerning learning and teaching quality. Thus, this supports Rodríguez de Céspedes's (2017) argument that the Bologna process's main goal was to implement a quality education system connected with research and lifelong learning to ensure graduates' employability across Europe. As far as the HE curriculum is concerned, the purpose was to design, develop, and evaluate a process for quality improvement of HE programmes, based on assessing students' competence with constructive alignment and achieving acceptance from academic staff (Lucander & Christersson, 2020). Hence, addressing the issue of quality might have one or more meanings, depending on the particular stakeholder, the relevant goals and objectives, and the institution's mission (Der Horst & Gerhardus, 2007).

Accruing from the above, the best practice in curriculum reviews, development, and implementation requires that discipline-based standards or requirements embody both curricular and programme scopes and sequences. Ensuring these are present and aligned in the course/programme content, activities, and assessments to support student success requires formalised and systematised review and development processes aiming at quality HE. To this effect, HEIs need to ensure that their faculty are familiar with sound curriculum practice. As a result, the underlying principles of constructive curriculum alignment, globalization, and quality assurance (QA) need to be integrated into their programmes would be met (McDonald et al., 2007).

Therefore, HEIs are responsible for designing curricula that prepare workforce-ready graduates to embrace quality (Lasrado & Kaul, 2021). In this case, documents serving as syllabi provide evidence of the intended curriculum through their representation of what was designed, planned, and to be communicated to students, as well as the actual artefacts or observations necessary to determine what was enacted, experienced, and assessed need to be dealt with to help to account for personal growth are the evidence that helps one to understand the learned curriculum. In other words, the HE curriculum must include components of knowing how to solve problems, work collaboratively, and think innovatively (Mojarradi & Karamidehkordi, 2016).

Past studies relating internal quality curricular review mechanisms to the quality of teaching and learning include Mensah (2016), Bornman (2004), and Hurst et al. (2017). For example, Mensah (2016), in a study aimed at investigating the implementation of QA in polytechnics in Ghana, found that focusing on the availability of internal QA policies including curriculum design evaluation and review increases the quality of students' learning. Meanwhile, Bornman (2004) found that QA in HE influences learning assessment, putting more emphasis on the theme of sustainability as integrated into HE programmes as criteria that should determine quality in education programmes in his study on programme review for QA in HE in South Africa. In a survey about sustainable employability in HE, Hurst et al. (2017) discovered that career programmes outside the curriculum allow students to practice, improve, and apply their employability skills in realistic environments.

In general, the literature cited herein reveals that internal quality curricular reviews play a significant role in enhancing the quality of teaching and learning. If any institution of higher learning is to attain its core functions, among which teaching and learning are, it must embrace the former. It is against this background that current researchers should give much attention to the quality of curriculum reviews to help policymakers and other related stakeholders address issues of poor graduates by filling the gaps left behind when dealing with detailed studies that are intended to deal with issues related to HE graduate production. Nevertheless, it is hard to guarantee that the way curriculum issues are handled elsewhere in Africa and Europe could be the same; therefore, the contextual knowledge gap will be bridged by undertaking this study.

4   Curricular Review Mechanisms and the Sustainable Development Goals

Novel trends within HE over the last decade have seen the emergence of innovative learning initiatives that involve the application of new and emerging technology tools, delivery platforms, and/or new business models and pedagogy (Kim, 2015). This has been in response to the Sustainable Development Goals (SDGs) that recognize the importance of education to sustainable development through SDG 4, which calls for providing "inclusive and equitable quality education and promote lifelong learning opportunities for all" by universities (Kestin et al., 2017). Noted is that universities are seen as engines of societal transformation that can nurture future citizens and navigate them towards sustainability through their educational programmes (Kioupi & Voulvoulis, 2020). To this effect, Education for Sustainable Development (ESD) is largely synonymous with quality education programmes but requires far-reaching changes to the way education functions in modern society, though the structure and implementation of quality education programmes for sustainable development is a key challenge (Didham & Ofei-Manu, 2014). The foregoing is not limited to the many contributions Universities can make to sustainability, education has the greatest potential, and this is reflected in SDG 4. For example, HEIs are mentioned in target 4.3, which aims to, "by 2030, ensure equal access for all women and men to affordable and quality technical, vocational, and tertiary education, including university". Further, HE also forms an important part of other goals related to poverty (SDG1), health and well-being (SDG3), gender equality (SDG5) governance, decent work and economic growth (SDG8), responsible consumption and production (SDG12), climate change (SDG13), and peace, justice, and strong institutions (SDG16) (Kioupi & Voulvoulis, 2020). This, therefore, advocates for the substantial efforts required to reinforce the responsibilities of HE to the SDGs (Shiel et al., 2020).

HEIs have introduced sustainability concepts into their curriculum through stand-alone courses, embedding sustainability, or offering minors/degree programmes/certificates on sustainability. These courses appear as part of the school strategy to better prepare change in agents who care about the world and their impact on it. As a result, three pillars of sustainability (environmental, social, and economic) should be addressed in HE curricula since literature shows that many, HE institutions predominantly highlight their community engagement projects on their websites as "proof" that they embrace sustainability (Zizka, 2019). The implication hereby is that to prepare today's students to be tomorrow's leaders, HE institutions must find more effective means of teaching sustainability principles and concepts that resonate with students and create authentic engagement with sustainability practices which will be continued upon graduation (Zizka, 2019). This factor and others consequently implement the SD Agenda across HEIs and continue to draw attention from the wider society because it is increasingly being looked up to for leadership in this regard (Awuzie & Emuze, 2017).

HE for SD is being significantly shaped by the global sustainability agenda. Many HEIs are responsible for equipping the next generation of sustainability leaders with the knowledge and essential skills, proactively try to action the SDGs in their curriculum and practice through scattered and isolated initiatives (Franco et al., 2019). This means that achieving a future SD agenda requires a strong centrality of education in enabling human well-being. Such an education goal needs to simultaneously address both the advancement of quality education and education for sustainable development (Didham & Ofei-Manu, 2014). Thus, faculty staff who wish to foster competencies for SD, therefore, need help with contextualizing and operationalizing competencies in their curricula. This could be often enforced through pedagogic-didactic understanding needed to implement competence orientation in their teaching, in an institutional context where knowledge transmission is traditionally rated higher than competence development (Wilhelm et al., 2019). At this point, university programmes should focus on learning outcomes (LOs) that define what graduates should know and be able to do at the end of their studies (Kioupi & Voulvoulis, 2020). The latter is well integrated into HE curricula, the ongoing discourse about sustainability and the realization of the 2030 Agenda of SDGs would appear to appreciate the balancing of economic growth, social equity, and environmental protection inclusively for developed and developing countries, leaving no one behind (Kioupi & Voulvoulis, 2020).

Besides, elsewhere in the world, the HEIs community has signalled its desire to support the national government's SD aspirations through their core activities of teaching and learning, research and operations (Didham & Ofei-Manu, 2014). According to the same source, a noticeable increase in the rates of adoption of SD-centred strategies amongst these HEIs has been observed. Although some of these strategies have been enunciated at the strategic level in many African HEIs in the form of policy documents, vision, and mission statements, not a lot has been reported on their implementation. This aligns with Awuzie and Emuze (2017) who observe that although several studies are quick to identify various factors which have driven the adoption of sustainable practices in HEIs, the paucity of studies seeking to identify the drivers for SD implementation remains glaring. Yet, on the contrary, it is pertinent to add that some HEIs have articulated holistic SD implementation frameworks for achieving their SD objectives within the African context (Awuzie & Emuze, 2017).

Moreover, the sober analysis of HE contributes to sustainable development by Kioupi and Voulvoulis (2020) and argues that developing an assessment framework for educational institutions to evaluate the contribution of their educational programmes to sustainability is elementary in finding out whether the intended learning outcomes of the proposed programmes have already established the competences they targeted, in which case those can be used in the assessment. According to the immediate-quoted coauthors, the framework takes a holistic and systemic approach based on the sustainability attributes required for the SDGs to be realized, avoiding the perils of having to evaluate the integration of each SDG in the programmes' intended learning outcomes separately. The trio do all gear towards integrating competencies in tertiary and university education for sustainable development (ESD) (Wilhelm et al., 2019) through curriculum development as well as an analysis of how twenty-first-century competencies might benchmark QA in curriculum development for HE for sustainability.

5   Research Methodologies

The study employed a non-experimental, descriptive, cross-sectional survey design. It was cross-sectional where the researchers visited respondents at once during the data-collecting process, as Amin (2005) would recommend. The cross-sectional survey was appropriate as it is friendly in both time and cost and as the study involved a considerable number of respondents (Kisige & Neema-Abooki, 2017). The study was descriptive as it described the situation of internal quality curriculum review mechanisms in institutions of higher learning. Data collection was approached quantitatively, where variables were measured using numbers. Data were collected from 253 academic staff of HEIs in Somalia. Due to the large population, 108 academic staff (response rate = 73%) were selected using Krejcie and Morgan's (1970) sample size determination table. The questionnaire was disseminated to academic staff that was nominated purposively and were requested to rate themselves following a five-point Likert scale: 1 = strongly disagree, 2 = disagree, 3 = undecided, 4 = agree, 5 = strongly agree. Data were analysed using frequencies, percentages, means, and simple linear regression.

QUALITY OF TEACHING AND LEARNING THROUGH INTERNAL QUALITY... 81

6 FINDINGS

6.1 Background of Respondents

Of the 85 respondents sampled, a little above average (45.9%) had teaching experience below five years, 45.9% were aged 30 but below 40, while males (84.7%) dominated the sample on the issue of academic qualification. The majority of the respondents (64.7%) had Master's qualifications, while holders of both the first degree and PhD stood at 35.2%. Regarding Academic ranks, a good many (58.8%) of the lecturers in the sampled Universities were at the rank of lecturer.

This study aimed to test the hypothesis that internal quality curricular review mechanisms significantly affect the quality of teaching and learning in private HEIs in post-conflict Somalia. Internal quality review mechanisms were operationalized into eight quantitative items. Using the eight quantitative items, lecturers were requested to do their self-rating based on a Likert scale ranging from: "strongly disagree", "disagree", "undecided", "agree", and "strongly agree". Results are depicted in the table below:

In Table 1, apart from the third, sixth, and seventh statements where a significant number of the respondents (45%, 42%, and 50%), respectively, expressed negative sentiments on whether the academic programmes are reviewed every 3 years by involving professionals and other stakeholders, the rest of the statements received a positive rating. In particular, according to the pattern of the responses, most of the academic staff asserted and agreed that the academic programmes are reviewed every 3 years by involving professionals and other stakeholders. For example, at one of the private institutions sampled, academic staff asserted that they have managed to form a departmental curriculum review committee responsible for examining the curriculum followed. In support of the preceding, a substantial number of the participants in the study credited their Deans, Heads of Departments, and Subject Unit Coordinators for encouraging them to engage in activities that yield good teaching practices. Such activities included generating departmental subject content for the respective subject units, benchmarking to see how other universities are conducting the same business, and writing research-project proposals to support teaching and learning. This, in one way, helps in creating a cordial responsibility of minding the curriculum offered to students.

82 A. M. OMAR AND A. KISIGE

Table 1 Internal quality review curriculum reviews

Indicator	Strongly disagree	Disagree	Undecided	Agree	Strongly Agree	Mean
Academic programmes in my department are designed to enhance teaching and learning		1 (1.2%)	0	64 (75.3%)	20 (23.5%)	4.21
The courses in my department are designed to enable students to acquire subject-specific and transferable skills		1 (1.2%)	4 (4.7%)	44 (51.8%)	36 (42.4%)	4.35
We review the academic programmes in my department after every three years		10 (11.8%)	35 (41.2%)	25 (29.5%)	15 (17.6%)	3.53
We evaluate the relevance of the academic programmes in my department regularly	2 (2.4%)	3 (3.5%)	10 (11.8%)	43 (50.6%)	27 (31.8%)	4.06
We involve students in my department in evaluating our academic programmes	9 (10.6%)	15 (17.6%)	28 (32.9%)	16 (18.8%)	17 (20.0%)	3.20
We involve other professionals in evaluating and reviewing the academic programmes of my department	6 (7.1%)	36 (42.4%)	22 (25.9%)	16 (18.8%)	5 (5.9%)	2.74
We involve other stakeholders in evaluating and reviewing the academic programmes of my department	8 (9.4%)	42 (49.4%)	24 (28.2%)	9 (10.6%)	2 (2.4%)	2.47
Regular curricular reviews are enhancing the quality of teaching and learning in my department	2 (2.4%)	4 (4.7%)	8 (9.4%)	59 (69.4%)	12 (14.1%)	3.88

In the same vein, 67% of the academic staff further agreed that the quality of teaching and learning had been enhanced in institution departments operating teaching and learning activities due to regular curricular reviews. In carrying out this activity, most of the staff maintained that institutions have always exhibited a commitment to supporting

QUALITY OF TEACHING AND LEARNING THROUGH INTERNAL QUALITY... 83

curriculum review processes with an emphasis on checking on how best to do their work; and this has enabled members of respective institutions the courage to be committed and innovative in whatever task deemed apt to bring quality teaching and learning. The previous truism does reveal how committed the academic staff is at the threshold of post-conflict HEIs in Somalia.

Unlike what is experienced elsewhere regarding course programmes designed without the intended objectives of transferring desirable skills and knowledge, the respondents agreed unanimously that despite being affected by the conflict for a long time, courses in most of the private HEIs in Somalia are designed to enable students to acquire subject-specific and transferable skills where 80% of the lectures affirmed that the courses are designed to enable students to acquire subject-specific and transferable skills. Presupposed, therefore, is that without quality academic programmes, it might be hard for higher education institutions to produce quality graduates, meaning failure to execute one of its core duties which is teaching.

The rest of the items in Table 1 that are not discussed here all scored "Agree", implying that internal quality curriculum review practices abound within the majority of higher institutions in Somalia.

Having obtained the responses from the questionnaire, the researchers endeavoured to establish whether the ratings on internal quality curriculum review mechanisms had any association with the responses on the quality of teaching and learning. A simple linear regression analysis was conducted to test the relationship between internal quality curriculum review mechanisms and the quality of teaching and learning. The results are as in Table 2.

Table 2 results show that internal quality curriculum review mechanisms explained a 24% variation in quality teaching and learning (Adjusted R2 = 0.24). This means that 76% of the variation was accounted for by extraneous variables, other factors not considered in the study. The regression model was good as $F = 4.410$, $p = 0.011$, $p > 0.05$. The null

Table 2 Internal quality review curriculum reviews

Model	Coefficients	Significance (p)
ICRM	−0.024	0.011
Adjust R^2 =-0.024		
F= 4.410		

hypothesis was rejected in favour of the research hypothesis that internal quality curriculum review mechanisms significantly determine the quality of teaching and learning amongst institutions of higher learning in Somalia. This counsels that internal quality curriculum review mechanisms significantly determined the quality of teaching and learning. Hence, through effective and efficient curriculum review mechanisms where professionals are involved in the evaluation and reviewing of the academic programmes to design courses that enable students to acquire subject-specific and transferable skills, the quality of teaching and learning is enhanced.

7 Discussion

Results obtained from the study are discussed based on the study findings and showed that internal quality curriculum review mechanisms significantly determine the quality of teaching and learning. This is supported by Mensah (2016) according to whom HEIs should have formal mechanisms for the periodic review or evaluation of the courses and the curriculum. The implication is that effective internal quality curriculum review mechanisms improve the quality of teaching and learning. The findings further render credence to one of the earlier studies, in the same manner as that of Saint, Hartnett, and Strassner (2003), which found that today's global competition for knowledge requires many HE managers to update their curriculum for both the institution and students, hence a significant permanent undertaking.

Habituated by these results, it is interesting that university management in private HEIs in Somalia reviews academic programmes every three years. This was revealed when the majority of the academic staff in the study institutions asserted and agreed that their academic programmes are reviewed every three years by involving professionals and other stakeholders. The results corroborated such studies as Clark (2001) and Kisige et al. (2021), who concluded that HEIs need to continuously review their curriculums to weed out outdated content in consonance with the regulatory body. In this way, the findings of the study further rhyme with El-Khawas (2001), who established that the curriculum review enables university administrators and academic staff inclusive of coming up with pedagogy innovations that could accommodate the increasingly diverse student population that has as a result of the massification of HE due to the policy of accessibility of HE. Hence, internal quality curriculum review mechanisms determine the quality of teaching and learning in private higher education institutions in Somalia.

Regarding the relevance of the academic programmes, the study spelt out the due strictness put up by the management and administration of private universities in evaluating the relevance of the academic programmes in all university departments, regularly checking their relevance to the community. This was revealed when most academic staff in the various departments in studied private universities ascertained that their administration reviews academic programmes by involving different professionals and other stakeholders. The finding thus strengthens and acts as a confirmatory to earlier studies equally with Kahsay (2012), who holds that the best practices of internal QA must focus on core educational processes involving coherence in the design, delivery, and relevance of the curriculum. More concise to the previous validation are Mensah (2016) and Kasozi (2006), as they observe that focusing on the availability of internal QA policies, including curriculum design, evaluation, and review, increases the quality of students' learning.

It was therefore apparent that once professionalism is involved in evaluating and reviewing academic programmes, the quality of teaching and learning is likely to improve.

8 LESSONS FOR OTHER PRIVATE INSTITUTIONS OF HIGHER LEARNING

8.1 Global Context

There have been various HE policy reforms at the global level to overcome the challenges and impacts of globalization in the current knowledge-based global economy. Universities have already been involved in various internationalization processes establishing both bilateral and multilateral cooperation across borders (Woldegiorgis, 2013). For example, the International Network of Quality Assurance Agencies in Higher Education established in 1991 recommends guidelines of good practice by HE QA agencies that need substantial revision before they can be considered adequate amongst stakeholders in any national HE system. Likewise, agencies in HE also debate that the adoption of its guidelines of good practice has international significance in that the decisions about HE quality made by agencies that comply with them can be accepted at face value by universities (Blackmur, 2008), which increases the academic attention to the QA teaching and learning (Steinhardt et al., 2017).

Similarly, QA of teaching and learning as part of universities' governance and quality management has become a major subject in HE and HE politics worldwide (Steinhardt et al., 2017). To this effect, various integration schemes, regional organizations like the European Union (EU) and its Commission, and the African Union (AU) and its Commission are engaging in HE policy harmonization processes to foster more integration and provide regional remedies for the common challenges of globalization in their respective regions (Woldegiorgis, 2013). Propagated, therefore, is that QA in teaching and learning is considered a common meta-goal that different countries pursue via sub-goals, comparatively with the "improvement of quality and public accountability: for standards achieved" (Billing, 2004). Adjacently, it is a truism that developing countries should assess their academic quality by working with parameters that are globally acceptable and transparent to all stakeholders if they are to operate and survive for long.

8.2   Regional Context

In recent decades, quality in HE has been at the top of the global agenda in many countries across Europe and beyond and has also—perhaps as a consequence—received a considerable amount of academic attention (Bloch et al., 2021). It is, therefore, indubitable that the idea has become extremely useful to academics in that; in hunting the scholarship of quality teaching and learning, HEIs and universities need to interpret internal curriculum review mechanisms as correspondence to quality teaching and learning processes where HEI managers and administrators need to continuously review their curriculum to weed out outdated content. The belief here is that curriculum reform—or curriculum "reviews" as it is sometimes referred to—should be on the agenda of HEIs. This would be responding to the global shift of "knowledge-based economies that have led to national and institutional curriculum debates about how best to prepare graduates for a knowledge economy" (Shay, 2015). This, therefore, calls for HE administrators to make programme provisions adequate curriculum reviews to emphasize quality teaching for enhanced student learning. The implication is that the generated information about internal quality curriculum review mechanisms and quality teaching and learning would benefit the region's managers and administrators of other private HEIs.

9 Conclusion and Recommendation

In this study, the researchers sought to establish the effect of internal quality curriculum review mechanisms on the quality of teaching and learning. The findings have shown that internal quality curriculum review mechanisms determine the quality of teaching and learning in private institutions of higher learning in Somalia, hence enabling graduates to be equipped with subject-specific and transferable skills of the twenty-first century relevant to the demands and needs of the global competition.

Higher learning institutions' managers are recommended to continuously review their curricula to weed out obsolete content. This should be done following regulatory body guidelines that provide periodic curriculum reviews —which guidelines should be adhered to by all HEIs.

References

Ainebyona, G. (2016). *The Fragility of Higher Education in the Post-Conflict Somaliland: A Dialogue*. Available at SSRN 2832711.

Amin, M. E. (2005). *Social science research: Conception, methodology, and analysis.* Makerere University Printery.

Asiyai, R. I. (2020). Best practices for quality assurance in higher education: Implications for educational administration. *International Journal of Leadership in Education*, 1–12.

Awuzie, B., & Emuze, F. (2017). Promoting sustainable development implementation in higher education: Universities in South Africa. *International Journal of Sustainability in Higher Education.*, *18*, 1176.

Aynte, A. (2013). *The state of higher education in Somalia: Privatization, rapid growth, and the need for regulation.* Heritage Policy Institute.

Babury, M. O., & Hayward, F. M. (2013). A lifetime of trauma: Mental health challenges for higher education in a conflict environment in Afghanistan. *Education policy analysis archives*, *21*(68), 68.

Billing, D. (2004). International comparisons and trends in external quality assurance of higher education: Commonality or diversity? *Higher Education*, *47*(1), 113–137.

Blackmur, D. (2008). A critical analysis of the INQAAHE guidelines of good practice for higher education quality assurance agencies. *Higher Education*, *56*(6), 723–734.

Bloch, C., Degn, L., Nygaard, S., & Haase, S. (2021). Does quality work? A systematic review of academic literature on quality initiatives in higher education. *Assessment & Evaluation in Higher Education*, *46*(5), 701–718.

Boitshwarelo, B., & Vemuri, S. (2017). Conceptualizing strategic alignment between curriculum and pedagogy through a learning design framework. *International Journal for Academic Development, 22*(4), 278–292.

Bornman, G. M. (2004). Programme review guidelines for quality assurance in higher education: A South African perspective. *International Journal of Sustainability in Higher Education, 5*(4), 372–383.

Clark, S. C. (2001). Work cultures and work/family balance. *Journal of Vocational behavior, 58*(3), 348–365.

der Horst, V., & Gerhardus, J. (2007). *Radiation tolerant implementation of a soft-core processor for space applications.* Doctoral dissertation, Stellenbosch: University of Stellenbosch.

Didham, R. J., & Ofei-Manu, P. (2014). Quality education for sustainable development.

El-Khawas, E. (2001). Who's in charge of quality? The governance issues in quality assurance. *Tertiary Education & Management, 7*(2), 111–119.

Endut, A. S. (2014). Enhancing internal quality assurance mechanism at HEI through responsive program evaluation. *Procedia-Social and Behavioral Sciences, 123*, 5–11.

Eno, M., Mweseli, W. N., & Eno, O. (2015). The revival of higher education in Somalia: Challenges and prospects. *Journal of Somali Studies, 2*(1&2), 9–44.

Fabiyi, A., & Uzoka, N. (2009). State of physical facilities in Nigerian universities: Implication for repositioning tertiary institutions for global competition. Proceedings of Towards Quality in African Higher Education. Higher Education Research and Policy Network (HERPNET), 180–187.

Franco, I., Saito, O., Vaughter, P., Whereat, J., Kanie, N., & Takemoto, K. (2019). Higher education for sustainable development: Actioning the global goals in policy, curriculum and practice. *Sustainability Science, 14*, 1621–1642.

Grainger, P., Crimmins, G., & Burton, K. (2019). Assuring the quality of curriculum, pedagogy, and assessment across satellite campuses. *Journal of Further and Higher Education, 43*(5), 589–600.

Hurst, C., Fowler, J., & Scapens, G. (2017). Sustainable employability in higher education: Career development outside of the curriculum. In *Success in higher education* (pp. 217–228). Springer Singapore.

IIEP. (2010). *Guidebook for planning education in emergencies and reconstruction.* International Institute for Educational Planning.

Kahsay, M. N. (2012). *Quality and quality assurance in Ethiopian higher education: Critical issues and practical implications.*

Kasozi, A. B. K. (2006). The politics of fees in Uganda. *International Higher Education,* (43).

Kestin, T., van den Belt, M., Denby, L., Ross, K., Thwaites, J., & Hawkes, M. (2017). Getting started with the SDGs in universities: A guide for universities, higher education institutions, and the academic sector.

Kim, J. (2015). Competency-based curriculum: An effective approach to digital curation education. *Journal of Education for Library and Information Science, 56*(4), 283–297.

Kioupi, V., & Voulvoulis, N. (2020). Sustainable development goals (SDGs): Assessing the contribution of higher education programmes. *Sustainability, 12*(17), 6701.

Kisige, A., & Neema-Abooki, P. (2017). Financial resource mobilization projects and their relationship to academic staff commitment in Uganda Martyrs University, .

Kisige, A., Ezati, B. A., & Kagoda, A. M. (2021). Teacher preparation by universities: Internal stakeholders' perception of teacher education curriculum content in Makerere and Kyambogo universities. *Education Quarterly Reviews, 4*(1).

Krejcie, R. V., & Morgan, D. W. (1970). Determining sample size for research activities. *Educational and Psychological Measurement, 30*(3), 607–610.

Lasrado, F., & Kaul, N. (2021). Designing a curriculum in light of constructive alignment: A case study analysis. *Journal of Education for Business, 96*(1), 60–68.

Lucander, H., & Christersson, C. (2020). Engagement for quality development in higher education: A process for quality assurance of assessment. *Quality in Higher Education, 26*(2), 135–155.

McDonald, J., Drysdale, R., Hill, D., Chisari, R., & Wong, H. (2007). The hydrochemical response of cave drip waters to sub-annual and inter-annual climate variability, Wombeyan Caves, SE Australia. *Chemical Geology, 244*(3–4), 605–623.

Mensah, M. A. (2016). Implementation of internal quality Assurance in Polytechnics: Evidence from Ghana. *European Scientific Journal, ESJ, 12*(19), 221.

Milton, S., & Barakat, S. (2016). Higher education as the catalyst of recovery in conflict-affected societies. *Globalisation, Societies, and Education, 14*(3), 403–421.

Mojarradi, G., & Karamidehkordi, E. (2016). Factors influencing practical training quality in Iranian agricultural higher education. *Journal of Higher Education Policy and Management, 38*(2), 183–195.

Ndirangu, M., & Udoto, M. O. (2011). Quality of learning facilities and learning environment: Challenges for teaching and learning in Kenya's public universities. *Quality Assurance in Education, 19*(3), 208–223.

Reimers, F. M., & Chung, C. K. (2010). Education for human rights in times of peace and conflict. *Development, 53*(4), 504–510.

Rodríguez de Céspedes, B. (2017). Addressing employability and enterprise responsibilities in the translation curriculum. *The Interpreter and Translator Trainer, 11*(2–3), 107–122.

Saint, W., Hartnett, T. A., & Strassner, E. (2003). Higher education in Nigeria: A status report. *Higher education policy, 16*(3), 259–281.

Shay, S. (2015). Curriculum reform in higher education: A contested space. *Teaching in Higher Education, 20*(4), 431–441.

Shiel, C., Smith, N., & Cantarello, E. (2020). *Aligning campus strategy with the SDGs: An institutional case study.* Universities as living labs for sustainable development: Supporting the implementation of the sustainable development goals, 11–27.

Somali National University. (2014). *University history*. Accessed Apr 17, 2017, from http://snu.edu.so/overview/

Steinhardt, I., Schneijderberg, C., Götze, N., Baumann, J., & Krücken, G. (2017). Mapping the quality assurance of teaching and learning in higher education: The emergence of a speciality? *Higher Education, 74*(2), 221–237.

Wilhelm, S., Förster, R., & Zimmermann, A. B. (2019). Implementing competence orientation: Towards constructively aligned education for sustainable development in university-level teaching-and-learning. *Sustainability, 11*(7), 1891.

Woldegiorgis, E. T. (2013). Conceptualizing harmonization of higher education systems: The application of regional integration theories on higher education studies. *Higher Education Studies, 3*(2), 12–23.

Zizka, L. (2019). Sustainability in higher education: Aligning sustainable development goals (SDGs) with curriculum/campus/community. In *EDULEARN19 proceedings* (pp. 2116–2123). IATED.

Students Satisfaction and Virtual Learning Service Delivery at Bugema University

Kulthum Nabunya [iD]

1 INTRODUCTION

In the past two decades, the educational system has progressed towards digitisation and online learning (Pande & Mythili, 2021). People's lives, and their ways of working, living, playing, and learning have been severely affected by the coronavirus disease (COVID-19) pandemic. Since classroom learning was adjourned and later suspended due to infection concerns, many national governments requested that educational institutions migrate online to circumvent interruptions (Faturoti, 2022). Using the development facilitated by technology, the world has become a small global village with instantly changing methods of education. Online learning is now an alternative form of course delivery used by colleges and universities worldwide (Al-Omairi & Hew, 2022). This has resulted in most university courses being made available online due to the demand for

K. Nabunya (✉)
Bugema University, Kampala, Uganda

Victoria University, Kampala, Uganda
e-mail: knabunya@kcca.go.ug

© The Author(s), under exclusive license to Springer Nature Switzerland AG 2024
P. Neema-Abooki (ed.), *The Sustainability of Higher Education in Sub-Saharan Africa*, Sustainable Development Goals Series, https://doi.org/10.1007/978-3-031-46242-9_5

online learning opportunities. Pande and Mythili (2021) define online learning as the transfer of skills and knowledge through the use of computers and other technological devices, like smartphones. This content is delivered via the internet or audio/video or satellite or CD-ROM and in a virtual learning environment (VLE). This is also done through other digital teaching platforms like Zoom, Google Meet, Google Classroom, and many others (Chen et al., 2020). E-learning defined as the conveyance of educational content via a computer network relies on the motivation and dedication of students and as defined by Puriwat and Tripopsakul (2021), it is the communication of learning materials to information seekers via technologies, VLEs, and mainly the internet.

The COVID-19 pandemic worldwide compelled all educational establishments to embrace online education without regard to how ready they were for it (Faturoti, 2022). As a result of this shift, a debate has been sparked about the quality of e-learning, student satisfaction, and future usage intentions of e-learning platforms and VLEs (Puriwat & Tripopsakul, 2021). Whereas enhancing student satisfaction by improving e-learning quality results in greater efficacy of students' online learning performance and outcomes. Alhumaid et al. (2020) applied the technology acceptance model to investigate the teachers' perceptions concerning online learning as a substitute for formal education during the COVID-19 outbreak in Pakistan and found a relationship between knowledge sharing, communication facility, motivation and usage, and e-learning acceptance among instructors. Even more, during the COVID-19 pandemic in Indonesia, Tj and Tanuraharjo (2020) analysed to examine the link between e-learning service quality and student satisfaction and found a positive relationship between e-learning service quality and student satisfaction. However, Means and Neisler's (2020) study with American undergraduates rated their ability to remain focused during an online session as worse or much worse compared to face-to-face learning.

Globally, universities are responsible for upholding ideal and excellent services that provide students with quality educational courses and experiences whether they are attending physically or online. However, since online learning is dependent on good internet access and adequate equipment for non-face-to-face classes, these factors have been major determinants or the spread of online learning throughout the world (Palau et al., 2021). The Western world speeds towards 4G and 5G internet connections which made the transition to online classes during the COVID-19 pandemic more seamless (Reed & Thompson, 2021). Africa's internet penetration has to some extent improved in the past decade; however, the continent has had difficulty integrating the internet into its educational systems (Faturoti, 2022); this is coupled with disparate access to information and computer technologies used in online education, data and internet connectivity. This has been made worse by the COVID-19 pandemic, which has widened the wealth gap and exposed society's weaknesses.

Internet access, connectivity, and speed have a significant and positive effect on student satisfaction in virtual learning environments. The results of a study conducted by Hassan et al. (2020) in Scotland propose that internet access has relatively major effects on students' performance. The researchers observed that not only were e-learning platforms difficult for the majority of students due to limited internet access and knowledge of ICT devices, but teachers also encountered difficulties when working with these platforms. In the United States, Means and Neisler (2020) point out that many students in virtual learning environments reported experiencing poor internet connections that prevented them from regularly attending synchronous sessions of online teaching.

Online learning is considered to be the best option for continuing education by Saxena et al. (2021); however, its affordability, in terms of ICT functioning as laptops, mobile phones, and data bundles, as well as the availability of a good and stable internet connection and the required ICT infrastructure, is a matter of debate among educators and policymakers. It has been observed that when these limitations are addressed, e-learning improves as seen in Kosovo, where a high percentage of citizens who are well acquainted with ICT are considered to be part of the success of distance or online learning in the country (Duraku & Hoxha, 2020).

A variety of factors, including the COVID-19 pandemic, widespread technological advances, higher internet usage, and students seeking courses that are convenient and fit their schedules, have contributed to the growth of online courses and online programmes, which continues to outpace the growth of traditional HE as a whole (Graham, 2021). In this sense, e-learning educational effectiveness and learning are not only international but national issues as well (Kaisara & Bwalya, 2021). Distance education has become a major topic in the last decade, as Institutions of Higher Education (IHEs) shift their focus towards online instruction. Online learning, as opposed to face-to-face learning, offers limited chances for student collaboration, with the probability of collaboration being worse in a virtual learning environment than in a traditional classroom environment (Ann & Aziz, 2022).

With distractions arising from a continuous barrage of advertisements (ads) appearing during online sessions, it becomes more difficult for an individual to stay focused. The added COVID-19 anxiety, the change in financial circumstances, and the difficulty in finding a quiet place to study within the house during the lockdown were also found to negatively influence the performance of students during e-learning (Hassan et al., 2020). In the traditional learning environment, instructors have used pedagogical approaches that encouraged the active participation and motivation of students. In online learning, on the other hand, talking to students and motivating them is not always evident, especially with limited individual sessions carried out (Duraku & Hoxha, 2020).

Even though these are different perspectives of a learning process, learning achievement and faculty perspectives, students' perspectives, and their satisfaction are particularly indispensable in an educational venture (Van Wart et al., 2020). Having a satisfied customer base is elementary to any business scheme, and online education is no different (Wengrowicz et al., 2018). A learner's satisfaction relates to their feeling about their accomplishments and their experiences in learning (Pande & Mythili, 2021). Learner satisfaction reflects students' perception of their learning experience and is defined as a student's overall positive assessment of his or her learning experience (Rabin et al., 2019). Studies on e-learning also confirm that student satisfaction is very likely to lead to improved student loyalty (Kilburn et al., 2016). According to Puriwat and Tripopsakul (2021), the learning performance and outcomes of online students improve significantly due to improved e-learning quality. However, despite attempts to provide empirical evidence to evaluate e-learning quality and its implications for improving the quality of online teaching, there are different results due to the diversity of contexts for e-learning across countries.

Considering the lack of empirical evidence regarding the quality of e-learning in third-world countries like Uganda, this study aimed to contribute both academically and practically to the discussion on how to improve student satisfaction in hybrid or online learning. Wu et al. (2010) used social-cognitive theory (Bandura, 1986) as a basis for their study of student satisfaction in a virtual learning environment. In this study, student interaction provides the greatest contribution to performance expectations which provides the greatest contribution to learner satisfaction. Student interaction could involve online interactions between students and teachers, between students and individual students, or between students and their content.

In a report by Means and Neisler (2020), students' satisfaction declined sharply after schools converted to all-online courses during the COVID-19 pandemic, with undergraduates struggling to stay motivated and some even missed receiving feedback from instructors and collaborating with their peers. In addition, the authors note that a number of students experienced issues with their internet connections, software and computing devices, some of which were bad enough to prevent them from attending classes. In Uganda, for instance, strategies were put in place to assuage the challenges which included training teachers to use technology, providing ICT tools, improving internet connectivity, and ensuring a stable power supply.

2   Virtual Learning and the Sustainable Development Goals

Sustainable Development Goal (SDG) 4, which aims to ensure inclusive and equitable quality education and promote lifelong learning opportunities for all, can benefit from virtual learning. This is because online education expands educational opportunities, particularly for people living in rural or remote areas, as well as those with disabilities (Secretariat, 2017). Through virtual learning, students can get to access quality learning material which initially weren't available to them. An example was the use of digital tools during the COVID-19 pandemic that made it easier for students to have access to online material. Students can also benefit from virtual learning because it allows them to balance work, family, and education; this applies to adult education students at tertiary institutions.

SDG 5, which aims at achieving gender equality and empowering all women and girls, can also be advanced through virtual learning (Febro et al., 2020). Virtual learning can assist in breaking down traditional gender barriers to education, allowing women and girls to gain access to education that they may have previously been denied (Prior & Woodward, 2017). Like the above, SDG 9 aspires to construct resilient infrastructure, promote inclusive and sustainable industrialisation, and support innovation (Denoncourt, 2020). Individuals and businesses can utilise virtual learning to gain new skills and knowledge that will help encourage innovation and long-term growth. Best practices from other regions of the world can be taught to people all over the world on virtual platforms, as has been the case with numerous webinars that have had professionals in many fields discuss virtually to thousands of people all over the world without having to travel to those areas.

SDG 10, which is anchored on the aim to decrease disparities within and between countries, can also be a beneficiary of virtual learning (Otto & Becker, 2019). People and communities can benefit from virtual learning by acquiring skills and knowledge that can lead to improved career opportunities and economic mobility. Overall, virtual learning has the potential to be a powerful tool for promoting sustainable development and progress towards the SDGs (Otto & Becker, 2019). Virtual learning can help to create a more equitable and sustainable society by increasing access to education, breaking down traditional barriers, and stimulating innovation and economic mobility. For example, it is now necessary for a university student to have a good internet connection and a computer device in order to virtually obtain information from professors from around the world, indicating that the disparities between countries, while still evident, are diminishing albeit marginally.

3   Sustainability of Education: The Ugandan Scenario

Education sustainability in the Ugandan context can be viewed from different perspectives (Altinyelken, 2010). This can include the sustainability of the education system itself, the sustainability of the education outcomes, and the sustainability of the education process (Altinyelken, 2010). Firstly, the sustainability of the education system in Uganda can be seen in terms of its ability to maintain quality education and provide access to education for all citizens in a constant manner over time. This requires continued investment in education infrastructure, human resources, and educational policies to ensure that the education system can continue to function effectively (UNESCO, 2017).

Secondly, the sustainability of education outcomes refers to the ability of the education system to provide graduates with the necessary knowledge, skills, and attitudes to succeed in their future careers and contribute positively to society (Otto & Becker, 2019). This requires an education system that is responsive to the needs of the economy and society, and that provides learners with practical and relevant skills that they can apply in their daily lives (UNESCO, 2017). Lastly, the sustainability of the education process involves ensuring that learners are motivated and engaged in the learning process and that they are equipped with the necessary tools and resources to learn effectively. This requires a focus on pedagogical approaches that foster critical thinking, problem-solving, and lifelong learning (Rieser, 2012).

There are a few important variables to take into account in the Ugandan setting when it comes to the sustainability of education. The first concern is financing. Uganda, like many developing nations, has great difficulty financing its educational system, especially at the postsecondary level. Although the government has made attempts to boost funding for higher education (HE), there is still a sizable resource gap, which may have an impact on educational quality and accessibility. The second concern is accessibility. Although Uganda has made great strides in recent years to increase access to HE, there are still significant disparities between urban and rural areas as well as between various socioeconomic categories. This may hinder HE's ability to support social and economic progress.

Next is the matter of relevance. The needs of the labour market and the broader community must be taken into consideration by Uganda's HEIs when designing their programmes. To achieve this, it is important to guarantee that graduates possess the abilities and information necessary to significantly advance the nation. The sustainability of HE in Uganda is showing some encouraging signs despite these obstacles. To increase access and improve quality, for instance, the government recently unveiled a new HE policy. The new HE policy in Uganda was launched in 2020 and is aimed at improving the quality of HE, expanding access, and promoting relevance. There has been a rise in private higher education institutions which may contribute to increased competition and access. Overall, several variables, in the same degree with adequate funding, increased access, and the relevance of programmes to the needs of the nation will determine the viability of HE in Uganda. Uganda can contribute to ensuring that HE plays a necessary role in fostering sustainable development and economic growth by addressing these issues.

Universities in Uganda have had the responsibility to continue delivering education following the closure of educational institutions as the lockdown and restrictions were extended for longer periods. In a learning period defined by significant changes in the mode of classroom delivery, where traditional classroom teaching and learning was replaced by online learning, understanding the perspectives of academic staff and students is necessary (Ouma, 2021). Coupled with the rapid development of technology paramount for online learning and the outbreak of the COVID-19 pandemic, many institutions of higher learning have used online learning as a solution to the closure of schools and interruption of learning (Means & Neisler, 2020). However, this sudden shift to complete online learning has had a considerable impact on students in regard to their satisfaction which is paramount to their academic performance. Therefore, maintaining student satisfaction with their virtual learning experience is a significant issue for stakeholders in IHE like Bugema University (Namboodiri, 2022). The university has had online courses for some of its programmes like Moodle, however, before the lockdown, not all faculties conducted a remote learning management system as many of these methods were implemented for the first time in the university. As student satisfaction is so important to their success, it is necessary to discover the factors that influence student satisfaction so that the university can develop methods to ensure the quality of their virtual learning environment.

4   VIRTUAL LEARNING SERVICE DELIVERY

Worldwide, educational systems were impacted by the closure of educational institutions as a precautionary measure against the spread of COVID-19 (Duraku & Hoxha, 2020). COVID-19 spread over the world, affecting over 850 million classroom children and interrupting instructional systems in all countries. As a result, institutions in many countries created virtual learning environments and began offering students online instruction via Zoom, Google Meet, Microsoft Teams, and other similar platforms (Chen et al., 2020). While many educational institutions throughout the world had already begun to transition their curriculum to an online format (Duraku & Hoxha, 2020), most African educational systems were unprepared for the abrupt shift to e-learning (Faturoti, 2022). E-learning is described as the delivery of educational content over a computer network (Zeng & Wang, 2021). However, the sort of e-learning used in an emergency is not often identified to the high-quality, carefully designed, web-based online learning, and virtual learning environments that have been used in IHEs for decades.

Both academic staff and students must be digitally literate to conduct successful virtual learning services. The capacity to identify, analyse, choose, utilise, and develop technology for lifelong learning in an effective and safe manner is referred to as digital literacy (Bugema University, 2020). Students must be digitally literate to recognise, engage in, relate to, generate, trade, and communicate with a range of learning technologies to obtain topic content, participate in learning activities, and pool resources with peers to accomplish assessment tasks (Dhillon & Murray, 2021). Regardless, employees must be digitally literate to connect with, support, encourage, and excite students in a virtual learning environment (Bugema University, 2020).

As both professors and students had very little time to prepare for the new virtual type of teaching and learning, the transition to online learning at Ugandan universities during the pandemic was swift and abrupt, as it had been in other areas of the world (Faturoti, 2022). However, Mpungose (2021) declares that for innovative virtual teaching strategies to grow, lecturers must focus on developing new online learning activities and supporting students in gaining specialised information in their VLEs (Adarkwah, 2021). This is achieved by employing and combining the most effective distance learning methods and technologies to inspire and motivate students to actively engage in the virtual teaching and learning process. According to Jung (2011), certain aspects of e-learning essentially differentiated communication, open access to multiple resources in the form of the internet, and the availability of software and hardware devices must be considered when evaluating the quality of virtual learning. And because e-learning is heavily reliant on student motivation and dedication, it can be difficult to measure and ensure the quality of e-learning.

According to Ouma (2021), virtual learning was found to be advantageous to students and academic staff in Uganda since it saved money on transportation and reduced the risk of infection with COVID-19 owing to reduced physical movement. Students and lecturers also became more resourceful, imaginative, and careful in their use of time throughout the virtual teaching-learning process as a result of the rising usage of ICT. However, the problems of limited data, poor internet connection, and inability to record lectures, as well as few Zoom connections at the faculty level, restricted class control, and unsteady student attendance, were identified by Ouma (2021).

In the African continent, online learning will always achieve less than expected due to a lack of Information and Communication Technologies (ICT) infrastructure to support and sustain it (Adarkwah, 2021). Without appreciating the need for a paradigm shift and acquiring the necessary ICT infrastructure, online learning will always achieve less than expected (Ouma, 2021). In support of its objective to create a comprehensive and supportive virtual learning environment for its students, Bugema University has made gains in its investment in ICTs for learning and teaching over the years. This policy supports the institution's virtual learning and teaching initiatives while building on and improving previous accomplishments and encouraging schools to use technology to enhance and promote virtual learning (Bugema University, 2020). The QA team at Bugema University identified various challenges associated with online education in the university in their survey conducted to evaluate their virtual learning management system, some of which included the poor internet connection, stressful nature of e-learning, difficulty in online assessments/exams, inadequate faculty-student interaction, and inadequate access to devices, which are reported as common challenges associated with online learning around the world as well (Zeng & Wang, 2021).

Aslam and Sonkar (2021) emphasise the importance of internet connectivity and a large amount of internet bandwidth for digital learning activities like videoconferencing, which are widely used in many courses. This is because even in privileged households with high-speed internet, internet connectivity issues arise when several individuals in the household are online at the same time, as was the case during the pandemic's peak. Internet connectivity problems are indeed significant enough to interfere with students' ability to attend or participate in their classes, which leads to frustration in this manner lowering students' satisfaction which is drawn from their inability to frequently and regularly attend sessions.

5   Students' Satisfaction

Even though IHEs were early adopters of remote learning using virtual learning environments during the pandemic, the use of this medium has received mixed reviews, with teachers and institutions in general finding it difficult to engage online learners to their satisfaction (Namboodiri, 2022). Student satisfaction has emerged as a pressing component of the virtual learning experience, with significant implications for students' willingness to continue their education and academic achievement (Perez, 2022). As a result, IHEs and universities have prioritised student satisfaction with their educational experience, which is regarded as one of the most important aspects in determining the quality of academic programmes (Al-Omairi & Hew, 2022). Student satisfaction is one of the metrics of an educational programme's effectiveness, whether it occurs in a VLE or a traditional classroom (Zeng & Wang, 2021).

Student satisfaction, according to Magni and Sestino (2021) is a short-term attitude coming from an appraisal of students' educational experience, services, and facilities. Student satisfaction is one of the most significant aspects to consider when assessing accomplishment in the deployment of a system in the context of e-learning (Tran & Nguyen, 2022). It is an irreplaceable subject in HE, and rigorous research into it may lead to greater student performance, changes in online teaching practices, and student retention in academic programmes (Tuan & Tram, 2022). It is also an important factor that may be utilised to assess the success of online learning.

Student satisfaction, according to Spencer and Temple (2021), is one of the most essential parts of continuing education since there is evidence that it is positively related to retention and students' desire to attend one or more further courses. Student contentment, on the other hand, is vital because happy students improve the public image of a college or institution (Rofingatun & Larasati, 2021). When a university treats students like consumers, their pleasure becomes critical to the university's recruiting efforts, and highly happy students become brand loyalists (Campbell, 2021). As a result, a more in-depth understanding of the elements influencing student satisfaction with online learning is necessary. Yunusa and Umar (2021) classified student satisfaction in e-learning into four categories: communication dynamics (interaction and information quality), e-learning-environmental elements (course structure and content), and organisational factors (technological support and service quality). Zeng and Wang (2021) gave a comprehensive evaluation of studies on online learning during the COVID-19 pandemic, adding that the same parameters impact the level of student satisfaction during emergency remote teaching.

Interaction and participation have become emerging issues and are still evolving in the virtual learning environment (Tuma, 2021). Many students would rather attend a virtual session where they are neither seen nor heard, arguing that they are 'too comfortable' to be seen by their classmates and teacher (Peace, 2021). Faculty, on the other side, may request that students be observed since some professors think that to have a relevant class, they must constantly examine students' reactions (Subekti, 2021). Some students, on the other hand, may want to attend lessons without video, preventing professors from seeing their faces or what they are doing. In an online setting, then, the challenge comes from managing the varied expectations of teachers and students regarding student involvement. In a similar vein, Sher and Toor (2021) found that online professor-student involvement, as well as student-student contact and mentoring support, all had an impact on perceived learning quality and student satisfaction.

Zeng and Wang (2021) discovered that teacher-learner interaction is an important component of student satisfaction during online learning. Students may feel as if they do not belong to a scholarly community in the absence of a good teacher-learner connection, resulting in student dissatisfaction. According to Sahu (2020), it is difficult to teach disciplines similar to nursing, sports, and music online due to the practical aspect and hands-on sessions. Ouma (2021) in Uganda agreed, citing the difficulty of teaching courses like engineering, fashion and textiles, and science laboratory-based subjects online. Aside from the difficulty of giving practical and laboratory-based examinations, many students are distracted by the barrier of internet connectivity during the assessment, leaving them disillusioned.

Students' motivation is an important component in determining student satisfaction and completion rates in web-based courses, and a lack of motivation is associated with high drop-out rates (James, 2021). Meanwhile, Zamzami (2021) discovered that highly motivated students were more productive and learnt more in web-based courses than students who were less motivated in a study done in Saudi Arabia. Students' active participation and motivation are increased when instructors use pedagogical approaches in the virtual classroom. Online learning, on the other hand, loses aspects of physically engaging with students and inspiring them in various courses (Duraku & Hoxha, 2020). Instructor feedback strives to improve and motivate student performance by informing students of their progress and guiding their learning efforts. Instructors provide students with basic cognitive feedback, diagnostic feedback, and prescriptive feedback through e-mail responses, graded work with comments, and online grade books (Cifuentes, 2021).

Some of the prerequisites for adopting online assessment, according to Tuah and Naing (2021), are developing an alignment of the assessment technique with the learning objectives; Using online assessments to motivate students to study; considering the style, layout, and timing of assessments; ensuring that students are aware of assessment challenges and that the online assessment is of good quality. Cifuentes (2021) also emphasises the importance of emphasising that online learning should be student-centred, with students' characteristics understood to identify potential learning barriers like motivation, costs, learning feedback, teacher communication, student support and services, and isolation.

Kizza et al. (2021) identified the delivery technique as one of the characteristics that either frustrate or excite virtual learners. Whereas effective virtual delivery techniques are more engaging and student-centred, poor virtual delivery methods delay student input and foster increased social isolation, resulting in student dissatisfaction. E-learning is becoming increasingly popular and acceptable to both students and lecturers due to the usage of numerous e-tools that promote enhanced engagement, kindred with e-chats, video conferencing, and discussion forums (Kharbat & Abu Daabes, 2021). Kizza et al. (2021) go on to explain that, while students and instructors at private colleges in Uganda support e-learning, they are not convinced of its utility. In both private and public universities, the status of preparation in terms of prior training and ICT infrastructure support continues to influence perceptions of e-learning performance. As a result, Kizza et al. (2021) recommend that universities train their staff and students in e-learning methodologies to improve their readiness to adopt e-learning, which would include acquiring user-friendly learner management systems (LMS) and seeking subsidies to assist in the acquisition of basic ICT infrastructure connatural with computers and smartphones, as well as ensuring easy access to reliable internet and power.

6 Purpose of the Study

The main purpose of the research was to determine the association between student satisfaction and service delivery at Bugema University. The specific goal of this study was to determine the association between students' virtual learning satisfaction and academic staff's virtual teaching service delivery at Bugema University.

7 Research Hypotheses

The following hypotheses were formulated to guide the study:

H1: There is a positive relationship between students' virtual learning satisfaction and the virtual teaching service delivery of academic staff at Bugema University.

H0: There is no relationship between students' virtual learning satisfaction and the virtual teaching service delivery of academic staff at Bugema University.

8 METHODOLOGY

Desk research, otherwise known as secondary research, was used to guide this investigation. Desk research, also known as secondary research, is a research method that incorporates the use of previously collected data, with the fascinating material summarised and compiled to boost the overall usefulness of the research. As a result of the ongoing lockdown, IHEs across the country used online learning platforms and tools for accessing online educational resources. The secondary data for the study was obtained from Bugema University. Data mining was performed on the data using tables for enhanced visualisation and to conduct research on the relationship between students' virtual learning satisfaction and academic staff's virtual teaching service delivery at Bugema University.

9 FINDINGS

The results from student questionnaires were recorded in percentage form in the categories of poor, average, good, and excellent.

To excel in e-learning, universities must provide adequate support in the form of e-learning administration and technical provision. Students were questioned on a variety of topics related to their satisfaction with the virtual learning system, comfortability with the e-learning system, notifications on the Learning Management System (LMS), clarity of instruction, and accessibility to technical support and University internet services, among others. Accordingly, 30% of the 64 participants were uncomfortable with the e-learning system, 22% mentioned that the e-learning system was excellent, and 69% rated the online LMS notifications as good or excellent.

The proportion of respondents who were not satisfied with the instructions on how to use the online system was 20% as opposed to 53% who had positive feelings to the status quo. Meanwhile, 23% of respondents revealed that learning materials were not always easily accessible, unlike the 49% according to whom the materials were always easily accessible and only 28% who rated accessibility as average.

The importance of consistent, efficient, and dependable internet access is the foundation for effective and productive e-learning experiences. However, nearly half of the respondents (47%) criticised the university's internet services (Eduroam). A minimum of 13% commended the stance.

The technical support received from the university in regard to e-learning was scored by 31% of respondents as poor, 23% as good, and 16% as excellent. Training on how to use the online learning system was regarded by half of the respondents as satisfactory, third of the respondents as average, and 27% as poor.

Equidistantly, 47% of respondents credited the lecturers as resolute in their teaching during the virtual learning programme while 22% and 31% pronounced average and poor rating, respectively. Meanwhile, 39% of the respondents judged the coverage of topics taught on the virtual platform as poor and an equal number of the respondents regarded the scenario as either good or excellent.

In regard to the phenomenon that online professor-student interaction, as well as student-student contact and mentoring support, all have an impact on perceived learning quality and student satisfaction, 27% of the respondents regarded the level of multimedia usage and interactivity of lecturers on the virtual platform while according to 51% it was good or excellent. Asked about the use of the university's e-learning resources, 32% said it was poor, while 40% upheld it as good or excellent.

Meantime, more than half of the respondents (56.3%) agreed that the online examination procedure used by their lecturers was good or excellent while only 25% had a different view.

Commonplace knowledge appreciates the use of online library learning resources as conducive to student performance during e-learning. When asked if they had been oriented and trained on how to use Bugema University Library e-resources, an above-average percentage enunciated the training as good or excellent, while the 30% and 25% perceived it as bad and average, respectively. In a related vein, nearly half of the respondents (49%) agreed that the University's E-resources for the Study Programmes were adequate, while 29% and 23%, respectively, construed them as poor or average. This could be attributed to the low utilisation of Bugema University's e-learning resources.

10 Discussion

The findings of this study were in consonance with the findings of Van Wart et al. (2020), who shed light on the potential ramifications of the COVID-19 aftermath as many universities around the world jumped from relatively low levels of online instruction at the beginning of 2020 to nearly 100% as the pandemic progressed. As a result, the university was unprepared to hold full virtual classes, as evidenced by students' dissatisfaction with the university's online learning system. The findings were in consonance with Puriwat and Tripopsakul (2021) who surmised that the lack of communication opportunities and difficulties in the interaction between learners and instructors online must be compensated for by thorough course and content design, congeneric with appealing page designs, clear course structures, and easily usable engaging content. Puriwat and Tripopsakul (2021) remarked that one of the most common mistakes university instructors made was failing to change or redesign the instructional materials they regularly use for offline learning to better suit the needs of e-learning. This was also reflected in research conducted at Bugema University, which called into question the adequacy of e-resources for the programme of study.

Puriwat and Tripopsakul (2021) conjecture that universities sought to provide adequate support in the form of e-learning administration and technical provision, not only to students but also to lecturers who may be unfamiliar with e-learning platforms, in line with the findings of poor technical support and university internet services (Eduroam) at Bugema University. They point out that without the proper training and materials for both hardware and software, instructors' and students' capacity to use e-learning methods may be jeopardised. Universities should make sure that students can always get help, and they should also provide free internet access to both lecturers and students.

This line of thought is congruent with Means and Neisler (2020) that many students in online learning environments reported experiencing bad internet connectivity, making it impossible for them to attend synchronous sessions of online teaching regularly. These findings are also in harmony with the effect that learners faced teacher-related, psychosocial, socioeconomic, management, and technical challenges, and as such tabled that both teachers and learners be trained in the use of technology, including the provision of technology tools, internet connectivity, and ensuring a stable power supply.

Spelt out also was inadequacy of e-resources for the programmes of study in the university with a low degree of engagement with the loaded content, and more interactivity between lecturers and students in Bugema University. This consorts with Means and Neisler (2020) that during online learning, in contrast to face-to-face learning, there are limited opportunities for student collaboration with their peers and lecturers, with the opportunity for collaboration worse in an online learning environment compared to the traditional classroom environment. In a related manner, Ouma (2021) discovered that the efficiency of virtual learning was hampered by the challenges of limited data, unpredictable internet connection, failure to record lectures, few Zoom links at the faculty level, limited class control, and unstable student participation.

Meanwhile, the findings on the need for improved technical expertise in virtual learning do augur with the observation of Ouma (2021) that insufficient skills and technical support are major barriers to the provision of online learning services in many universities, particularly in the developing world. He also points out that while many teachers use technology in basic teaching activities analogous with PowerPoint and emails, there is minimal technical assistance for employing ICTs to supplement virtual learning.

This survey discovered that the majority of students had difficulty using the university's LMS. This was in agreement with Sahu (2020) that many students and a few academic staff were unable to access and use institutional learning management systems due to a shortage of computers and reliable internet connections.

Querying the future of high education distance learning in Canada, the United States, and France, Wotto (2020) sought insights from and before COVID-19 and found that evolving information and communication technology creates new spaces, learning materials, and demands in training institutions concerning virtual service delivery. This is related to the study at Bugema University where various learning materials and training have been carried out to enhance their virtual service delivery. Wotto (2020) also found that HE distance learning responses to these transformations are miscellaneous and its development strategies vary from one country to another.

In China, Zeng and Wang (2021) examined *college student satisfaction with online learning, with a focus on those studies investigating the elements of the online courses designed by the instructors who moved face-to-face courses to distance during the COVID-19 pandemic and found that due to individual differences, some students can do well in online courses while other students may not be able to do well in online courses. Their review identified elements in the online course design that contribute to student online learning satisfaction and evinced that instructors could proactively help student online learning by modifying online course components. This review is quite similar to the submissions in this study where lectures play a key role in the satisfaction of students during the virtual learning service delivery. Hence the need for universities to invest in their academic staff training in virtual service delivery.*

Bawa'aneh (2021) carried out a study investigating students' satisfaction, attitudes, and challenges in the United Arab Emirates public schools during the virtual learning phase of the academic year due to the COVID-19 pandemic and found that student satisfaction level, attitudes, and challenges were found to lie within the 'strong' category indicating high satisfaction level, positive attitude, with minimal challenges faced during the virtual learning. Bawa'aneh (2021) attributed this positive result to the fact that students of the United Arab Emirates public schools were partially exposed to electronic before the pandemic, the tremendous efforts of the ministry of education to guarantee smooth transfer to full distance learning after the outbreak of the pandemic, and the well-established infrastructure of the country. However, unlike the UAE which had partially exposed its learners to virtual learning technologies before the pandemic, Uganda like most sub-Saharan countries was exposing its students to these facilities during the pandemic, albeit ineffectively. This study helps cement the findings of this study on the need for proper investment by the ministry of education and other educational stakeholders for virtual learning to be of great benefit to learners.

Muftahu (2020) explored the implications of the COVID-19 pandemic to HE by identifying matters arising and the challenges of sustaining academic programmes with specific attention to developing universities in the African context and found that the COVID-19 pandemic pushed universities in different nations beyond their limits towards developing appropriate and creative alternatives congenial to transitioning to remote learning, training of academic staff in the use of online instructional materials and tools, and encouraging students to complete their educational requirements through online learning in response to the COVID-19 pandemic. This case was found to be true in a Ugandan context as well as universities were involuntarily forced into the virtual learning mode rather than voluntary.

Mbiydzenyuy (2020) carried out a study on Teaching and Learning in resource-limited settings in the face of the COVID-19 pandemic. The closure of schools did not only interfere with the traditional patterns of socialisation but intercepted the academic sequence and plunged ill-prepared educational systems reminiscent of African universities into a cosmos of teaching and learning uncertainties. African universities were ill-prepared to cope with the threats to their educational systems and needed to adapt and or influence the transition. Despite having a teeming youth birth into technology, most African universities weren't cognitively and technically adept to teach or be taught within the accredited online e-learning platforms, sadly this was the case with this study.

11 Conclusion and Recommendation

From this study, it is imperative to note that for African universities and other institutions of HE, to harness the usage of e-learning platforms as a method of increasing their learning and student happiness, must learn vital lessons on the importance of electronic infrastructure. Universities and stakeholders should invest in educating their teachers on the most up-to-date instructional tools to avoid falling behind. Most importantly, as a result of the increase in the usage of online platforms as a result of the COVID-19 pandemic, universities may now have the most prominent professors from the greatest universities all over the world educate their students from thousands of miles away.

According to the findings, designing effective e-learning courses is just as important for teaching and learning quality as it is for face-to-face classrooms. Second, to sustain the quality of teaching and learning throughout the COVID-19 pandemic, e-learning instructors should adjust their content to match the virtual learning needs rather than simply using instructional materials previously employed in face-to-face classrooms. The failure of lecturers to change or rethink the teaching materials they frequently use for offline learning to better suit the needs of e-learning was detrimental.

There are numerous lessons to be learned for African countries, particularly Uganda, in terms of virtual learning, as IHE must create skilled, available technical teams to completely support their professors and students during and after virtual learning sessions. Training courses on the use of the various hardware and software components that are required for efficient virtual learning experiences are required. The availability of consistent, efficient, and dependable internet services is the foundation for effective and productive e-learning experiences. As a result, countries must invest in lowering the cost of the internet by providing subsidies to institutions.

Universities and educational stakeholders will need to pay close attention to the cost of internet bandwidth to enable online teaching and learning. Students may be required to pay additional fees in addition to their tuition fees, or universities may be required to charge an online fee to provide stable and reliable internet data for use; however, the high data costs are likely to prevent learners from fully participating in online learning. It is vital to emphasise that even the best internet infrastructure would be useless without a reliable power supply and adequate infrastructure, both of which require government investment.

Universities ought to work with their governments to obtain ICT devices that will allow students to participate more fully. This may have been accomplished through the distribution of radios and the sponsorship of televised lessons in secondary and primary schools, but more needs to be done for tertiary and IHEs. Finally, universities must establish specific specialised teams of specialists to give routine guidance and support to students and other users of all e-learning platforms in use, as digital illiteracy is a barrier to effective virtual learning.

Future studies examining student satisfaction with online learning should aim for a multi-university-based distribution, as well as collect other demographic characteristics related to socioeconomic level, internet connection availability, and online learning technologies. In addition, because this study was undertaken in the early phases of the pandemic, consideration should be designated so as to carrying out a similar study in the later stages of the pandemic.

REFERENCES

Adarkwah, M. A. (2021). An outbreak of online learning in the COVID-19 outbreak in sub-Saharan Africa: Prospects and challenges. *Global Journal of Computer Science and Technology.*

Alhumaid, K., Ali, S., Waheed, A., Zahid, E., & Habes, M. (2020). COVID-19 & eLearning: Perceptions & Attitudes of teachers towards E-learning acceptance in the developing countries. *Multicultural Education, 6*(2), 100–115.

Al-Omairi, A. R. A., & Hew, S. H. (2022). Students satisfaction of online learning in Oman. *F1000 Research, 11*(101), 101.

Altinyelken, H. K. (2010). Pedagogical renewal in sub-Saharan Africa: The case of Uganda. *Comparative Education, 46*(2), 151–171.

Ann, L., & Aziz, Z. (2022). Avatars meet face-to-face: Learning leadership online: A thematic analysis of East-African perspectives. *Journal of Leadership Education, 21*(1).

Aslam, S., & Sonkar, S. K. (2021). *Platforms and tools used for online learning all over the world during Covid-19: A study. Library philosophy and practice* (pp. 1-18).

Bandura, A. (1986). Social foundations of thought and action (pp. 23–28). Englewood Cliffs, NJ.

Bawa'aneh, M. S. (2021). Distance learning during COVID-19 pandemic in UAE public schools: Student satisfaction, attitudes and challenges. *Contemporary Educational Technology, 13*(3).

Bugema_University. (2020). *Blended learning policy for Bugema (BU-BL POLICY version 1). POLICY NO. BU/BLP/1.0 (pp. 1–12).* Bugema University.

Campbell, J. D. (2021). *Understanding the recruitment, admissions, and enrolment experiences of non-traditional students across generations.* Doctoral dissertation,. North-eastern University.

Chen, T., Peng, L., Yin, X., Rong, J., Yang, J., & Cong, G. (2020, September). Analysis of user satisfaction with online education platforms in China during the COVID-19 pandemic. In *Healthcare* (Vol. 8, No. 3, p. 200). Multidisciplinary Digital Publishing Institute.

Cifuentes, L. (2021). *Course designs for distance teaching and learning.* In A guide to administering distance learning (pp. 174-205).

Denoncourt, J. (2020). Companies and UN 2030 sustainable development goal 9 industry, innovation and infrastructure. *Journal of Corporate Law Studies, 20*(1), 199–235.

Dhillon, S., & Murray, N. (2021). An investigation of EAP teachers' views and experiences of e-learning technology. *Education Sciences, 11*(2), 54.

Duraku, Z. H., & Hoxha, L. (2020). *The impact of COVID-19 on education and the well-being of teachers, parents, and students: Challenges related to remote (online) learning and opportunities for advancing the quality of education.* Retrieved online from https://www.researchgate. Net/publication/ 341297812.

Faturoti, B. (2022). Online learning during COVID-19 and beyond: A human right based approach to internet access in Africa. *International Review of Law, Computers & Technology, 36*, 1–23.

Febro, J., Catindig, M., & Caparida, L. (2020). Development of E-learning module for ICT skills of marginalized women and girls for ICT4D. *International Journal of Emerging Technologies in Learning (iJET), 15*(16), 94–105.

Graham, M. (2021). *Impact of online education on student success outcomes and institutional Effectiveness: Study of Florida State University System.*

Hassan, A., Alazzeh, D., Leung, D., Sidhva, D., & Obasi, C. (2020). Investigating students' support for learning experience during COVID-19 & the way forward.

James, P. C. (2021). What determines student satisfaction in an E-learning environment? A comprehensive literature review of key success factors. *Higher Education Studies, 11*(3), 1–9.

Jung, I. (2011). The dimensions of e-learning quality: From the learner's perspective. *Educational Technology Research and Development, 59*(4), 445–464.

Kaisara, G., & Bwalya, K. J. (2021). Investigating the E-learning challenges faced by students during COVID-19 in Namibia. *International Journal of Higher Education, 10*(1), 308–318.

Kharbat, F. F., & Abu Daabes, A. S. (2021). E-proctored exams during the COVID-19 pandemic: A close understanding. *Education and Information Technologies, 26*(6), 6589–6605.

Kilburn, B., Kilburn, A., & Davis, D. (2016). Building collegiate e-loyalty: The role of perceived value in the quality-loyalty linkage in online higher education. *Contemporary Issues in Education Research, 9*(3), 95–102.

Kizza, J., Kasule, W., Amonya, D., Nakimuli, L., & Komugabe, A. (2021). Perceptions towards the effectiveness of E-learning in private and public universities in Uganda: A comparative study. *East African Journal of Arts and Social Sciences, 3*(1), 156–169.

Magni, D., & Sestino, A. (2021). Students learning outcomes and satisfaction. An investigation of knowledge transfer during social distancing policies. *International Journal of Learning and Intellectual Capital, 18*(4), 339–351.

Mbiydzenyuy, N. E. (2020). Teaching and learning in resource-limited settings in the face of the COVID-19 pandemic. *Journal of Educational Technology and Online Learning, 3*(3), 211–223.

Means, B., & Neisler, J. (2020). *Suddenly online: A national survey of undergraduates during the COVID-19 pandemic.* Digital Promise.

Mpungose, C. B. (2021). Students' reflections on the use of the Zoom video conferencing technology for online learning at a South African university. *International Journal of African Higher Education, 8*(1), 159–178.

Muftahu, M. (2020). Higher education and Covid-19 pandemic: Matters arising and the challenges of sustaining academic programs in developing African universities. *International Journal of Educational Research Review, 5*(4), 417–423.

Namboodiri, S. (2022). Zooming past "the new normal"? Understanding students' engagement with online learning in higher education during the covid-19 pandemic. In *Re-imagining educational futures in developing countries* (pp. 139–158). Palgrave Macmillan.

Otto, D., & Becker, S. (2019). E-learning and sustainable development. *Encyclopedia of sustainability in higher education*, 475–482.

Ouma, R. (2021). Beyond "carrots" and "sticks" of online learning during the COVID-19 pandemic: A case of Uganda Martyrs University. *Cogent Education, 8*(1), 1974326.

Palau, R., Fuentes, M., Mogas, J., & Cebrián, G. (2021). Analysis of the implementation of teaching and learning processes at Catalan schools during the Covid-19 lockdown. *Technology, Pedagogy and Education, 30*(1), 183–199.

Pande, J., & Mythili, G. (2021). Investigating student satisfaction with online courses: A case study of Uttarakhand Open University. *International Journal of Information and Communication Technology Education (IJICTE), 17*(3), 12–28.

Peace, T. A. (2021). COVID 19 Pandemic and Lockdown: Examining the Challenges and Opportunities of Online Teaching of Religious Education in Institutions of Higher Education in Uganda. *Interdisciplinary Journal of Virtual Learning in Medical Sciences, 11*(2), 1–3.

Perez, M. L. (2022). *Strategies to improve student satisfaction and Foster belonging among students in online programs.* In quality in online programs (pp. 137-151).

Prior, J., & Woodward, T. (2017). *Sustainable development goal 5: Achieve gender equality and empower all women and girls.* Integrating global issues in the creative English language classroom: With reference to the United Nations sustainable development goals, *57.*

Puriwat, W., & Tripopsakul, S. (2021). The impact of e-learning quality on student satisfaction and continuance usage intentions during covid-19. *International Journal of Information and Education Technology, 11*(8), 368–374.

Rabin, E., Kalman, Y. M., & Kalz, M. (2019). An empirical investigation of the antecedents of learner-centred outcome measures in MOOCs. *International Journal of Educational Technology in Higher Education, 16*(1), 1–20.

Reed, A., & Thompson, K. M. (2021). Never waste a crisis: Digital inclusion for sustainable development in the context of the COVID pandemic. *Library Journal, 40*(2), 14.

Rieser, R. (2012). *Implementing inclusive education: A commonwealth guide to implementing Article 24 of the UN Convention on the Rights of Persons with Disabilities.* Commonwealth Secretariat.

Rofingatun, S., & Larasati, R. (2021). The effect of service quality and reputation on student satisfaction using service value as intervening variables. *The International Journal of Social Sciences World (TIJOSSW), 3*(01), 37–50.

Sahu, P. (2020). Closure of universities due to coronavirus disease 2019 (COVID-19): Impact on education and mental health of students and online. *Cures, 12.*

Saxena, C., Baber, H., & Kumar, P. (2021). Examining the moderating effect of perceived benefits of maintaining social distance on e-learning quality during the COVID-19 pandemic. *Journal of Educational Technology Systems, 49*(4), 532–554.

Secretariat, C. (2017). *Ensure inclusively and equitable quality education and promote lifelong learning opportunities for all* (SDG 4).

Sher, A., & Toor, A. (2021). Achieving success in online teaching and learning: Strategies. In *Teaching in the post COVID-19 era* (pp. 393–401). Springer.

Spencer, D., & Temple, T. (2021). Examining Students' online course perceptions and comparing student performance outcomes in online and face-to-face classrooms. *Online Learning, 25*(2), 233–261.

Subekti, A. S. (2021). Covid-19-triggered online learning implementation: Preservice English teachers' beliefs. *Metathesis: Journal of English Language, Literature, and Teaching, 4*(3), 232–248.

Tj, H. W., & Tanuraharjo, H. H. (2020). The effect of online learning service quality on student satisfaction during the COVID-19 pandemic in 2020. *Journal Management Indonesia, 20*(3), 240.

Tran, Q. H., & Nguyen, T. M. (2022). Determinants in student satisfaction with online learning: A survey study of second-year students at private universities in HCMC. *International Journal of TESOL & Education, 2*(1), 63–80.

Tuah, N. A. A., & Naing, L. (2021). Is online assessment in higher education institutions during the COVID-19 pandemic reliable? *Siriraj Medical Journal, 73*(1), 61–68.

Tuan, L., & Tram, N. (2022). Examining student satisfaction with online learning. *International Journal of Data and Network Science, 6*(1), 273–280.

Tuma, F. (2021). The use of educational technology for interactive teaching in lectures. *Annals of Medicine and Surgery, 62*, 231–235.

UNESCO. (2017). *Education for sustainable development in Uganda: A situational analysis.*

Van Wart, M., Ni, A., Medina, P., Canelon, J., Kordrostami, M., Zhang, J., & Liu, Y. (2020). Integrating students' perspectives about online learning: A hierarchy of factors. *International Journal of Educational Technology in Higher Education, 17*(1), 1–22.

Wengrowicz, N., Swart, W., Paul, R., Macleod, K., Dori, D., & Dori, Y. J. (2018). Students' collaborative learning attitudes and their satisfaction with online collaborative case-based courses. *American Journal of Distance Education, 32*(4), 283–300.

Wotto, M. (2020). The future high education distance learning in Canada, the United States, and France: Insights from before COVID-19 secondary data analysis. *Journal of Educational Technology Systems, 49*(2), 262–281.

Wu, J. H., Tennyson, R. D., & Hsia, T. L. (2010). A study of student satisfaction in a blended e-learning system environment. *Computers & Education, 55*(1), 155–164.

Yunusa, A. A., & Umar, I. N. (2021). A scoping review of critical predictive factors (CPFs) of satisfaction and perceived learning outcomes in E-learning environments. *Education and Information Technologies, 26*(1), 1223–1270.

Zamzami, I. (2021). The factors influencing the acceptance of web-based E-learning systems among academic staffs of Saudi Arabia. *Future Computing and Informatics Journal, 6*(2), 5.

Zeng, X., & Wang, T. (2021). College student satisfaction with online learning during COVID-19: A review and implications. *International Journal of Multidisciplinary Perspectives in Higher Education, 6*(1), 182–195.

Sustainability and the Triple Mission of the University: *Uganda Martyrs University in Perspective*

Christopher Mukidi Acaali ⓘ

1 INTRODUCTION

This chapter, structured into seven sections, focuses on the sustainability of higher education institutions (HEIs), emphasizing their triple mission with Uganda Martyrs University (UMU) as a case in point. The chapter considers that the four pillars of sustainability as well as the triple mission of teaching and learning, research and community engagement can propel HEIs to implement United Nations (UN) sustainable development goals (SDGs). In light of this, the chapter highlights the pillars of sustainability, the relation between higher education (HE) and SDGs, the triple mission of HEIs (with UMU in perspective), the way forward for UMU and UMU's success stories as benchmarks for other HEIs. The chapter pleads that HEIs, through the sustainability agenda and triple mission of teaching, learning and community engagement, can play a pivotal role in

---

C. M. Acaali (✉)
Uganda Martyrs University, Nkozi, Uganda
e-mail: acaalimukidi@gmail.com

© The Author(s), under exclusive license to Springer Nature
Switzerland AG 2024
P. Neema-Abooki (ed.), *The Sustainability of Higher Education in Sub-Saharan Africa*, Sustainable Development Goals Series,
https://doi.org/10.1007/978-3-031-46242-9_6

the implementation of SDGs and thus contribute meaningfully to the social and economic development of society.

2 PILLARS OF SUSTAINABILITY

The term "sustainability" refers to four distinct but interrelated areas, namely, human, social, economic and environmental—known as the four pillars of sustainability. Benn et al. (2014) contend that human sustainability is the preservation and improvement of human capital in society. To promote human sustainability, and thus preserve and improve human capital, investments have to be made in those areas that contribute to its perpetuity or continuity, namely, health, nutrition, knowledge and skills as well as the education system. In the context of HE, human sustainability consists in investing heavily in the human capital that is directly involved with the various facets of education. Borrowing from Benn et al. (2014), this implies that any human capital directly or indirectly involved in the teaching and learning of students and eventual churning out of graduates from HEIs or the services they render to society would be an important focus for human sustainability. Looking at the end product (the graduate), the focus would therefore be placed on the development of skills and human capacity to support the functions and promotion of the well-being of society (Benn et al., 2014).

Unlike human sustainability which focuses on human capital, social sustainability instead pays attention to capitalizing and generating services that constitute the framework of society, the concept visualizes the preservation and protection of humanity's future generations with the acknowledgement that people's actions can impact others in particular and the world in general. Weighing this concept of impact, the concern of social sustainability is to maintain and improve the quality of life in society with an emphasis on cohesion, equality, reciprocity, honesty and relationships (Diesendorf, 2000). Deriving from the concept of social sustainability, HE focuses on investing and generating services (cognate to Quality Assurance) that would contribute to the well-being of society. Education, as a public good, promotes to some extent social sustainability as its impact is intended to improve the well-being of the present as well as of future generations. Social sustainability determines that what is done in and through HE can create an impact not only on individuals but also on the whole world. In other words, HE, using its arm of social sustainability, can contribute to the well-being of the people and the world in these present and future times.

The third pillar of sustainability is economic sustainability which some scholars have understood to mean the improvement of people's standards of living (Anand & Sen, 2000). Based on the concept of economic sustainability, the impact of HE would be realized when it positively affects people's standards of living. Besides improving people's standards of living, the United Kingdom (UK) Government in its Annual Report understood economic sustainability in the business sense to mean "the efficient use of assets to maintain company profitability over time." It observed: *"Maintaining high and stable levels of economic growth is one of the key objectives of sustainable development. Abandoning economic growth is not an option. But sustainable development is more than just economic growth. The quality of growth matters as well as the quantity"* (Annual Report, 2000 and January 2001). Based on the UK's assessment of the economic sustainability of the business, the following can be observed concerning HE. To achieving its objectives, HE could use a business model such that in the final analysis it would contribute to improving people's economic standards of living. Like business, education would efficiently utilize all its assets to bring this about. Educational assets include things like curricula, quality assurance, human capital (human resource), students who are its key stakeholders, Alumni, parents, and Government and non-government organizations especially those that have a bearing on education. It is through the efficient use of these "assets" that HE could be said to maintain "profitability" over time. Based on the above, abandoning *or* ignoring these educational assets in HE, therefore, is not an option.

The fourth pillar is environmental sustainability which is the maintenance and improvement of the well-being of humanity through the protection and responsible use of natural resources, ditto land, forests, water, air, animals and minerals. The focus of this pillar is to ensure that the current generation's needs are not met at the expense of future generations. This pillar is marked with signs of "responsible use of the earth's resources" sparingly for the future generations (Dunphy & Griffiths, 2007). The same emphasis is made by Pope Francis in his reference to humanity's promotion of pollution of the earth caused by industrial and transport fumes, harmful wastes and residues, loss of biodiversity due to ill use of natural wetlands, forests and woodlands, species of plants and animals in the name of economic development (Encyclical Letter Laudato Si of the Holy Father Pope Francis, 2015). In the context of HE, this fourth pillar does not directly affect human capital. But since the actions of the human person on the environment when used irresponsibly can affect others and especially those of future generations, one can say that indirectly HE would be affected since its sole purpose is to serve, maintain and improve the human capital and all its environs. Moreover, there would be no future human beings to serve in HE, when the environment and the entire eco-system that sustains them are destroyed.

To sum up, the thread that holds together the four pillars—that is, human, social, economic and environmental—sustainability is the overall focus on the well-being of individuals and society. Along with this focus, emphasis through all four pillars is the concern for the well-being of the present as well as future generations. Environmental sustainability, for example, advocates for the responsible use of natural resources by the present generation because of the future generations. Interestingly, these objectives are also the primary concern of HE. The well-being of individuals and society is the primary objective of HE. While HE focuses on the present needs of the present generation, its aspirations and planning are aimed at making the future better than it is today. Finally, the four pillars of sustainability contribute to the well-being of individuals and society. HE, through these pillars, can contribute to the well-being of individuals and society, especially through the implementation of SDGs.

3   HIGHER EDUCATION AND SUSTAINABILITY (SUSTAINABLE DEVELOPMENT GOALS)

The question to consider is whether or not there is a relationship between HEIs and Sustainability (and/or the UN) SDGs. Though the response is evidently in the affirmative, the subsequent challenge is to describe the role HEIs can play in the implementation of the SDGs. This section therefore shall attempt to show the relationship between HEIs and SDGs and also highlight the role HEIs can play in the implementation of SDGs.

The 17 SDGs of the UN 2030 agenda that were promulgated and adopted by member states in 2015 cover the social, economic, environmental and technological aspects of the world's development programme (Chankseliani & McCowan, 2021). In the SDGs, the needs of the educational domain from the primary and secondary sectors as contained in the Millennium Development Goals (MDGs) and Education for All (EFA) were expanded to include Tertiary education of which HEIs are a key part. The point of intersection between HE and SDGs is in SDG 4 where the emphasis is put on equal accessibility of learning opportunities for all. The relationship, therefore, is that HEIs are part of the SDGs agenda of the United Nations. Without HEIs, it would be impossible to implement the SDGs. Hence, HEIs are not only part of the SDGs agenda but rather play a fundamental and vital role in the implementation of SDGs.

The relation between HEIs and SDGs ultimately leads to the role HEIs can play in the implementation of SDGs. Worldwide and particularly in Africa, HEIs until recently were perceived to be ivory towers purposely intended for the elites and forming them for the religious, professional and administrative positions. Education was so specific that those who went through it became skilled to perform specific jobs. For the rest of the population whose ambitions were not directed at getting these jobs, HE remained mostly unrelated to the solving of their societal issues. HEIs, for their part, saw themselves as only meant for the generation of knowledge which for the most part was meant to be used by other scholars. Indeed, scholars of various levels used this research to complete their research projects leading to completion and graduation. Until recently, some of the research done in some HEIs, especially in the arts section, tended to be too theoretical to be used meaningfully for the development of society. Amidst such developments in HEIs, communities surrounding such HEIs languished in all forms of underdevelopment manifested, for instance, in abject poverty, hunger and disease.

However, recently HE, particularly through its generation of knowledge in research and innovations, took on a new and advanced role of contributing to the development agenda of society (Ekene & Oluoch-Suleh, 2015). This was and is especially exposed through the churning out of an increased number of professionals connate with teachers, doctors, nurses and engineers who have in many different ways impacted positively on the well-being of society. HEIs have not only impacted positively on society by producing such professionals but rather some of their research particularly in social sciences have contributed to the understanding of life and the place of the human person in society.

Recently, several HEIs have jumped on the bandwagon of directing their research towards solving society's challenges. Maia et al. give the example of the University of Pretoria in South Africa which directed its research energies towards solving the continents' food security. The other university is Ahfad for Women in Sudan whose programmes of teaching, research and community engagement are directed at supporting women as change agents in their communities (Maia et al.). Interestingly, the UN has identified these universities as Hubs for the implementation of SDGs.

By so doing, the UN has implicitly admitted that HEIs are key partners in the implementation of SDGs. In addition to these initiatives, the formation of The Times Higher Education (THE) University Impact Rankings in 2019 raised the stakes higher by gauging universities' rankings based on their implementation of SDGs. This impact ranking measure recognized more than 700 universities throughout the world based on their research targeted at addressing the social and economic challenges of the world. Maia et al. cite three universities as examples, namely, the University of Auckland in New Zealand for its research on the utilization of earthly ecological systems, the Tongji University in China for its research on inexpensive clean energy and the University of Sao Paulo in Brazil for its research on poverty reduction (Maia et al.).

Basing on this new thinking and the desire to be relevant to society, HEIs can no longer afford to remain in ivory towers and indifferent to society's contemporary needs under the guise of being conscious thinkers and generators of knowledge. With the emerging trend of wanting to provide solutions to the social and economic challenges of society, HEIs are beginning to see themselves having a role to play in implementing SDG 4 in the first place but also in being drivers for the implementation and achievement of the other 16 SDG goals as well. Scholars like Serafini et al. (2022) contest that HEIs not only have a role but are strategic partners in the achievement of the SDGs through their key tripartite mission of teaching, research and community engagement (Serafini et al., 2022).

Besides carrying out this mission, Garcia-Feijoo et al. (2020) uphold that HEIs have the noble duty of making the population aware of the SDGs but also assist them to acquire the skills and competencies needed to respond to "the challenges of sustainable development" (Garcia-Feijoo et al., 2020). HEIs, therefore, are charged with the duty of knowing and understanding the 17 SDGs so that they can repackage them for implementation by the population. To sum up, HEIs not only have a role to play but are key partners in the implementation of all SDGs. None of the SDGs would be implemented without their direct role (SDSN Australia/Pacific, 2017). For HEIs to act as vehicles for the implementation of SDGs, serious attention, therefore, should be directed to the implementation of the triple mission of teaching and learning, research and community engagement.

4 THE TRIPLE MISSION OF HIGHER EDUCATION INSTITUTIONS

The three-fold phenomenon of teaching, learning and community engagement contributes to the HEIs' implementation of SDGs. Teaching is understood as the exchange of knowledge through a lecture room context, and research is the art of using knowledge to discover more and particularly to intervene in and engage with the community's or society's activities.

4.1 Teaching and Learning

Worldwide, HEIs exist within societies or communities. Graduates from these institutions end up going back to live and work in society. This implies, therefore, that the graduate should be equipped with the skills and expertise to enable him or her to live and work within society; in other words, the graduate should be employable. In addition, the graduate should be formed by a curriculum that has been influenced by what is happening in society. The challenges and general issues in society to some extent should influence or even be determinants of the curriculum developed, taught and learnt in HEIs. In that way what is taught and learnt by students will positively impact the social and economic well-being of society.

Contrary to the above, some curricula in HEIs are developed without the input of society as a key stakeholder. Developers of curricula in HEIs, even though they may claim to have benchmarked with the potential users of the programme, usually do not have in mind the social, economic and other needs of society. Instead of advancing the approach of addressing society's challenges and providing solutions, HEIs, for the most part, move around with made-up solutions to address problems that they have not identified in society. As opposed to the curriculum being influenced by society, the curriculum in HEIs is more anchored on the possible numbers of students the new programme may attract than on the social and economic impact the programme may create on society. The same can be said of reviewed programmes. Inasmuch as stakeholders relatable to current students, alumni and parents may be consulted on the review of a certain programme, usually, the interest of some reviewers is not so much the impact the programme may create on society but rather the possible clients the programme may attract; plus, the fulfilment of the requirements for accreditation by the National Council for Higher Education (NCHE). Thus, to some extent, the needs of society may not necessarily inform the curricula when it is being designed, developed and reviewed. This may partly explain why some graduates from HEIs fail to get employment. For excellent curricula to be developed, Jure Erjavec recommends that various stakeholders be involved in its development and/or review (Jure Erjavec, 2021).

With curricula aside, the subsequent focus should be turned to the mode of delivery of knowledge in HEIs. Until recently, many HEIs in Uganda, Africa and the world at large were using the traditional methods of teaching and learning where the lecturer was understood as the "possessor of knowledge" and the students as "the receiver of that knowledge." While the lecturer did most of the talking, the students only listened. As Paulo Freire long ago put it, it is assumed that the students have nothing to contribute and their minds are empty—tabula rasa (Paulo Freire, 1993). At the time of assessment or the end of the lecture, the lecturer will expect students to reproduce exactly what they were taught. Such a lecture is accompanied by little interaction. Some graduands on the day of graduation look forward to being employed by other people. Once the euphoria of graduation has cooled down, many of them begin the endless treks to cities and towns searching for employment opportunities. This kind of education, unfortunately, tunes students to be job seekers for white-collar jobs as opposed to acquiring skills and expertise that would make them job creators and hence positively impact society. In reversing this type of education and thus creating job makers as opposed to job seekers, Garcia-Feijoo et al. (2020) advocated for High Impact Entrepreneurship Education Practices (HIEEP) that comprise Entrepreneurship Internship Programme, Business Incubation Programme and Entrepreneurial Supportive Environment, entrepreneurial self-employment and entrepreneurial education knowledge. In their study, where these practices were sustained, graduates tended to be more thinkers but above all creators of their employment (Subramaniam Sri Ramalu et al., 2020).

UMU, while implementing the project "Transforming Employability for Social Change in East Africa" (TESCEA), funded by Strategic Partnerships for Higher Education Innovation and Reform (SPHEIR) programme, emphasized the need to categorically reflect on how teaching and learning at the university were done. TESCEA called for a paradigm shift from teacher-centred to student-centred teaching where the student, through critical thinking, problem-solving and gender-sensitive approaches, is empowered to navigate his/her learning right from the admission to graduation. In the language of TESCEA, students are taught how to think and not what to think. Consequently, in assessments, students are required not to reproduce verbatim what their teachers have taught them but rather use the ideas and problem-solving skills that address challenges identified in their communities (Ngabirano, Maximiliano on TESCEA, 2017).

Since the inception of this new approach to learning, several positive changes have occurred concerning the teaching experience of the lecturers and the learning experience of the learners. Majorly due to the emphasis on competencies of aspects like problem-solving skills, and gender-sensitive approaches plus the involvement of business and communities in the learning process of students, particularly through guest lecturers, both learners and lecturers have found this new approach interesting and relevant. In addition, these new approaches have bridged the gap between academia and business/industry. The formation of the Joint Advisory Group (JAG) in UMU but also other partner universities, for example, is a clear case in point. JAG, which is an assembly of people representing the community, businesses, industry, the non-government organization (NGO) world and government, has assisted UMU, especially in the redesigning of its programmes to make them more relevant to the needs of society. By so doing, through JAG, the relationship between academia (UMU) and society was enhanced. But even more importantly, JAG assisted UMU to come up with redesigned programmes that were practical and user-friendly for students (Ngabirano, ibid.).

4.2   Research

Research, as a key component of HEIs, should play a key role in the socio-economic development of society. Scholars like A.B.K. Kasozi have made a causal link between HE and economic development (A.B.K. Kasozi, 2016). Scholars like Carita Lillian Snellman (2015) have grappled that for Europe to develop and compete favourably with her competitors in the socio-economic development of her people, the way to go is to invest massively in HE and particularly in the areas of research and development. In her perspective, HEIs can no longer afford to remain as ivory towers, insisting on only being knowledge producers and churning out a class of great thinkers, but rather they have to develop a new mindset that is aimed at developing a crop of researchers and people of skills who will meet the ever-increasing challenges and demands of society.

She states further: that if Europe [is to aim] at being competitive on the global stage, HE will have to assume a central role in creating facilities for meeting the demands of a knowledge society.

With the onset of independence of most countries in Africa in the 1960s, African leaders committed to investing in HEIs as a way to prepare the human capital that would fill the spaces of the civil servants and professionals of the departing expatriates. Investments in HEIs were understood to be the key to Africa's social and economic development. In light of this, the United Nations Educational Scientific and Cultural Organization (UNESCO) in 1962 and the Association of African Universities (AAU) in 1972, both underscored the role HE was going to play in the development of Africa (Cloete et al., 2011). African leaders' mindset on the significance of investing in HE was well captured in the Accra Declaration's statement that insisted that "all universities must be development universities" (Yesufu, as cited by Nico Cloete et al., 2011). Unfortunately, this noble route to a great extent of the period after independence was abandoned in favour of the World Bank and other international bodies' priorities of funding socio-economic activities directly without using the arm of research of HEIs in Africa (Nico Cloete et al.).

In tandem with the new teaching pedagogies, plus the new thinking of understanding the role of HEIs as contributing to the socio-economic development of society, UMU undertook initiatives that related research to development. One such initiative is the programme Entrepreneurship for Impact (E-4 Impact) in Masters of Business Administration where applicants are admitted to the programme and have identified a research project that addresses a specific challenge within one's community. An example is a student who identified fistula as a problem that was affecting young women in her Eastern Region whose private parts were torn at the time of birth of their children. The mothers not only suffered physically but also emotionally, especially arising from the stigmatization of society. The student, though not a medical doctor nor a nurse, through her research that sought to advocate dignity for her patients, was able to concretize her research project into a fistula hospital in Soroti City that is now serving the women of Soroti and Uganda as a whole. The research of the E-4 Impact Programme orients its students to address the social and economic challenges of their communities.

The other example is where UMU students and staff are engaged in studying the Black Soldier Fly Larvae as a source of high-value feeds for both animals and human beings. The Black Soldier Fly Larvae, which students and staff have used as an alternative to maize and soya feeds mixed with fish, is anchored on the premise that because of the ever-increasing human population, humans and animals in the future will compete for the same foods (Nina et al., 2023). This is also against the understanding that because of the increasing human population, the environment is increasingly being overused which affects soya production. The misuse of the environment, for instance, by polluting rivers and lakes with plastic bags and other synthetic materials has also contributed to the reduction of some fish species worldwide.

UMU's research projects through students of Agroecology have focused on identifying the Black Soldier Fly, its diet and its feeding formula that boosts the hatching of eggs as well as "larval biomass gain." In addition, students, working with selected communities, have been able to identify the bio-waste forms that feed the flies (Nina et al., 2023.). The studies so far show evidence that animal growth based on Black Soldier Fly larvae diets compares favourably with conventional feeds. This research helps students of Agroecology to learn new knowledge and basic techniques of farming but even more importantly enables the UMU community to share their research knowledge with farmers in the community. About 20,000 smallholder farmers in the nearby communities who have learnt the Black Soldier Fly Larvae technology boast of now using cheap quality feeds as well as reducing production costs in maintaining their animal farms (Nina et al.). These researches at UMU are some of the interventions that HEIs can make in contributing to the socio-economic development of their neighbouring communities.

4.3 Community Engagement

The third mission consists of HEIs engaging in activities that address the social and economic challenges of society; ably captured by the study of Nabaho et al., in their description of what they called the third mission of African Higher Education Institutions (AHEI) (Nabaho et al., 2022).

Community engagement is one of the three pillars, besides teaching and research, that define Uganda Martyrs University's Vision, Mission and the other core values and activities that describe the University's essence and existence. In tandem with its Charter and the community's expectations, UMU's community engagements started in 1996 with the creation of the Centre for Extra-Mural Studies (CEMS). The purpose of this centre originally was to run Distance Learning Programmes and also act as an incubator for new programmes and short courses. Another unit was created in 1999 to liaise community activities of the university. Later, this unit was converted into a Directorate of Outreach with the sole mandate of coordinating community outreach programmes. In the restructuring exercise of 2015, the Directorate was changed into the Department of Education for Sustainable Development and Community Engagement. Part of the work of this Department was to mainstream education for Sustainable Development in the university and coordinate the Regional Centre of Expertise Greater Masaka.

So, how has UMU used this Department to reach out to the community? Between 1999 and 2002, UMU implemented a "Nutrition Project" for the undernourished children of the sub-counties of Buwama, Kituntu and Nkozi. UMU did not just provide nutritious soya flour but also provided extension workers to guide households on their farms. It was out of this that the University Farm was borne (Education for Sustainable Development and Community Engagement-ESD & CE, 2022).

With the support of AP-TRIAS, a Belgian Funding organization, UMU started another initiative between 2003 and 2008 whose purpose was to assist households in having sustainable farm productions. UMU trained farmers on the formation of farmers' groups, business management and entrepreneurship skills, post-harvesting and handling of crops, processing and marketing techniques and collaborations among farmer groups and communities (ESD & CE, 2022). By the end of this intervention, the lives of farmers across Nkozi Sub-County, Mpigi District had greatly changed. Farmers were able to feed themselves on nutritious foods but also had a surplus to sell. UMU's intervention improved the socio and economic development of the Sub-County.

Between 2009 and 2010, UMU, with the support of the SIEBEN Foundation, Belgium, and the University of Notre Dame, USA, embarked on the promotion of commercial agriculture through cooperatives and the improvement of nutrition in households in Nkozi Sub-County, Mpigi District (ESD & CE, 2022). By the end of the implementation of the project, food security and nutrition had increasingly improved in households. But more pleasing was that farmers not only knew what nutritional foods to feed their households but also their farming outputs had considerably increased.

The most successful UMU community engagement activity was with the "University Partnership for Research and Development" (UPFORD) implemented in Nindye, Nkozi Sub-County, Mpigi District between 2009 and 2015 (ESD & CE, 2022). This project, implemented in partnership with the University of Notre Dame, USA, saw farmers improve their agricultural production as well as access better health services, water sanitation and hygiene. With improved incomes, UMU facilitated the formation of micro-saving groups known as the "Savings and Internal Lending Communities" (SILC). The primary purpose of these groups was to encourage a saving culture for the people but also create possible financial opportunities for their businesses (ESD & CE, 2022). The Nindye project not only created an improved learning environment for the children in the community but also empowered the community. For UMU, the Nindye community became a hub for their research and students' internship, especially for the Faculties of Agriculture, Business Administration and Management and School of Arts and Social Studies.

As projects of agriculture and business were implemented, one missing link was connected with people's health. Between 2012 and 2013, UMU, with the support of VEROZON, implemented a health project with Nindye Health Centre. The project's essence was to help the community access health information and treatment through the use of mobile phones (ESD & CE, 2022). The project augmented the efforts of the university in improving the health aspects of the households in the Nindye community.

The Social Enterprise Project (SEP), implemented in Nindye between 2015 and 2019 with the support of UPFORD, was a continuation of the SILC project. Its purpose is "aimed at increasing access to finance for starting up locally-run enterprises; providing longer loan repayment periods; and, on-site business/entrepreneurial skills training for the participating SILC groups' members and other entrepreneurs" (ESD & CE, 2022).

UMU, with support from UPFORD, implemented the Community Lending and Outside Capital (CLOC) research project in Nkozi Sub-County between 2020 and 2021. The CLOC research project was a result of wanting to address some of the challenges faced during the implementation of SILC and SEP projects (ESD & CE, 2022). One of the key challenges SILC groups faced was that funds for loans were not sufficient to be accessed by members. This was because most of the members seeking loans were farmers and hence, they tended to seek loans within the same period due to agricultural seasons. CLOC intervened to temporarily put money into SILC groups at the beginning of the agricultural season. CLOC was able to establish that SILC members' livelihoods changed for the better as a result of the project's intervention. Members were able to invest loan funds into poultry, businesses and agricultural projects.

To sum up, the three pillars of teaching, research and community engagement, as validated by UMU, are pathways through which HEIs can implement the SDGs and consequently contribute to the social and economic development of society. As in Europe and other developed countries, HEIs should be seen to contribute to Africa's growth and development. By becoming active participants in the implementation of SDGs using the three pillars, HEIs will also benefit by overcoming the general perception within society that labels them as ivory towers. The implementation of SDGs will make HEIs more relevant to the communities they serve.

5   What Next for Uganda Martyrs University?

One of the key drivers for HEIs to be sustainable is their resolve and determination to develop their human capital. In UMU's context, its ability as an institution to carry out its mandate of teaching and learning, research and community engagement will greatly depend on the skills and competencies of its academic staff. The new pedagogies that require lecturers to engage their students in skills will require training and re-tooling to prepare the ideal graduate. As experienced in the training carried out by TESCEA, the work of re-tooling staff requires a team of dedicated staff with the proper skills and competencies to training others. UMU should invest more in these training sessions and activities because of preparing academic staff for the task. Thus, the development of academic staff in new pedagogies is not an option but a must for UMU.

Since due to the aftermath of COVID-19 UMU has opted for blended learning, more investments should be made in the ICT infrastructures that support teaching and learning. Investments should be made in the availability of computers and their accessories and the availability and accessibility of fast internet on all UMU campuses. With the state-of-the-art ICT infrastructure, students at all UMU campuses can be able to access online materials for their learning. More importantly, students will be able to access the UMU library from wherever they are. Heavy investments in the ICT infrastructure will greatly enhance teaching and learning and research at all UMU campuses. Long after COVID-19, UMU will continue with the blended-learning approach as learners have appreciated it more due to its suitability for those who are near and far.

The development and review of UMU's curricula should be driven by the Government's National Development Plan and the UN's SDGs agenda for the social and economic development of the country. The deliberate move to develop and/or review programmes in tandem with international and national benchmarks will make UMU's programmes attractive and responsive to society's needs. Curriculum development and review, accompanied by new pedagogies that are anchored on skills that require deliberate investments and taking staff through the appropriate methodologies. Going forward, all academic staff should be required to go through a course on methodology and encouraged to attend periodical short courses on the same to equip them with the required skills.

UMU should dedicate time and resources to initiate and develop high-impact curricula proportionate to the E-4 Impact that would make UMU contribute to the social and economic development of the country. Since UMU has taken a position that all curricula to be developed should be entrepreneurial in nature, resources should be put into this development such that UMU's future graduates should not be job seekers but job creators. Resources should be invested not only in the development of curricula but also in the teaching and learning of such curricula and the training of staff.

UMU's interventions in community engagements, albeit funded by external budgets, should be rolled out to other communities so that the accomplishments of such initiatives can be replicated. There is a need for UMU to deliberately plan for these community engagements in line with the SDGs and the Government's plan for the social and economic transformation of Uganda. UMU's research agenda should be in tandem with SDGs and Government's National Development Plan for economic development. The totality of such initiatives should create synergies that would inform and contribute to the country's social and economic development. However, the implementation of such noble initiatives should depend mostly on UMU's internal budgets to avoid scenarios of initiatives or projects closing after the funding has ended.

UMU can use students across different Faculties in identifying problems and challenges arising from various communities. By collecting data from these communities, the university will learn more about the communities and vice versa; and consequently, a relationship will be established. After the study, findings and recommendations towards addressing the challenges can be shared for improving life in the community. This engagement, to some extent, will contribute to the transformation of society and also make UMU not only visible but also relevant to the community. Through such engagements, UMU will have done away with the perception of the *Ivory Tower* mentality that tends to portray HEIs as places of learning but cut off from the realities of life.

In addition to community engagements, UMU should endeavour to establish working collaborations with industries. Working with industries will create a lot of benefits for UMU. One of the benefits will be that UMU's students will have the opportunity to go through industrial (internship) training where they will acquire experience and thus be prepared to have skills for employability. Industries too will gain by having free labour but also by students passing on their theoretical knowledge to their co-workers. More importantly, the real beneficiary is society as it will receive in its midst graduates who will not learn on the job as they will already have industrial experience. But even more importantly, graduates will utilize their learned skills and competencies to contribute to the well-being of society. In this symbiotic relationship, UMU, industries and society will all win.

There is a need for UMU to invest in the development of start-ups and incubation centres. This development presupposes that out of the action, and research carried out, some innovative ideas, especially those based on the problems and challenges identified by different communities, will be selected, leading to further development into either products or services. While start-ups are described as rudimentary innovations that can be turned into sellable products and services; incubation centres are spaces or offices created to support students who have innovative ideas that can be fully developed into start-ups. UMU needs to come up with spaces or offices where students' new ideas can be developed. Students of Engineering and Applied Sciences, for example, can come up with innovations geared at solving issues of irrigation, water collection at the family level and renewable energy. Those of Agri-ecology, for instance, can innovate around the development of organic pesticides and fertilizers that can mitigate against the various crop diseases within the communities. Such innovations can be UMU's initial start-ups that can be patented and rolled out nationally. This will not only attract funds to the university but will also create jobs for the innovators. Students involved in these innovations will not have to look for jobs but rather will be employed by the businesses they have created out of the research and innovations did.

6 Benchmarks for Other Universities (Drawn from the Successes of UMU)

6.1 Staff Development

There are five areas that HEIs can learn from UMU's experience. Human capital is an important factor in the running of HEIs and in impacting the learners and society. Without developing the human capital, no development in the university and the communities surrounding the university can be achieved. UMU was greatly assisted by the TESCEA project staff who had been trained on the type of learning that would bring about the social and economic transformation of the communities.

6.2 Lining Curricula with the National Development Plan and SDGs

With the re-designing of curricula in the implementation of TESCEA projects at UMU, many of the programmes were reviewed in view of addressing the new pedagogies that are anchored on requisite skills and gender-sensitive approaches. These approaches are meant to help universities develop curricula that can be responsive to society's needs. Universities wanting to make their curricula effective should develop these new approaches. Existing curricula should be redesigned to bring out the desired goal of churning out a self-employing graduate.

6.3 Applying New Pedagogies in Teaching

Though many universities embraced online teaching and learning, especially during lockdowns following COVID-19, some of them were limited only to conducting classes on Zoom. Universities should invest in the training of their staff so that they can be re-tooled with new pedagogies of teaching and learning. Universities need to create a team among their staff that can re-tool others. These new pedagogies should be accompanied by the acquisition of ICT skills that will assist the lecturer to facilitate students in their learning in a more efficient and efficacious manner.

6.4 Enhancing Community Engagements

Some universities have continued to operate under the traditional mantra of perceiving themselves as ivory towers with little connection to communities. Some of their research is not informed by the problems and challenges arising from communities. For universities to be relevant, they should position themselves to address the social and economic challenges of their communities. Universities ought to create centres or offices that are directly concerned with staff and students' community engagements.

6.5 Linkages with Industries

If universities are to leap the contribution towards the social and economic development of their communities, one of the key ways should be investing in developing a working relationship with industries. Industries will create opportunities for interns but also will expose avenues through which staff and students' innovations can be facilitated and developed. Through these links, universities can directly play a role in the implementation of the UN's SDGs.

7 Conclusion

This chapter has highlighted the four pillars of sustainability, namely, human, social, economic and environmental sustainability. These interrelated pillars are harmonized by their common task of contributing to the present and future well-being of individuals and society. The UN and other world bodies have had different programmes to accentuate these pillars; the Millennium Development Goals (MDG) and lately the SDGs are some of the interventions made by these bodies to improve the well-being of society. HEIs contribute to the improvement of individuals and society through the implementation of the four pillars and as recently captured by the SDGs. SDG 4 is for that matter dedicated to HE. The chapter believes that HEIs are at the heart of implementing the SDGs through its four pillars. Without HEIs, none of the pillars and the SDGs would be implemented. HEIs, therefore, need to recognize this role that their active participation in the implementation of the SDGs will contribute to the social and economic improvement of society. In view of this, universities, as part of HEIs, should prioritize teaching and learning, research and community engagement in their core activities.

REFERENCES

Anand, S., & Sen, A. (2000). *Human development and economic sustainability in world development* Vo. 28, Issue 12. Accessed Apr 4, 2023, from https://www.sciencedirect.com/science/article/abs/pii/S0305750X000007

Benn, S., Dunphy, D., & Andrew, G. (2014). *Organizational change for corporate sustainability.* Routledge.

Chankseliani, M., & McCowan, T. (2021). *Higher education and the sustainable development goals in higher education.* https://doi.org/10.1007/s10734-020-00652-w.

Cloete, N., Bailey, T., Pillay, P., Bunting, I., & Maassen, P. (2011). *Universities and economic development in Africa.* Published by the centre for higher education transformation (CHET).

Diesendorf, M. (2000). Sustainability and sustainable development. In D. Dunphy, J. Benveniste, A. Griffiths, & P. Sutton (Eds.), *Sustainability: The corporate challenge of the 21st century* (pp. 19–37). Allen & Unwin.

Dunphy, S. B., & Griffiths, A. (2007). *Integrating human and ecological factors: A systematic approach to corporate sustainability.* Accessed Mar 14, 2023, from file:///C:/Users/ICT/Downloads/Integrating_human_and_ecological_factors_A_systema.pdf

Education for Sustainable Development and Community Engagement-ESD & CE. (2022). *A brief to the vice chancellor.* Uganda Martyrs University.

Ekene, O. G., & Oluoch-Suleh, E. (2015). Role of institutions of higher learning in enhancing sustainable development in Kenya. *Journal of Education and Practice, 6.* www.iiste.org. Accessed Apr 4, 2023, from https://files.eric.ed.gov/fulltext/EJ1079963.pdf

Encyclical Letter *Laudato Si* of the Holy Father Francis on Care for our Common Home. (2015). Copyright © Dicastero per la Comunicazione - Libreria Editrice Vaticana.

Erjavec, J. (2021). *Stakeholders in curriculum development—case of supply chain and logistics programme in the 7th International Conference on Higher Education Advances (HEAd'21).* Accessed Mar 22, 2023, from file:///C:/Users/ICT/Downloads/Stakeholders_in_curriculum_development_-_case_of_S.pdf.

Freire, P. (1993). *The pedagogy of the oppressed. The continuum international publishing group Incb15 east 26th street* (Vol. 10010).

Garcia-Feijoo, M., Eizaguirre, A., & Aspiunza, A. R. (2020). *Systematic review of sustainable development goal deployment in business schools.* https://doi.org/10.3390/su12010440.

Kasozi, A. B. K. (2016). *The National Council for higher education and the growth of the university sub-sector in Uganda, 2002–2012, Council for the Development of Social Science Research in Africa (CODESTRIA).* Dakar.

134   C. M. ACAALI

Maximiliano Ngabirano on TESCEA. (2017). Accessed Mar 23, 2023, from www.umu.ac.ug

Nabaho, L., Turyasingura, W., Twinomuhwezi, I., & Nabukenya, M. (2022). The third Mission of universities on the African continent: Conceptualisation and operationalisation. *Higher Learning Research Communications, 12*(1).

Nina P. M., Kiambati, K., & Mbalassa, M. J. (2023). *Using black soldier fly bio-waste conversion to produce commercial animal feed rations.* A grant proposal submitted to bio-innovate Africa. Accessed May 30, 2023, from https://bioinnovate-africa.org/call-for-proposals-for-early-stage-bio-incubation-projects/

Ramalu, S. S., Nadarajah, G., & Aremu, A. Y. (2020). Turning students from job seekers into job creators: The role of high impact entrepreneurship educational practices. *International Journal of Innovation, Creativity and Change, 13*(4) Accessed Mar 17, 2023, from www.ijicc.net. file:///C:/Users/ICT/Downloads/IJICC2020.aspx.pdf

SDSN Australia/Pacific. (2017). *Getting started with the SDGs in universities: A guide for universities, higher education institutions, and the academic sector.* Australia, New Zealand and Pacific Edition. Sustainable Development Solutions Network. Accessed Mar 29, 2023, from https://reds-sdsn.es/wp-content/uploads/2017/09/University-SDG-Guide_web.pdf

Serafini, P. G., Morais, J., de Moura, M., de Almeida, R., & de Rezende, J. F. D. (2022). Sustainable development goals in higher education institutions: A systematic literature review. *Journal of Cleaner Production, 370,* 133473. Accessed Mar 16, 2023, from https://www.sciencedirect.com/science/article/abs/pii/S0959652622030542

Snellman, C. L. (2015). University in knowledge society: Role and challenges. *Journal of System and Management Sciences, 5*(4), 84–113.

UK Annual Report. (2000 and January 2001). *In "The four pillars of sustainability introducing the four pillars of sustainability; human, social, Economic and Environmental".* Accessed Mar 14, 2023, from https://www.futurelearn.com/info/courses/sustainable-business/0/steps/78337

A Framework for Embedding Graduate Employability: *Attributes of Higher Education Curriculum in Africa*

Romanus Shivoro 🆔, *Rakel Kavena Shalyefu* 🆔,
and Ngepathimo Kadhila 🆔

1 INTRODUCTION

The subject of graduate employability continues to be increasingly popular in the higher education (HE) sector across the globe including Africa. Employers view graduates with employability skills to be assets since they are dynamic and readily adaptable to today's work environment. As a result, HE throughout the world is under pressure to produce graduates who are employable and can continue studying while working (Ngulube, 2020). However, the embedded generic skills approach has been widely critiqued, and there is endless debate on the extent to which generic skills and competencies can be taught effectively in the classroom and then transferred to the workplace. Others champion that there is very little

---

R. Shivoro • R. K. Shalyefu • N. Kadhila (✉)
University of Namibia, Windhoek, Namibia
e-mail: rshivoro@unam.na; rkshalyefu@unam.ha; nkadhila@unam.na

© The Author(s), under exclusive license to Springer Nature
Switzerland AG 2024
P. Neema-Abooki (ed.), *The Sustainability of Higher Education in Sub-Saharan Africa*, Sustainable Development Goals Series,
https://doi.org/10.1007/978-3-031-46242-9_7

evidence that these attributes have a direct bearing on graduate employment success and that factors equivalent to social class, gender, ethnicity, social networks, and university status are more salient to job search outcomes (Clarke, 2018). In addition, Clarke holds that when HEIs adopt graduate employability attributes, they fail to recognise differences between different professions, different organisations, and even different cultural contexts.

In line with the focus on graduate employability, HEIs have come under pressure from government bodies to measure graduate employability by providing hard data to graduate outcomes. In addition, in many HE systems, a driver for graduate employability measurements has been the pressure to meet learning outcome standards required by external accrediting bodies. However, measuring graduate employability using graduate employment status has been under scrutiny and its reliability has been questioned. Research points to that graduate employment data generally measure current employment status, not employability over the longer term. Realistically, employability can only be evaluated using a time dimension which considers movement in, and between, organisations, not just initial employment post-graduation (Yorke, 2006; Clarke, 2018).

Tran (2016) states that this is mostly attributed to the effects of globalisation, job competitiveness, and increasingly changing patterns of graduate recruitment. This calls on HEIs to produce employable graduates (Hill et al., 2017). Many HEIs have begun working towards the inclusion of explicit employability attributes in undergraduate curricula. This is evident by varied lists of graduate attributes that HEIs deem important for their graduates to possess. These efforts continue to proliferate, in many cases, driven by policy imperatives for increased industry involvement in HE curriculum development (Wilton, 2014). In addition to determining graduate employability attributes, HEIs and government agencies have gone further to develop employability frameworks to guide the production of employable graduates.

Commonwealth of Learning (COL) attests that there have been several concerns from industry, HE pundits, and critics that HEIs create half-backed graduates who lack certain competencies that employers want (COL, 2019). In the African context, although employers have been lamenting the lack of work-readiness of graduates, no attempt has been made to interrogate the issue of graduate employability in Africa to inform the development of strategies to address the challenge expressed by employers. Therefore, this chapter attempts to analyse the issue of graduate employability attributes in African HE and develop a framework for integrating graduate employability attributes into the curriculum to enhance graduate readiness for the ever-changing labour market requirements.

2   CONCEPTUAL FRAMEWORK

The labour market conditions have evolved from prior decades whereby employees required cognitive skills and routine and non-routine manual abilities, therefore possessing a high school certificate was a strong display of human capital accumulation to perform a job. At present, an individual graduate's success on the job depends more on non-routine analytic and interactive skills (Levy & Murnane, 1996). As such, employers have expressed a need for well-rounded graduates, who are aware of the labour market and business practices and expectations. Thus, instilling graduate employability attributes at HE levels in Africa will ensure graduate preparedness for the labour market. These attributes should be instilled through curriculum design and implementation and appropriate pedagogical methods.

Knight and Yorke (2003a, 2003b) and Yorke and Knight (2006) present some models of embedding graduate employability attributes in the curriculum. Among these models, there is a USEM Model that stands for Understanding, Skilful practices, Efficacy beliefs, and Metacognition. These elements match the underlying assumptions of the Human Capital Theory.

The model proposes four inter-related components of employability (Yorke & Knight, 2006):

- Understanding (of disciplinary subject matter and how organisations work);
- Skilful practices (academic, employment, and life in general);
- Efficacy beliefs (reflects the learner's notion of self, their self-belief, and the possibility for self-improvement and development); and
- Metacognition (complements efficacy, and embraces self-awareness, how to learn and reflect. It encompasses knowledge of strategies for learning, thinking, and problem-solving and supports and promotes continued learning/lifelong learning).

The USEM Model is a framework for thinking about how to integrate employability into the curriculum, and it recognises the need for considering the requirements of students, employers, and other stakeholders. It also challenges us to consider how curricula incorporate assessment that fosters student efficacy and metacognition, as well as how this relates to the development of subject knowledge and professional skills that may be used in the workplace. It claims that there is a link between employability and effective learning, and it emphasises that employability is the consequence of a combination of successes in four main categories (Yorke & Knight, 2006). Good curriculum designs will continue to assist students to establish an understanding of the subject matter and maintain the more recent interest in developing a variety of skilled practices, or skills, in line with the implications of the USEM Model for the curriculum. They will, however, exhibit concern for the development of positive efficacy beliefs, metacognition, and other complex achievements valued by employers. Therefore, this model was used as the lens for the development of a framework for integrating graduate employability attributes in HE curricula in Africa.

3   Definition of Graduate Employability

Employability is a nebulous term. Pool and Swell (2007) assert that for many people, employability is merely about having a job; and the phrase is increasingly being conflated with "enterprise," which is then associated with "entrepreneurship." Employability is now being interpreted in a variety of ways, ranging from simple metrics like whether or not a graduate has found work (through graduate first-destination surveys) to in-depth scholarly publications on the issue. Employability is simply a very vague and imprecise indicator of what a student has gained if it is judged in basic terms of whether or not a graduate has been able to get a job within six months after graduation. Questions must be raised about whether the graduate is applying their degree-related skills, knowledge, and understanding in a "graduate level employment," which in turn raises a fresh discussion about what a "graduate level job" involves. There's a lot more to employability than just getting a job, and first-destination figures don't account for the reality that some graduates may have selected lower-paying professions to alleviate financial strains, especially after accruing loans during their studies (Pool & Swell, 2007).

Much research has been conducted to establish the general abilities that companies require. There is a large amount of research on the notion of graduate employability, and several frameworks specify lists of knowledge, abilities, and traits that graduates should possess. However, there are still significant gaps between employer, graduate, student, and staff expectations for what, when, and where required student learning should take place. According to the literature, students must gain graduate employability skills while attending HEIs. Employability is defined more generally as the ability of a graduate to find a job or employment regardless of economic conditions. One often-used definition is a collection of accomplishments—talents, understandings, and personal characteristics—that make graduates more likely to find a job and succeed in their chosen fields, benefiting themselves, the workforce, the community, and the economy (Yorke, 2006).

The immediate foregoing scholar explicates that employability is a set of achievements—skills, understandings, and personal attributes—that makes graduates more likely to gain employment and be successful in their chosen occupations, which benefits themselves, the workforce, the community, and the economy. In the same vein, graduate employability refers to the process by which individuals transition from university to the workforce. Building on skills and attributes acquired in gaining a degree, individuals begin to develop a graduate identity which they present to "gatekeepers" (recruiters and managers). That is, they must display the graduateness that sets them apart from nongraduates, as well as from other graduates within the labour market, and thus offer a future employer the type of inputs and outcomes they expect (Clarke, 2018).

Pool and Swell (2007) define employability as the ability to find and sustain a satisfying job. Employability, in a broader sense, is the ability to move independently within the labour market to realise one's potential through long-term work. Pool and Swell observe that employability is made up of four basic components. A person's "employability assets," or knowledge, skills, and attitudes, are the first of these. The second, "deployment," comprises job search skills as well as career management abilities. Finally, "presentation" refers to "job-seeking abilities," commensurate with CV writing, work experience, and interview strategies. Clarke (2018) argues that a person's capacity to maximise their "employability assets" is dependent on both personal circumstances (such as identical to family duties) and external ones (e.g., the current level of opportunity within the labour market).

140   R. SHIVORO ET AL.

Employability is therefore a lifelong process that applies to all students regardless of their situation, course, or mode of study; it is complex and involves several interconnected areas; it is about assisting students in developing a range of knowledge, skills, behaviours, attributes, and attitudes that will enable them to be successful not only in employment but in life; it is a university-wide responsibility; and it is about making a difference (Yorke & Knight, 2006). The qualities, skills, knowledge, and abilities of HE graduates, beyond disciplinary content knowledge, that apply to a variety of contexts and are acquired as a result of completing any undergraduate degree, have been adopted as a working definition of graduate employability attributes in this chapter, and they should represent the core achievements of a HEIs graduate (Barrie, 2006; Yorke & Knight, 2006).

4   GRADUATE EMPLOYABILITY MODELS

In light of the working definition, scholarly contributions have identified different graduate employability attributes which are viewed as important for the world of work. Bennett et al. (1999) developed a model of HE course-offering that includes five components: (1) disciplinary subject knowledge; (2) disciplinary skills; (3) workplace awareness; (4) workplace experience; and (5) generic skills. This model comes close to integrating all of the required factors to assure a graduate's maximum employability, but it still lacks a few key components. The most well-known and recognised model in this field is the USEM Model of employability (Yorke & Knight, 2006). The term USEM stands for four interconnected components of employability: (1) understanding; (2) skills; (3) efficacy beliefs; and (4) metacognition.

Rosenberg et al. (2012) identified eight graduate employability attributes that employers usually look out for in HE graduates, namely basic literacy and numeracy and other employability attributes that include critical thinking skills, management skills, leadership skills, interpersonal skills, information technology, systems thinking skills, and work ethic. These employability attributes seem to be congruent, less complex, easy to explain, and more generally agreed upon by various authors. Below is a brief description of each attribute:

– *Basic literacy and numeracy*—the ability to read, write, speak, listen, and perform basic mathematical procedures. Reading includes the ability to interpret written information, writing includes the ability to communicate thoughts in letters and reports and mathematical skills include the ability to solve practical problems through the use of a variety of mathematical techniques.

- *Critical thinking skills*—the use of cognitive skills or strategies that increase the probability of a desirable outcome. Critical thinking is used to describe thinking that is purposeful, reasoned, and goal-directed—the kind of thinking involved in solving problems, formulating inferences, calculating likelihoods, and making decisions when the thinker is using skills that are thoughtful and effective for the particular context and type of thinking task.
- *Management skills*—include the activities of planning, organising, leading, and controlling to meet organisational goals.
- *Leadership skills*—the ability to take control of a situation and empower peers and motivate and positively influence others to achieve organisational goals. Graduates should make plans for their own or group activities, identify and mitigate risks, and deal with change and uncertainty.
- *Interpersonal skills*—include the ability to work in teams, help others to learn, provide customer service, negotiate agreements, resolve differences, and work in a multicultural organisation; also, the ability to relate to and feel comfortable with people at all levels and to maintain relationships as circumstances change.
- *Information technology*—refers to the use of technology that contributes to the effective execution of tasks. This includes the ability to select procedures and the equipment and tools to acquire and evaluate data.
- *Systems thinking skills*—including the ability to understand and operate within social, organisational, and technological systems, as well as designing and encouraging modifications to systems and explaining the interaction of systems in the context of the global economy.
- *Work ethic*—refers to the individual's disposition towards work and includes attendance, punctuality, motivation, the ability to meet deadlines, patience, a positive attitude, dependability, honesty, professionalism, and realistic expectations of job requirements and career advancement (Rosenberg et al., 2012).

The triple authors point out that these attributes are essential for job performance and that more efforts should be done to emphasise these in curricula. In addition, Pool and Swell (2007) created a model that represents the statement that each component is necessary and that one missing component will significantly impair a graduate's employability. Some overlap between the components is acknowledged, and this is reflected in the model's visual presentation. However, this does not imply that these are the only sites where overlap occurs, as it does so at numerous points. Work experience, for example, is not only a life-or-death aspect of career development learning, but it may also directly feed topic learning related to the degree programme being studied in some situations.

The model depicted in Fig. 1 illustrates the essential components of employability and also suggests the direction of interaction between the various elements. The acronym "CareerEDGE" is used to help recall the five components of the model's lowest layer. It is proposed that providing students with opportunities to access and develop everything on the lower tier, as well as opportunities to reflect on and evaluate these experiences,

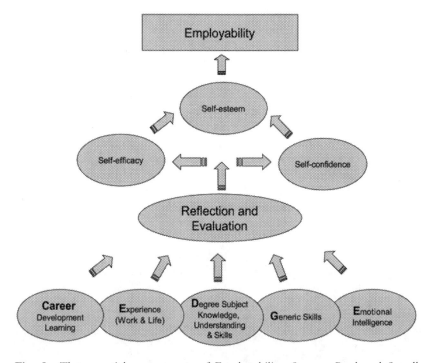

Fig. 1 The essential components of Employability. Source: Pool and Sewell (2007, p. 280)

will result in the development of higher levels of self-efficacy, self-confidence, and self-esteem—all of which are important links to employability. As a result, according to the CareerEDGE, employers should expect graduates to have gained—as sung by Pool and Swell (2007) general abilities that include: imagination/creativity; adaptability/flexibility; willingness to learn; independent working/autonomy; working in a team; ability to manage others; ability to work under pressure; good oral communication; communication in writing for varied purposes/audiences; numeracy; attention to detail; time management; assumption of responsibility and for making decisions; planning, coordinating, and organising ability; and ability to use new technologies (not listed herewith but mentioned in many others and an important element).

Suffice to emphasise the enterprise and entrepreneurial abilities, which are frequently discussed in the literature on employability. An entrepreneurial graduate is likely to be recognised in any organisation, whether profit-making or non-profit-making, large or small. For the sake of this model, it is assumed that an entrepreneurial graduate would be inventive, creative, flexible, and eager to learn—in fact, they would possess the majority of the abilities proposed under the "generic" category. Entrepreneurial skills, on the other hand, may be a desirable complement for certain graduates, but not for all. Not everyone aspires to own their own successful business. Entrepreneurial skills were not included in the model since they are not considered a required component.

5   Embedding Graduate Employability Attributes in the Higher Education Curriculum

Mason et al. (2009) assert that the major purpose of teaching graduate attributes is to enhance graduates' skill sets to boost their competitiveness in the job market. Some researchers have intimated approaches through which graduate employability can be enhanced. Saunders and Zuzel (2010) write that the current trend of enhancing graduate employability is to incorporate opportunities for employability skills alongside subject-specific knowledge and skills. This approach separates the employability intervention from the programme of study, and it is not clear if the employability intervention is credit-bearing. In contrast, Mason et al. (2009) state that some universities have adopted strategies of integrating graduate employability into their programmes of study. Here, employability is built into the programme alongside specific learning outcomes of the degree programme. These strategies include offering work experience, work-related learning and employability skills modules, and "ready for work" events, as well as involving employers in course design and delivery. Another hint is offered by Gardner (1998) advocating for faculty to incorporate employability attributes coincident with leadership, team work, and communication in their different courses of study and not as a stand-alone module or once-off experience. To Gardner, this should be accompanied by internships or work-based placements and should be contextual to discipline. However, Gardner acknowledges the challenge to incorporate such opportunities in incremental stages for the intellectual, professional, and personal development of students.

Literature has proposed the following levels of the curriculum at which graduate employability can be enhanced (Muldoon, 2009; Quality Assurance Agency for Higher Education, 2009; Yorke & Knight, 2006). Firstly, employability *through the whole curriculum* where individual students are expected to evidence competency in abilities conforming to teamwork, problem-solving, and communication, in a progressive manner; secondly, *employability in the core curriculum* where a university designates a module or two as vehicles for enhancing employability attributes. The modules seek to enable students to acquire a variety of employability skills through practical volunteering experience and reflection (Quality Assurance Agency for Higher Education, 2009); thirdly, *employability-related modules within the curriculum* whereby students are required to take a theoretical employability module during an undergraduate degree; fourthly, *work-based or work-related learning interspersed within the curriculum* which includes short and long-term placements and internships; and finally, *work-based or work-related learning in parallel with the curriculum* whereby students are employed part-time in parallel with their studies (Muldoon, 2009). To the above levels, Fallows and Steven (2000) point out that for a HEI to ensure that every student is fully equipped with employability attributes necessary for the workplace, such institution should make decisions regarding the attributes that should be highlighted; how the institution recognises meaningful skills progression from the first to the final year of studies; and how the institution ensures that each student is fully exposed to each attribute.

6   GRADUATE EMPLOYABILITY ATTRIBUTES AND THE SUSTAINABLE DEVELOPMENT GOALS

Despite the disruptions to education caused by COVID-19, the role of academia in advancing Sustainable Development Goals (SDGs) on college campuses is more important than ever. To adequately address graduate employability attributes, this chapter adduces that HEIs must adopt the SDGs. Chankseliani and McCowan (2022) introspect that the 17 SDGs, which were adopted by all UN member states in 2015, address a wide range of socioeconomic, environmental, and technological development-related challenges and apply to all countries around the globe. The SDGs broadened the focus beyond primary and secondary education to include HE as part of its comprehensive mandate (Chankseliani & McCowan, 2022). It is indisputable that governments must collaborate across policy sectors to make progress towards the SDGs; nevertheless, it is also evident that political commitment alone will not be sufficient without precise frameworks to guide their implementation. To achieve the SDGs, policy decisions must be supported by evidence that is relevant to that policy, co-designed and co-produced with the appropriate stakeholders, and considering the local and political context (Turner & El-Jardali, 2017).

HEIs are in a unique position to lead the multi-sectoral fulfilment of the SDGs since they are largely seen as neutral and influential stakeholders, as well as an invaluable source of expertise in research and education across all sectors of the SDGs. It is recognised that, in addition to universities, think tanks, and other organisations that produce and disseminate information play a significant role in furthering the SDG agenda (El-Jardali et al., 2018). Despite being included in the SDG agenda, HEIs in middle-income and low-income countries, particularly in Africa, participate less actively than anticipated (Blessinger, 2023). Therefore, such institutions are encouraged to incorporate SDGs in their activities to adequately address graduate employability attributes. Hence, HEIs are judged on how well they build students' global competency and capacity for international collaboration and leadership, teaching is a basic component of SDG implementation (Blessinger, 2023; Grund, 2020). HEIs are also judged based on the SDGs' incorporation in their curricula and the extent to which students are aware of them. The ranking of such institutions considers important factors congenerous with academic staff knowledge and awareness of SDGs and their capacity to integrate particular SDG-related topics into their teaching. All of these factors strongly support the gradual implementation of SDGs in the HE sector—in such a way that students will develop graduate employability attributes that make a meaningful contribution to the attainment of SDGs through active engagement in social and economic transformation activities. Therefore, to attain the SDGs, students and academic staff should continue to make devoted efforts in education awareness and training to inform communities and society and to encourage their commitment to the attainment of SDGs. Incorporating the SDGs and sustainability in general into university courses and curricula, in Blessinger's (2023) opinion, involves at least two life-and-death elements. The first is through degree programmes in sustainability and the environment, whose students intend to follow these areas as a career and whose professors have already made these areas their area of expertise. The second, and more challenging, part is including SDGs in every course offered at universities to reach students who might not be interested in sustainability as a career.

7   An Analysis of Curricula for Graduate Employability Attributes at the University of Namibia

There is a range of levels of curricula in which employability is fostered, either explicitly or implicitly. As perceived by Yorke and Knight (2006) there is a spectrum of ways in which employability can be developed through curricula. The following represent "ideal types" (in the Weberian sense) whose differentiation is not clean-cut—indeed, they smudge into each other:

- Employability through the whole curriculum
- Employability in the core curriculum
- Work-based or work-related learning incorporated as one or more components within the curriculum
- Employability-related module(s) within the curriculum
- Work-based or work-related learning in parallel with the curriculum.

An analysis of six curriculum documents for management sciences degree programmes at three HEIs in Namibia was conducted in this chapter, based on Yorke and Knight's (2006) arguments, to see how and where employability-related learning is incorporated into curricula, and where there might be gaps.

7.1 Graduate Employability in the General Aims and Intended Outcomes of Degree Programmes

Three out of the six programmes studied contain an overarching statement in the aims of the programme that refers to graduate employability. Each of the three programmes, in addition to discipline-specific aims, asserted that "*students will have the opportunity to develop key transferrable skills and be able to apply these in a context of an organization.*" In addition, each programme has intended outcomes specified. In this case, a generic outcome of each of the three programmes states that students will gain the "*skills to work effectively as individuals and as team members, the ability to communicate effectively in the workplace, and analyse information from a variety of sources.*" Each of the three programmes has an additional emphasis on specific employability attributes, like "*responsible citizenship, and good communication, among others,*" and in another, "*display honesty, social responsibility, objectivity and professional demeanour,*" as programme outcomes.

Reference to graduate employability attributes has not been made explicit in the remaining three programmes. A business programme states that the programme is aimed at developing the "*intellectual ability, executive personality, management skills*" of students. Another programme states that it aims at developing "*quantitative and qualitative analysis, and critical thinking skills.*" One of the programmes has vaguely enumerated that "*the qualification will sharpen the students' analytical skills through integrating their knowledge of economic theory with real-life economic issues encountered in day-to-day economic decision-making processes, both in the government and private sector.*" One of the programmes has no reference to graduate employability. However, the analysis found implicit employability attributes within the curriculum.

From the above analysis, it is evident that some programmes have referred to key transferable skills conformable with good communication, work ethics, and professional responsibility. Some did not contain labels to refer to attributes that may not be discipline-specific, but random terms fungible to analytical skills. Therefore, there is a need to define graduate employability attributes. In addition, the programmes should articulate intent towards developing graduate employability attributes in the programmes. The intention and commitment must be known by the lecturers, students, and employers to enable all the stakeholders to play their part in this endeavour. Therefore, graduate employability should form part of the general aims of the programmes. The programmes specify all graduate employability attributes that will be developed as a result of participating in the degree programme.

7.2 Graduate Employability in the Core Curricula

Edith Cowan University (2013) highlights that regardless of the course content, there should be multiple opportunities at module levels to provide students with the chance to develop and exhibit their employability attributes. Analysis of graduate employability in core curricula has established that none of the core courses has referred to graduate employability in course outcomes. The descriptions of these modules have a strong emphasis on preparing students solely for the next levels of academic engagements. However, there appear implicit employability attributes in the course content in modules virtual to Computer User Skills, Computer Literacy, Information Competence, English Communication and Study Skills, Language in Practice, English for Academic Purposes, English in Practice, Contemporary Social Issues, and in some programmes, Business Mathematics.

7.3 Graduate Employability Through Work-Based or Work-Related Learning

The concept of work-integrated learning (WIL) has become a popular method of work-based or work-related learning (Jackson, 2014a). It is described as "the practice of combining traditional academic study, or formal learning, with student exposure to the world of work in their chosen profession, with a core aim of better-preparing undergraduates for entry into the workforce" (Jackson, 2014b). Similarly, Smith (2012) notes that WIL is a curriculum design in which students spend time in professional, work, or other practice settings relevant to their degrees of study and to their occupational futures. It has benefits for students, universities, and employers. It also builds students' confidence and an understanding of the standards of industry-required skills (Jackson & Wilton, 2016; Smith, 2012). Universities benefit through producing employable or work-ready graduates (Blackwell et al., 2001). Employers become informed about the university and the circumstance under which it operates, and it provides an employability signal for students that have participated in WIL experiences (Blackwell et al., 2001). WIL requires a close interaction between educators and practitioners to bridge the gap between capabilities acquired and competencies required in the industry (Leong & Kavanagh, 2013). Typologies of WIL include the following.

- *Sandwich course*: periods of work experience between years of a course usually 6 or 12 months;
- *Co-operative programmes*, which are periods of work experience that may be integrated into the overall curriculum, are designed both to integrate theory and practice and improve graduate employment.
- *Job shadowing*, which emphasises observation and absorption of organisational culture of the workplace;
- *Placement/Practicum*, which is an extended period in work settings to learn skills and gain experience of requirements of future work; and
- *Fieldwork*, is short periods, tantamounting to one day per week of fieldwork in an agency to observe and learn about the organisational culture of the workplace. (Gibson et al., 2002)

Analysis of management sciences degree programmes provided that one of the three HEIs in Namibia has a compulsory WIL module that is offered during either Semester 1 or 2 of Year 3 of the four-year degree programme. The module is customised to each of the degree programmes. For example, the statement for an accounting programme reads that the module enables "*students to apply various accounting and finance-related theories/ best practices to real-world accounting and finance situations/problems; developing capabilities equal to effective oral and written communication, teamwork, planning and organising, thinking creatively and problem-solving; enabling students to acquire general work experience and work ethics; and to manage time, communicate effectively, make presentations...,*" while the WIL module in economics states that students will "*apply economics theory to the various tasks that they are to do during WIL, manage time, communicate effectively, and prepare for business meetings; make presentations to large groups of people, conduct supervised practical mini researches on organisations of their choice; and, review some practical economic aspects of the Namibian economy and make appropriate recommendations.*" Literature provides that the most popular type of WIL is placements/internships whereby students spend a duration of three to six months at an organisation. In most cases, the student spends a whole semester and acquires 12–16 credits. For short WIL experiences parallel with shadowing and fieldwork, students are usually required to spend between 100 and 160 hours on WIL experience (Business/Higher Education Roundtable, 2017). WIL has benefits in terms of building student confidence in their workplace capabilities; providing students with a better understanding of the nature and standard of industry-required skills; and a better appreciation of the world of work; promoting certain elements of career self-management; providing education that responds to present and future needs; and providing learning that is useful to society and not just an addition to students' disciplinary knowledge base (Coll et al.; Freudenberg, Brimble, and Cameron, as cited in Jackson, 2014a).

In two of the programmes, the WIL module has 360 notional hours and 240 hours in another programme. The programme has not specified the type of WIL experience that students are required to undertake. The analysis established that the WIL module runs concurrently with other semester modules. If the WIL experience is the internship, it is not clear how students undertake this experience and at the same time attend to the academic demands of other semester modules. Other programmes do not have a work-integrated learning module or component. In light of the above, universities should specify the types of WIL experiences that students are required to undertake. In addition, universities should make appropriate adjustments to the curriculum to suit the types of WIL experiences. All universities should develop a structured WIL programme as part of degree offerings.

7.4 Graduate Employability Through Other Modules in the Degree Programme

Edith Cowan University (2013) proposes that while all employability attributes are important, there are likely to be some that are more appropriately developed within the context of a specific module. Therefore, to incorporate the attributes in the module, universities should ensure that there is a written outcome towards the attribute, design teaching and learning activities, consider WIL experience relevant to the attribute, and develop assessment tasks. An analysis of management sciences curricula found a strong relationship between graduate employability attributes and the learning outcomes of the modules. Modules uniform to Systems Thinking; Business Ethics; Innovation, Creativity, and Entrepreneurship; Business Ethics and Leadership; Business Research Methods; and Research Project contain objects that can be considered employability attributes. For example, the Systems Thinking module enables students to *"Apply knowledge of systems thinking to improve decision-making in the business administration and management domain."* Another example is for Innovation, Creativity, and Entrepreneurship module that enables students to *"create innovative and feasible business ideas and develop a comprehensive and functional business plan for a sustainable entrepreneurial business."* These examples can be considered to fall within the employability attributes dimension of Systems Thinking and Innovation.

Although graduate employability is not explicit in the aims of these modules, it is clear that these modules can be considered as modules enhancing graduate employability. Therefore, explicit reference to developing graduate employability attributes must be inserted in the intended outcomes of the modules. In addition, universities should identify modules that can be used as vehicles to enhance relevant attributes.

7.5 Embedding Graduate Employability Through a Stand-Alone Module

Owens and Tibby (2014) stress that employers are increasingly expecting graduates to be innovative, adaptable, resilient, and flexible and have an enterprising mindset. These attributes are needed by graduates to enable them to be successful in the ever-changing global economic environment. Concerning this, a report "Enterprise Education for All" was compiled and advocated that all universities in the UK should offer training on enterprise and entrepreneurship. Below are three examples of these stand-alone modules. Owens and Tibby (2014) present examples of ways through which graduate attributes can be enhanced using stand-alone modules.

- A stand-alone "Biotechnology and Business Module" for the School of Bioscience at Cardiff University. This module is offered during Year 2 of the four-year degree and is aimed at developing commercial awareness, enterprise core skills resemblant to communication, interpersonal and teamwork, and developing student's professional identity and interests of students before they begin industry placement.

- A university-wide "*Think it. Try it. Do it*" module offered at the University of Exeter aimed to raise students' awareness of enterprise and entrepreneurship, entrepreneurship and self-employment as alternative career paths, and the pervasiveness of enterprise skill requirements across all career paths. This module allows students to engage in enterprise workshops, engage with alumni and local entrepreneurs, and overall prepares them to start their businesses.
- A university-wide programme on "Enterprise, Entrepreneurialism, and Employability (3Es)" at Staffordshire University. The programme is aimed at improving the students' discipline expertise, professionalism and professional integrity, global citizenship and sustainability, communication and teamwork, reflective and meditative learning, and lifelong learning. This programme is delivered through different disciplines as a core module but customised to a specific discipline.

This study has established that employers, graduates, and HEIs in Namibia have unanimously expressed that graduates need training in "Innovation" and "Professional Responsibility." Therefore, in response to the expressed need, universities should consider developing specific training for graduates in "Innovation" and "Professional Responsibility" in all degree programmes.

8 A Proposed Framework for Embedding Graduate Employability Attributes in the Higher Education Curriculum in Africa

Cole and Tibby (2013) emphasise that the goal of the employability framework is to develop a defined, cohesive, and more comprehensive approach to employability. They urged comprehensive guidelines for developing graduate employability attributes framework for a university, faculty, or department. The guidelines are divided into four stages, firstly, "Discussion and Reflection" involves creating and defining a shared point of reference, which includes agreeing on a working definition of graduate employability, addressing employability responsibilities and activities in the curriculum, and discussion on work-based or work-integrated learning with employers. Secondly, "Reviewing and Mapping" specific employability attributes in the curriculum and determining how these can be audited.

Thirdly, "Action" involves addressing the gaps identified in the first two stages and determining how these can be prioritised, capacitating staff, and sharing the lessons learnt and best practices on enhancing graduate employability. Finally, "Evaluating" what success looks like, reflecting if all faculty members are engaged, and assessing impact.

Although the above guidelines can be further elaborated to fit a university, faculty, department, or course context, the overall employability approach for each university would differ depending on the outcomes of the above process. The UK Quality Assurance Agency for Higher Education (2009) advises that approaches to employability and personal development planning should allow for diversity in subject-based approaches while maintaining a coordinated institutional approach. In addition, the approach should ensure an even rate of progress in its implementation across faculties, schools, and departments, to bring about consistency in student experiences, and to enable more effective monitoring of progress made by students.

In light of the results of the study, this paper proposes the following framework for HEIs in Africa:

1. Building consensus on the definition of graduate employability; this should be understood by students, lecturers, and employers. The present understanding is that graduate employability attributes are qualities, skills, knowledge, and abilities of university graduates, beyond disciplinary content knowledge, which apply to a range of contexts and are acquired as a result of completing any undergraduate degree and should prepare graduates for the working lives in an ever-changing global economic environment.

2. The specific graduate employability attributes identified by the stakeholders are: numeracy and literacy, critical thinking, systems thinking, interpersonal, leadership, management, work ethic, information technology, innovation, and professional responsibility.

3. Key stakeholders in enhancing these attributes are the students, academic staff, employers, offices responsible for Work-Integrated Learning, Career Development Offices, alumni, and all other relevant offices. Employability is a university-wide responsibility; therefore, all members of the institution should be sensitised about the university's employability efforts.

There are many ways through which employability can be embedded into the curriculum. HEIs should develop strategies for embedding graduate employability attributes in programmes of study. This should be presided over by curriculum mapping to determine the present state and how to improve it. Therefore, efforts of incorporating graduate employability attributes in the curriculum can include:

(a) The general aims of all degree programmes, specific intended outcomes of individual programme modules, and units of learning. This should be done through discussions with all academic staff, to ensure that all staff understand enhancing graduate employability attributes, and they are motivated to plan relevant activities.

(b) Using foundational courses like English and computer literacy to enhance related attributes cognate with communication and information communication technology with a specific focus on graduate employability. Current academic foundational modules that are undertaken by all university students should be enhanced to equip students with relevant employability attributes. These should be across all degree programmes. Academic staff should be prepared to deliver such modules with an employability component.

(c) Development of appropriately structured WIL experiences in all degree programmes and specific modules identified to enhance each attribute. It has been acknowledged that WIL is an indispensable vehicle for combining theoretical knowledge and a practical work environment.

   (i) Every university student should participate in various forms of WIL experiences while at university.

   (ii) The curriculum should be adjusted to suit the envisaged WIL experiences. WIL calls for structured delivery specifying the intended outcomes, the attributes to be developed, the type of WIL, and how such experiences will be assessed.

   (iii) The time of WIL experience should be specified, for example, 6 months, 360 hours, etc.

   (iv) Credits earned from WIL experience should be specified, for example, 4, 8, or 16 credits.

(v) It is encouraged that WIL experiences are coupled with reflective journaling and portfolios to allow students to think about their professional growth and plan how to improve subsequent experiences.

(vi) The engagement of the industry in WIL planning is essential for the effective and efficient delivery of a WIL experience.

(vii) Typologies of WIL include job shadowing, field work, internships, placements, sandwich courses, cooperative learning, and service learning.

(d) Development of a special employability module to respond to the expressed need of the industry. For example, a module on innovation for all students.

(i) In addition to other modules in the programme, owing to the changing labour market, universities should develop programmes to enhance innovation and prepare students for professional working lives as responsible members of society.

(ii) These modules prepare students to be enterprising, to be creative, and at the same time maintain responsibility and accountability for their actions.

(iii) These programmes may be offered to students in varied disciplines.

4. Employability attributes should be reflected in module outcomes if such a module has been identified to enhance an attribute.

5. Develop a university-industry partnership.

(a) The industry is a major stakeholder in enhancing graduate employability.

(b) For universities to plan for relevant skills and experiences, the industry should be involved to ensure appropriateness and relevance to industry needs.

(c) The industry can serve as a resource to effectively prepare students for work lives. There should be a continuous collaboration between the university and the industry.

6. Assessment of university employability efforts should be conducted, both formative and impact assessment.
   (a) To celebrate and improve the practice.
   (b) To share success and good practice.
7. Platforms to share good employability practices should be shared among members of the university and other institutions as well. These can include seminars, scholarly papers, or conferences.

Figure 2 provides summarised components of the proposed framework for embedding graduate employability attributes in the HEIs curriculum in Africa. In consideration of the framework, HEIs have the task of determining specific actions required at all levels of the university, that is, university management, faculty, department, course, module, and unit levels. Although there might be a difference in the operationalisation of graduate employability attributes in different programmes, there should be

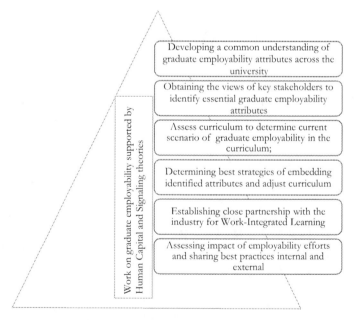

Fig. 2 A proposed framework for embedding graduate employability attributes in the higher education curriculum in Africa

consistency in the rigour of embedding graduate employability attributes across the university curriculum. This would ensure the equivalent benefit for all HE students. Thus, enhancing graduate employability would become an important part of curriculum offerings.

9 Lessons for Africa from the International Arena

Global trends homogeneous with increasing vocationalisation of HE education, mass education, marketisation, and increased competitiveness between HEIs have led to the focus on graduate employability more than ever before in the history of HE (Jonck, 2014a, 2014b). Clarke (2018) clarifies that a common approach adopted in many countries (e.g., UK, New Zealand, Australia, Europe, and North America) has been to incorporate generic skills into programme and course learning objectives based on the assumption that students will develop the requisite skills by the time they graduate and as a consequence achieve employment success.

A further assumption is that graduate employability attributes are directly correlated to the workplace skills demanded by employer groups and therefore relevant to a graduate's future career prospects. In some courses, skills are directly taught and assessed; in others, they are indirectly linked to assessment but not explicitly included as part of the curriculum. For example, many courses assess communication skills (written and oral) within assignment tasks but do not teach the theory or practice of effective communication. Accordingly, the Commonwealth of Learning (COL, 2019) has developed a graduate employability toolkit whose attributes may at first seem unremarkable, their generic-ness is in fact what makes them valuable because students with these skills, mindset, and attitudes can deal with challenges and adapt to new environments quickly and effectively. COL (2019) has developed an Employability Model for HEIs to enable them, especially those with minimal employability strategies, to explore diverse ways to assess, develop, implement, and evaluate employability within their home context. Fundamental to COL Employability Model is the need for tools and strategies to be contextualised to deeply reflect the realities of an institution's context, both social and institutional. With this in mind, the examples and templates provided are intended to generate thinking but are not designed to be used verbatim (COL, 2019). Therefore, African HEIs can learn from this COL Employability Model as a best practice in embedding graduate employability attributes in the curriculum.

10 Conclusion

Governments, HEIs, and companies in Africa should make embedding employability within the HE curriculum a top priority. This will benefit both the private and governmental sectors, proving HE's important role in economic growth as well as its importance in social and cultural development. HEIs in Africa must consider the following factors when developing and implementing effective employability strategies: their definition of employability, how it can be translated into practice, how students and staff can be involved, current practice and gaps in provision, and how to track progress. They will need a flexible framework to help them achieve this, one that includes a process for debate, contemplation, action, and assessment. Therefore, this chapter developed and proposed a framework for the integration of graduate employability attributes in curricula. The proposed framework was developed using best practices identified in the literature. The framework contains components that are considered important for HEIs in Africa to enhance graduate employability. These components include understanding what is graduate employability; identifying key stakeholders; determining employability attributes to be enhanced; determining strategies to best integrate employability in the curriculum; HE-industry partnership; assessing university employability efforts; sharing good practices through scholarly contributions and information-sharing platforms. Although the study used a case of management sciences curricula in Namibia, the framework may apply to any discipline in any country on the continent.

References

Barrie, S. C. (2006). *Understanding what we mean by the generic attributes of graduates* (Vol. 51, p. 215). University of Sydney.

Blessinger, P. (2023) *Advancing SDG goals in higher education institutions globally.* Accessed Jan 16, 2023, from https://www.linkedin.com/pulse/advancing-sdg-goals-higher-education-institutions-patrick-blessinger/?midToken=AQEznDxqUpLpZQ&midSig=1921SNatEMGWA1&trk=eml-email_series_follow_newsletter_01-newsletter_content_preview-0-title_&trkEmail=eml-email_series_follow_newsletter_01-newsletter_content_preview-0-title_-null-2yvbij~lcxjsk5j~zl-null-null&eid=2yvbij-lcxjsk5j-zl

Blackwell, A., Bowes, L., & Harvey, L. (2001) Transforming work experience in higher education. *British Educational Research Journal,* 27(3). https://doi.org/10.1080/0141192012004830

A FRAMEWORK FOR EMBEDDING GRADUATE EMPLOYABILITY... 159

Business/Higher education Roundtable. (2017). *Taking the pulse of work-integrated learning in Canada.* Accessed Jan 16, 2023, from https://www.bher.ca/sites/default/files/documents/2020-08/BHER-Academica-report-supplement.pdf

Chankseliani, M., & McCowan, T. (2022). Higher education and the sustainable development goals. *Higher Education, 81,* 1–8.

Clarke, M. (2018). Rethinking graduate employability: The role of capital, individual attributes and context. *Studies in Higher Education, 43*(11), 1923–1937.

COL. (2019). *Introducing: Employability: A toolkit for the commonwealth of Learning's employability. Model.* Canada.

Cole, D., & Tibby, M. (2013). *Defining and developing your approach to employability.* The Higher Education Academy. Accessed Jan 16, 2022, from https://www.advance-he.ac.uk/knowledge-hub/defining-and-developing-your-approach-employability-framework-higher-education

Edith Cowan University. (2013). *Employability: A good practice guide.* Edith Cowan University.

El-Jardali, F., Ataya, N., & Fadlallah, R. (2018). Changing roles of universities in the era of SDGs: Rising to the global challenge through institutionalizing partnerships with governments and communities. *Health Research Policy Systems, 16*(1), 1–5.

Fallows, S., & Steven, C. (2000). Building employability skills into the higher education curriculum: A university-wide initiative. *Education and Training, 42*(2), 75–82.

Gardner, P. (1998). Are college seniors prepared for work? In N. Gardner & G. van der Veer (Eds.), *The senior year experience. Facilitating integration, reflection, closure, and transition.* Jossey-Bass Inc Publishers.

Gibson, E., Brodie, S., Sharpe, S., Wong, D. K. Y., Deane, E., & Fraser, S. (2002) *Towards the development of a work integrated learning unit.* Accessed Apr 24, 2022, from https://researchers.mq.edu.au/en/publications/towards-the-development-of-a-work-integrated-learning-unit

Grund, L. (2020). *How can universities meaningfully and effectively use the SDGs?* Accessed Jan 16, 2023, from https://sdg.iisd.org/commentary/generation-2030/how-can-universities-meaningfully-and-effectively-use-the-sdgs/

Hill, J., Walkington, H., & France, D. (2017). Graduate attributes implications for higher education practice and policy. *Journal of Geography in Higher Education, 40*(2), 155–163.

Jackson, D. (2014a). Employability skill development in work-integrated learning barriers and best practice. *Studies in Higher Education, 40,* 1–18.

Jackson, D. (2014b) Modelling graduate skill transfer from university to the workplace. Journal of Education and Work. https://doi.org/10.1080/1363908 0.2014.907486

Jackson, D., & Wilton, N. (2016). Developing career management competencies among undergraduates and the role of work-integrated learning. *Teaching in Higher Education, 21*(3), 266–286.

Jonck, P. (2014a). A human capital evaluation of graduates from the Faculty of Management Sciences Employability Skills in South Africa. *Academic Journal of Interdisciplinary Studies, 3*(6), 265–274.

Jonck, P. (2014b). Human capital evaluation of graduates from the Faculty of Management Sciences Employability Skills in South Africa. *Academic Journal of Interdisciplinary Studies, 3*(6), 265–274.

Knight, P. T., & Yorke, M. (2003a). *Assessment, learning and employability.* SRHE and Open University Press.

Knight, P. T., & Yorke, M. (2003b). Employability and good learning in higher education. *Teaching in Higher Education, 8*(1), 3–16.

Leong, R., & Kavanagh, M. (2013). A work-integrated learning (WIL) framework to develop graduate skills and attributes in an Australian university's accounting programme. *Asia-Pacific Journal of Cooperative Education, 14*(1), 1–14.

Levy, F., & Murnane, R. J. (1996). With what skills are computers a complement? *American Economic Review Papers and Proceedings, LXXXVI,* 258–262.

Mason, G., Williams, G., & Cranmer, S. (2009) *Employability skills initiatives in higher education: What effects do they have on graduate labour market outcomes? 17*(1), 1–30. https://doi.org/10.1080/09645290802028315

Muldoon, R. (2009). Recognizing the enhancement of graduate attributes and employability through part-time work while at university. *Active Learning in Higher Education, 10*(3), 237–252.

Ngulube, B. (2020). Undergraduate economics curriculum and employability skills in South Africa. *Problems of Education in the 21st Century, 78*(6), 1000–1013. https://doi.org/10.33225/pec/20.78.1013

Owens, J., & Tibby, M. (2014) *Enhancing employability through enterprise education: Examples of good practice in higher education.* Accessed Apr 13, 2022, from https://www.makingthemostofmasters.ac.uk/media/microsites/mmm/documents/6.-enhancing_employability_through_enterprise_education_good_practice_guide.pdf

Pool, D. L., & Swell, P. (2007). The key to employability: Developing a practical model of graduate employability. *Education and Training, 49*(4), 277–289.

Quality Assurance Agency for Higher Education. (2009). *Learning from ELIR 2003–07 emerging approaches to employability and personal development planning sharing good practice.* QAA.

Rosenberg, S., Heimler, R., & Morote, E. S. (2012). Basic employability skills: A triangular design approach. *Education and Training, 54*(1), 7–20.

Saunders, V., & Zuzel, K. (2010). Evaluating employability skills: Employer and student perceptions. *Bioscience Education, 15*(2), 27–43.

Smith, C. (2012). Evaluating the quality of work-integrated learning curricula: A comprehensive framework. *Higher Education Research & Development, 31*(2), 247–262.

A FRAMEWORK FOR EMBEDDING GRADUATE EMPLOYABILITY... 161

Tran, T. (2016). Enhancing graduate employability and the need for university-enterprise collaboration. *Journal of Teaching and Learning for Graduate Employability, 7*(1), 58–71.

Turner, T., & El-Jardali, F. (2017). Building a bright, evidence-informed future: A conversation starter from the incoming editors. *Health Research Policy Systems, 15,* 88.

Wilton, N. (2014). Employability is in the eye of the beholder employer decision-making in the recruitment of work placement students. *Higher Education, Skills and Work-Based Learning, 4*(3), 242–255.

Yorke, M. (2006). *Employability in higher education: What it is–what it is not.* The Higher Education Academy.

Yorke, M., & Knight, P. (2006). *Employability embedding employability into the curriculum. Learning and employability.* The Higher Education Academy.

Addressing SDG 4 via Learners' Prior Numerical Cognition: *a Predictor of Educational Performance in a Developing Country in Sub-Saharan Africa*

Alexander Michael ⓘ *and Anass Bayaga* ⓘ

1   INTRODUCTION

A growing practical phenomenon is that mathematics or numeracy educators (SDG 4 describes) at all levels of educational systems in South Africa and globally have employed more progressively varied and intricate teaching/learning methods with the hope of improving learners' performance in numeracy skills at primary schools. From a theoretical disposition too, an overabundance of local and international studies articulate that the present digital-technology spaces of learners are not simply passive recipients of knowledge, in contrast, active participants and facilitators (Adams,

---

A. Michael (✉) • A. Bayaga
Nelson Mandela University, Gqeberha, South Africa
e-mail: alexander.michael@mandela.ac.za; Anass.Bayaga@mandela.ac.za

© The Author(s), under exclusive license to Springer Nature   163
Switzerland AG 2024
P. Neema-Abooki (ed.), *The Sustainability of Higher Education in Sub-Saharan Africa*, Sustainable Development Goals Series,
https://doi.org/10.1007/978-3-031-46242-9_8

2015; Can & Yetkin Özdemir, 2020; DBE, 2014; Egodawatte, 2011; Orakci et al., 2019; Pournara et al., 2016; Sasman, 2011). For example, Adams (2015) offered that in elementary school, learners' abilities of learning and knowing multiplication tables by rote require a qualitatively different kind of thinking compared to those using multiplication skills through solving word problems. In both cases, the teacher could evaluate learners' knowledge and skills by asking them to testify to those skills in action as supported by the sustainable development goals, particularly SDG 4. DBE (2014) reported that the 2014 South African learners' over-all mathematics results for Annual National Assessments (ANA) in grades 1–6 points towards an upward movement of test scores, while other higher grades approximating to grade 9 performance remained at a low level as the case in previous years. Adjacently, Pournara et al. (2016) opined that South African learners commit a wide variety of errors in simple algebra items. From the foregoing, a study needed to be conducted on learners' prior mathematical cognition as a predictor of performance at current educational levels. In this study educational levels refer to the hierarchical educational programmes that learners usually undertake to obtain a quali-fication, unvarying with primary school, high school, certificate, diploma, degree, and postgraduate studies.

Few inferences could be drawn so far from both practice and theory. For instance, in most school systems, subjects kindred with mathematics are structured in a way that could be inculcated into the minds of learners through basic skills in classrooms. Another ramification to the purpose is to equip learners with a certain level of cognitive skills which may include analytical and reflective thinking to make decisions and solve problems in mathematics (Orakci et al., 2019).

Despite such ongoing efforts, South African learners' performance in mathematics at higher educational levels (above primary school) gives a cause for concern (Egodawatte, 2011; Sasman, 2011; DBE, 2014; Can & Yetkin Özdemir, 2020; Pournara et al., 2016). For instance, Sasman's (2011) study in the Kwa-Zulu Natal province of South Africa reports that grade 9 learners' underperformance in mathematics is consequently due to insufficient prior knowledge of the subject and wrong application of the algorithm in problem-solving and algebra. A similar Canadian province of Ontario study by Egodawatte (2011) investigating errors and misconcep-tions in algebra at grade 11 found that learners' underperformance in mathematics was due to a lack of understanding of basic concepts of the variables in existential problems.

There is a dearth of studies on the actual happenings on the ground for how learners' prior mathematical cognition at the primary school level might or not impact their HE levels (Adams, 2015). Thus, mathematical cognition as conceived and expounded, while they refer to learners' more active ways of learning mathematics in classrooms, Adams (2015) quips that it is insufficiently examined.

The consensus from the works of Egodawatte (2011), Sasman (2011), DBE (2014), Can and Yetkin Özdemir (2020), and Pournara et al. (2016) are that in South Africa and internationally, learners' underperformance in mathematics at other levels above primary school level has become a reoccurring decimal. The additional highlight, however, is that it is worrisome to note that little is available in the literature on whether learners' underperformance at the primary level has a bearing on their performance at HE levels.

Based on the consensus, what sets this study apart from the previous is that these highlighted studies are from different contexts and so there might be variations like schools' systems and classrooms. However, it is important to note that the extant studies investigated learners' errors and misconceptions in mathematics topics kindred with algebra and problem-solving at the high school level with no regard to their prior knowledge of the subject. What is also unknown is the investigation of the relationship between learners' prior mathematical cognition at the primary school level and their performance in future educational levels; while persistent and worrisome, efforts towards such endeavours are minimal.

As a matter of contribution to the extant knowledge and taking into account the foregoing, it was compelling to assert whether a relationship existed or not between learners' prior mathematical cognition compared to their performance at current educational levels. It was pertinent therefore to examine such a dearth of research in the context of South African learners of rural settlement.

This chapter is structured as follows. In the next section, the authors reviewed some related research work around topics on students' mathematical cognition and their performance in the subject. For instance, the authors examined factors affecting learners' fluency and acquisition of mathematical skills through both the works of Pekrun et al. (2009) and supplemented by Linnenbrink (2007). Guided by the views of DBE (2014) and Can and Yetkin Özdemir (2020), the authors also explored scholarship on the effects of primary school mathematics instruction on high school learners' underperformance. Simultaneously, the authors also explored not just learners' underperformance in high school mathematics (through Sasman, 2011; Pournara et al., 2016), but also the examination of the practical application of mathematics topics in South African schools (through DoE, 2001; Mji & Makgato, 2006).

Guided by the revised Bloom's taxonomy related to cognitive learning and hinged upon Adams (2015) and Anderson et al. (2001), the theoretical framework section dealt with the description of the theory underpinning the study. In the Methodology section, we described the approach used and the procedures enjoyed in conducting the study. Based on the synopsis thus far, it is hypothesised that there is no significant relationship between learners' prior mathematical cognition compared to their performance at current educational levels.

2 Background

2.1 Sustainable Development Goals and Sustainability of Higher Education

HE has over the years played several important roles entirely with research, education, and social contribution. However, globalisation has impacted greatly on HE, thereby causing a gradual change, which led to the development of SDGs. SDG 4 in particular targeted an issue that has a direct bearing on the performance of HE in developing countries serving as South Africa. Generally, SDG 4 aims to ensure equal access for all women and men to affordable and quality technical vocational and HE (United Nations, 2015). It is therefore important to note the targeted goals as well as the quality of HE. Therefore, this chapter deals with the relationship between SDG 4 via Learners' Prior Numerical Cognition as a Predictor of Educational Performance in Developing Countries and HE by considering South Africa as an example. The nature and structure of the network in education in the past year have been multifaceted globally, with more universities and higher institutions collaborating for better and quality outcomes (United Nations Academic Impact, 2021). All the collaborations are aimed at supporting and contributing to the SDG 4 goals and mission of the UN. As reported by the World Bank (2020) and UNESCO (2021), HE is faced with both long- and short-term challenges. These challenges or problems as revealed included dwindling resources for a higher institution, academic, and social challenge for the learners and the institutions. Thus, the need for better facilities to support in-person and online learning models. HE, globally, is therefore facing difficult times as far as the need to addressing SDGs-related issues is concerned, while also dealing with the challenge of rural-urban learners' educational performance in a developing country and learners' fluency and acquisition of numeracy (mathematical) skills as noted in SDG 4, hence the need for this study.

2.2 Factors Affecting Learners' Fluency and Acquisition of Numeracy (Mathematical) Skills—SDG 4

Past studies had found several reasons why learners underperformed mathematically in their high schools (DBE, 2014; Can & Yetkin Özdemir, 2020). These factors include environmental, cognitive, and affective including changes in learners' mathematics learning. experts (e.g., Geary, 2011) in mathematics education had studied the relationship between learning mathematics and some cognitive factors. Geary (2011) reveals that working memory, intelligence, and processing speed have a high impact on mathematics achievement. Can and Yetkin Özdemir (2020) reported that learners' creativity in mathematics was seen to be related to fluent thinking skills, which were directly associated with how many ideas each learner produced. It follows that learners would only achieve meaningful mathematical skills and consequently acquire better mathematical fluency when mathematics teachers create an enabling environment for them to be emotionally and psychologically stable.

As discussed earlier, while South African learners' overall mathematics results for ANA at the primary school level (grades 1–6) point towards an upward movement of test scores, that of secondary school levels (higher grades) has remained at a low level as the case in previous years (DBE, 2014). Therefore, elaborating on the factors above in the South African context required an investigation to determine if there is or is no significant relationship between learners' prior mathematical cognition compared to their performance at current educational levels. Exploring other factors from past studies, for example, methods of instruction by teachers that contribute to learners' underperformance and non-acquisition of mathematical skills, could shed light on this study. The following subhead expounds on the effects of primary school mathematics instruction on high school learners. Therefore, this study proposed the following research question:

Research Question 1 (R1): What matters most to learners about ease of learning (fluency) and numeracy skills?

3   The Effects of Primary School Mathematics Instruction on High School Learners' Underperformance

As justified by Can and Yetkin Özdemir (2020), learners in elementary class (grade 4) underperformed in mathematics because of their inability to recognise number size. Can and Yetkin Özdemir (2020) underscored that as numbers increase in quantity, learners had difficulty recognising them. Two factors might be responsible for learners' underperformance in Can and Yetkin Özdemir's (2020) study. It may be due to the context that learners found themselves in or they were not taught the basic mathematical skills of number identification. Although achievement in mathematics is strongly related to intelligence, learners need motivation and guidance from their teachers on how to solve mathematical problems using cognitive approaches in other levels for them to perform better in their higher grades/high schools. As for Zan and Martino (2008), high school learners' attitude towards mathematics is key. Their likes and dislikes as well as their belief towards mathematics can affect their learning. Other external factors close to the accessibility of good mathematics textbooks might affect high school learners' performance as well (Zan & Martino, 2008). Above all lack of parental involvement and motivation from other family members could also have a significant effect on high school learners' performance in mathematics.

In any case, deriving from the discussion above, the effects of primary school learners' mathematical knowledge on their performance at high school/HE levels in the South African context need an in-depth investigation. Closely related to the above is the relationship between primary school learners' prior mathematical cognition compared to their performance at current educational levels (high school, certificate, diploma, first degree, and postgraduate studies). Thus, this study formulated the following research question:

Research Question 2 (R2): How related is learners' current underperformance in mathematics to their past years in primary school?

4   Learners' Underperformance in High School Mathematics Topics

Topics indistinguishable from fractions, algebra, analytics, and Euclidean geometric cover large sections of the mathematics syllabus in South Africa. Several researchers, including Sasman (2011) and Pournara et al. (2016), have revealed that learners' underperformance rate in mathematics at high schools and other HE levels in South Africa is high. According to Pournara, Hodgen, Sanders, and Adler (2016), learners commit a wide range of errors even in basic algebraic problems. Some of the errors of the learners as reported by Pournara et al. (2016), included conjoining, which is difficulties with negatives and brackets and a tendency to evaluate the algebraic expressions rather than leaving them in the required open forms. For instance, when grade 9 learners were asked to add 5 with 3 x, the learners produced conjoined answers coextensive with "5 plus 3 is 8 and then put x next to it to get 8". Again, when grade 11 learners were also asked to simplify algebraic expressions involving the subtraction of a negative term, the teacher asked, respond to the question: subtract 2b from 8. The learners answered the question as follows: "It's going to be 8". When the teacher asked why 2b was left, the learners answered, "cos it had to be subtracted from 8 and we couldn't". Pournara et al. (2016) concluded that learners' errors in simplifying algebraic expressions imply that they were not paying attention to signs and operations.

Similarly, Sasman's (2011) study reveals that some high school learners do not like mathematics. For example, when a grade 11 learner was asked why she underperformed in some mathematics topics self-same with algebra, fractions, and Euclidean geometric, her response was, "I hated algebra because I did not understand it". Sasman (2011) reported that learners committed errors in solving mathematical problems because they had a poor understanding of the basics and foundational competencies in some of the topics taught to them in their earlier grades. These errors discussed above often lead to underperformance and are also an indication that learners lack basic knowledge of some content areas in mathematics (DBE, 2014) at the primary school level.

This study investigated learners' prior mathematical skills compared to their performance at current educational levels in the rural settlement of Kwa-Zulu province. Again, there is an overlapping and interconnectedness between content areas and the practical applications of these content areas in mathematics classrooms. Practical applications of the content areas in every mathematics classroom play a key role in learners' better performance in the subject, the following subhead explores in depth this subject matter in the South African context. Hence, this study made the following research question:

Research Question 3 (R3): To what extent is learners' current underperformance in fractions, algebra, analytics, and Euclidean geometric associated with inappropriate practices?

5 Practical Application of Mathematics Topics in South African Schools

Mathematics topics as mentioned earlier often look quite abstract and far removed from real-life situations, making learners wonder why study this. Fortunately, connecting these mathematical content areas with their application can make a difference for learners. The practical application of mathematics topics in this study referred to how learners made use of these topics in real life. It involved letting the learners make sense of the mathematical content they learned to have a significant positive impact on their lives.

Mji and Makgato (2006) in examining the practical application of mathematics topics in South African schools reported that the significant increase in the number of underperforming mathematics learners was due to the traditional (teacher-centred) method of teaching/learning in classrooms and the lack of practical application of the mathematics topics. According to Mji and Makgato (2006), due to the lack of practical application of some mathematics topics, most learners could not effectively acquire the skills expected of them.

As triggered by Mji and Makgato (2006), mathematics teachers in schools needed to adopt a practical approach to their teaching/learning process of the subject to enhance learners' understanding. In this study, the practical application of mathematics topics was investigated concerning learners' performance. Table 1 shows a summary of the major themes of the reviewed kinds of literature, the research questions, all key gaps, and their sources. Consequently, this study proposed the following research question:

Research Question 4 (R4): How could making mathematics topics visible in real-world situations improve learners' achievement?

6 Theoretical Framework

As noted in the earlier sections, this chapter seeks to analyse learners' prior mathematical cognition as a predictor of performance at current educational levels in Kwa-Zulu province, South Africa. The study's hypothesis thus, there is no significant relationship between learners' prior mathematical cognition compared to their performance at current educational levels. The formulated four research questions (R1–4) centred on ease of learning (fluency) mathematics topics, relationships between learners'

Table 1 Summary details of major themes, questions, key gaps

Major themes	Questions	Key gaps	Sources
Factors affecting learners' fluency and acquisition of mathematical skills	R1. What matters most to learners in relation to ease of learning (fluency) mathematics topics?	Learners' prior mathematical cognition compared to their performance at current educational levels	DBE (2014); Can and Yetkin Özdemir (2020); Geary (2011)
The effects of primary school mathematics instruction on high school learners' underperformance	R2. How related is learners' current underperformance in mathematics to their past years in primary school?	Negative effects of primary school learners' mathematical knowledge on their performance at high school	Can and Yetkin Özdemir (2020); Zan and Martino (2008)
Learners' underperformance in high school mathematics	R3. To what extent is learners' current underperformance in fractions, algebra, analytics, and Euclidean geometric associated with inappropriate practices?	Learners' prior underperformance in mathematics compared to their performance at current educational levels	Sasman (2011), Pournara et al. (2016)
Practical application of mathematics topics in South African schools	R4. How could making mathematics topics visible in real-world situations improve learners' achievement?	Practical application of mathematics topics concerning learners' performance in specific topics congeneric with fractions, algebra, analytics, and Euclidean geometric	Mji and Makgato (2006)

Source: Authors' own design

current underperformance in mathematics to their past years in primary school, learners' current underperformance as per the status quo, as an associate with inappropriate practices, as well as how to improve learners' achievement by making mathematics topics visible in a real-world situation.

Again, the major gaps emanating from the reviewed literature were lack of study in the South African context on learners' prior mathematical cognition compared to their performance at current educational levels, a gap exists in understanding the negative effects of primary school learners' mathematical knowledge on their performance at high school. Learners' prior underperformance in mathematics compared to their performance at current educational levels needed further exploration as well as the practical application.

The available resources which informed the choice of the theory in the investigation of this research were kinds of literature relating to mathematical cognition. According to Adams (2015) mentioned earlier in the Introductory section, mathematical cognition refers to the more active ways of learning mathematics in classrooms by learners in their activities (Anderson et al., 2001). Thus, the selected theory should enable the analysis of learners' prior mathematical cognition as a predictor of performance at current educational levels. The theoretical perspective would help in the detailed analysis of the major themes emerging from the literature that addressed the collected data and the gaps identified.

From the foregoing that the authors needed to explore learners' prior mathematical cognition as a predictor of performance at current educational levels, this implied an approach from revised Bloom's taxonomy cognitive learning theory perspective (Adams, 2015; Anderson et al., 2001).

Revised Bloom's taxonomy, related to cognitive process dimension and hinged upon Adams (2015) and Anderson et al. (2001), is a more active approach to learning, where learners' answers are not just determined by correctness, but also on how learners remember, understand, apply, analyse, evaluate, and create before arriving at their answer. Revised Bloom's taxonomy theory also has a sub-dimension of knowledge called "Metacognition", which is somewhat related to thinking about learners' thinking. The implication is that if mathematics teachers employ revised Bloom's taxonomy and in particular the metacognition of the theory in the practical application concerning learners' performance; hence the urgency to examine how the learners arrived at their answers. The metacognition dimension of the theory is important in this study because it takes into thought that all learners think differently based on their previous knowledge and related information they have learned in the past. Understanding how learners think differently can help expand their knowledge. For example, looking at the gaps in Table 1, each theme from the literature and the research questions discussed, revised Bloom's taxonomy theory would expose the characteristics of each category of the educational levels for this study. The triangular representation (Fig. 1) adopted from Anderson et al. (2001) gives the discerption of the revised bloom's taxonomy as described in this study.

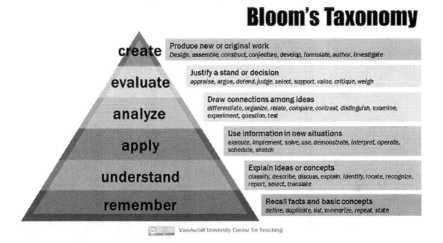

Fig. 1 Revised Bloom's taxonomy theory. Adopted from Anderson et al. (2001)

As articulated by DBE (2014), Can and Yetkin Özdemir (2020), and Geary (2011) as well as the revised Bloom's taxonomy above, the research question "what matters most to learners with ease of learning (fluency) mathematics topics?" was met. It counsels that learners would only achieve meaningful mathematical skills and consequently acquire better mathematical fluency when mathematics teachers create an enabling environment for them to be emotionally and psychologically stable.

Drawing from the work of Can and Yetkin Özdemir (2020) and Zan and Martino (2008), the second research question, how related is learners' current underperformance in mathematics to their past years in primary school, was responded to. It was evident from the foregoing that mathematics is strongly related to intelligence, but learners need motivation and guidance from their teachers on how to solve mathematical problems using cognitive approaches in other levels for them to perform better in their higher grades/high schools. While arising from the works of Sasman (2011) and Pournara et al. (2016) the research question, to what extent is learners' current underperformance in fractions, algebra, analytics, and Euclidean geometric associated with inappropriate practices, was also re-joined. It was put that basic knowledge on some content areas in mathematics namely, fractions, algebra, analytics, and Euclidean geometric needed to be reviewed and all grey areas making learners commit errors to be resolved to improve better performance.

Finally, the research question of "how could making mathematics topics visible in real-world situations improve learners' achievement" was solved using the work of Mji and Makgato (2006), who voiced that mathematics teachers in schools needed to adopt a practical approach to their teaching/learning process of the subject to enhance learners' understanding.

7 METHODOLOGY

7.1 Design

This study investigates the relationships between learners' prior mathematical cognition and their performance at current educational levels. The study employed a multivariate analysis research design. Multivariate analysis is a statistical design that is concerned with understanding the different aims/objectives and background of each of the several kinds of multiple dimensions while considering the effects of all variables on the responses of interest and how relevant/related they are to each other. The real implication of multivariate analysis to a specific research problem might involve a few forms of univariate and multivariate analyses to understand their relationships between variables as well as their relevance to the real problem under investigation. In this study, a stepwise discriminant analysis was conducted to find out if learners' prior mathematical cognition could be a predictor of their performance at current educational levels.

7.2 Characteristics of the Participants

Five educational levels within the rural settlement of Kwa-Zulu Natal province had a total population of 150 learners. The five educational levels as earlier enumerated in the Introductory section were postgraduate, degree, diploma, certificate, and high school with 150 learners. The participants were made up of mixed gender with a minimum age range of 21–23 years and a maximum age range of 30 and above years.

7.3 Sampling Procedure

The study adopted a quantitative approach and was conducted in a rural settlement of the Kwa-Zulu Natal province. Since the 150 of the study population was known, and since the numbers of learners/students were not the same across the various levels of education, proportionate stratified random sampling was used to sample the participant from each of the educational levels. According to Agyedu et al. (2011), proportionate stratified random sampling is used when the population for a study is finite. A sample size of 71 was used for the study, with high school learners represented by 1, while 2, 3, 4, and 5 represented certificate, diploma, degree, and postgraduate students, respectively.

7.4 Administration of the Questionnaire

A self-administered questionnaire was designed to assess learners' responses to their performance in mathematics and their prior learning experience. The questionnaire consisted of 34 items with 4 of them focusing on learners' mathematical cognition and performance characteristics and the remaining on other areas. The themes and their codes for mathematical cognition and performance characteristics are presented in Table 2.

The questionnaire items as above (see Table 2) were on a four-point nominal scale ranging from strongly disagree to strongly agree, with strongly disagree fitted into the score of 1. Whiles disagree, agree, and strongly agree were within the scores of 2, 3, and 4, respectively.

7.5 Data Analysis

To enable the analysis of data, the mean, standard deviation, Box's M, Levene's test, and Multivariate Analysis of Variance (MANOVA) were employed. MANOVA is the multivariate extension of the univariate techniques for assessing the differences between group means. In contrast to ANOVA, it can examine more than one dependent variable at the same time. The study made use of a four-point Likert scale (Boone & Boone, 2012). Common place understanding is that Likert-scale data are analysed at the nominal measurement scale and are generated by calculating a composite score (sum or mean) from four or more type Likert-type items; therefore, the composite score should be analysed at the nominal measurement scale. Descriptive statistics recommended for nominal scale items include the mean for central tendency and standard deviations (SD) for variability. The mean and SD were employed to answer the research questions for this study to determine learners' prior mathematical cognition compared to their current performance at the five educational levels

176   A. MICHAEL AND A. BAYAGA

Table 2 A sample of the administered questionnaire

Themes	Codes	Scale			
		1	*2*	*3*	*4*
Ease of learning (fluency) mathematics and acquisition of skills in any of the topics in mathematics and related programmes matters to me	MAS				
I associate my current underperformance with my past underperformance in my early (primary school) years of mathematics instruction	CUP				
I associate my current underperformance with inappropriate preparatory practices and a lack of understanding of topics like fractions, algebra, analytics, and Euclidean geometric	CUI				
Improving my understanding could have been achieved by making topics visible and presenting them in the real-world situation	IUP				

earlier mentioned. The Box's M and Levene's tests were employed to determine the normality and homogeneity of the data before the hypotheses (significant difference) testing. MANOVA was employed to determine the difference between the independent and dependent variables and which of the determinants significantly affects the dependent variables.

8   MULTIVARIATE ANALYSIS QUALITY ASSURANCE MEASURES

Multivariate analysis quality assurance (QA) in this study refers to all measures employed to achieve high-quality analysis of the data. In this study, QA was employed to ensure that the data generated met defined standards of quality. The chapter considers only the generalised structure of a QA system in the sampling, data collection, and analysis. The statistical analysis employed is the mean, SD, and MANOVA.

8.1   Reliability and Validity

A study ought to satisfy acceptable standards of a research inquiry through the provision of checks and balances. This section deals with the issues of reliability and validity that were relevant to the design of the questionnaire for this study. To reduce the shortcomings of the questionnaire, the following guidelines were adhered to: (1) there were clear and up to par instructions for participants for completion of questionnaires. (2) Items

ADDRESSING SDG 4 VIA LEARNERS' PRIOR NUMERICAL COGNITION... 177

were short and precise and devoid of double negatives to increase validity. (3) Questionnaire items were scrutinised by colleagues for content and ambiguity.

Internal consistency measures for assessing inter-item consistency are part of the reliability process designed to discover whether the items in the instrument measure the same thing. This study used the Likert-type instrument and the Cronbach alpha coefficient to measure the inter-item consistency. The Cronbach alpha is a reliable method to show the extent items positively correlate to each other (Sekaran & Bougie, 2010). For strongly correlated items with high internal consistency, the value of Cronbach's alpha is close to one.

In this study, the instrument used was a questionnaire of the Likert-type instrument. The Statistical Package for Social Science (SPSS) version 28.0 was employed to calculate the Cronbach alpha coefficient to determine the internal consistency of the questionnaire items. Cronbach alpha test conducted was greater than 0.70, which is dependable.

8.2 Ethical Consideration

Ethics is an important requirement in any research activity that involves humans. It shows sensitivity and respect on the researcher's part for the study participants (Opie, 2004). The authors applied for ethics clearance from the Faculty of Education and received approval.

9 RESULTS

As a way of recap, this study investigates if learners' prior mathematical cognition could be a predictor of their performance at current educational levels. Hence the research questions formulated to guide the study were: R1. What matters most to learners in relation to ease of learning (fluency) mathematics topics? R2. How related is learners' current underperformance in mathematics to their past years in primary school? R3. To what extent is learners' current underperformance in fractions, algebra, analytics, and Euclidean geometric associated with inappropriate practices? R4. How could making mathematics topics visible in real-world situations improve learners' achievement? The major gaps emanating from the reviewed literature thus far that needed further exploration consisted of learners' prior mathematical cognition compared to their performance at current educational levels. Negative effects of primary school learners' mathematical knowledge on their performance at high school. Learners' prior underperformance in mathematics compared to their performance at current educational levels. And practical application of mathematics topics for learners' performance in fractions, algebra, analytics, and Euclidean geometric.

Preceding the analysis of the data (71 items if the questionnaire was administered with a response rate of 100%), the questionnaire items were examined for accuracy and formatting, for purposes of entry into SPSS. The analyses, presentation, and discussion of the results are in three segments: descriptive analysis, the test of assumptions, and MANOVAs. The significance level was determined at a probability level of 0.05. The mean and SD were used to determine the level of learners' mathematical cognition on their performance at their current educational levels. And to also answer the four research questions. The test items on the questionnaire coded as MAS, CUP, CUI, and IUP were deliberately done to cover each of the research questions where MAS corresponds to research question 1 while CUP, CUI, and IUP correspond to 2, 3, and 4, respectively.

9.1 Descriptive Statistics

Descriptive statistics (mean and SD) were tested to answer the research questions and consequently determine the various levels of learners' prior mathematical cognition on their performance at their current educational levels.

9.2 Result for MAS

R1. What matters most to learners with ease of learning (fluency) mathematics topics?

A higher mean implies strongly agree/agree of learners at their current educational levels and a lower mean construes strongly disagree/disagree. A mean between one (1) and two (2) means that learners' answers were rated between strongly disagree and disagree on the average in that category. MAS result in Table 2 shows that learners at current certificate levels seem to agree that ease of learning (fluency) mathematics and acquisition of skills in any of the topics in mathematics and related programmes matter to them since their mean rating was 2.50. On the other hand, learners at current High School, Postgraduate, and Diploma mean ratings were

1.20, 1.39, and 1.76, respectively, indicating that it did not worry them much in terms of ease of learning (fluency) mathematics and acquisition of skills in any of the topics in mathematics and related programmes.

9.3 Result for CUP

R2. How related is learners' current underperformance in mathematics to their past years in primary school?

CUP result shows that learners at current postgraduate levels agree and seem to associate their current underperformance to past underperformance in their early (primary school) years of mathematics instruction since their mean rating was 2.22. However, learners at current certificate, Diploma, and High School mean ratings were 2.00, 1.87, and 1.40, respectively, indicating that they did not agree to associate their current underperformance to past underperformance in their early (primary school) years of mathematics instruction.

9.4 Result for CUI

R3. To what extent is learners' current underperformance in fractions, algebra, analytics, and Euclidean geometric associated with inappropriate practices?

According to CUI results, learners at current certificate levels agreed to associate their current underperformance to inappropriate preparatory practices and lack of understanding of the topics in question since their mean rating was 2.22. However, learners with current Postgraduate, High School, and Diploma mean ratings were 2.00, 1.87, and 1.40, respectively, indicating that they did not associate their current underperformance to inappropriate preparatory practices and lack of understanding in topics on focus.

9.5 Result for IUP

R4. How could making mathematics topics visible in real-world situations improve learners' achievement?

Looking at the IUP result above it is evident that learners at current certificate levels agreed that making topics visible and presenting them in the real-world situation by mathematics teachers would improve their achievement since their mean rating was 2.50. While learners at current High School, Diploma, and Postgraduate mean ratings were 1.20, 1.41, and 1.44, respectively, indicating that it did not matter much to learners at current High School, Diploma, and Postgraduate levels for mathematics teachers making topics visible and presenting them in the real-world situation.

9.6   Test of Assumptions

In using MANOVA, the following assumptions must be met. The data set must have multiple dependent variables which are continuous or Likert-scale. The data set must also have one independent variable which must be categorical (binary or nominal), and it is clear that the data set has met these assumptions mentioned above. Test procedures for analysis of MANOVA of homogeneity of covariance (HC), independence of observation (IO), homoscedasticity (LR), multicollinearity (MIV), significant outliers (CD), and normal P-P plots must also be conducted to ensure that the data set is valid for use in the multivariate analysis. Subsequent paragraphs present the screening process for these assumptions (Alexopoulos, 2010).

The box's test for HC (also called the box's test for equality of covariance matrices) was adopted. In this study, the Box's test was used to compare variation in the 71 sampled questionnaire data set for the five educational levels of learners. It checked to find out whether covariance matrices for the five educational levels were homogeneous (Table 3).

With a significant value of 0.373, it can be concluded that the assumption of homogeneity of covariance has been subjugated across the five educational levels. In simple terms, it means that the multivariate homogeneity assumption is met.

Table 3 Test for HC

Box's test of equality of covariance matrices	
Box's M	29.457
F	1.075
df1	20
df2	457.534
Sig.	0.373

On checking for the independence of observation using the Durban-Watson test statistics, the model summary gives the results as 1.849 which was within the generally accepted range of 1 < 1.849 < 3, which means the data is tenable because it has independence of observation.

A test was also conducted to check for LR of the data set in this study, hence partial regression plots of the scatter plots were used for the data of the five education levels of learners to conduct the test. All the scatter plots flagged up homoscedasticity (which means the data set collected for this study is similar, constant, and elliptical). It also means that the multivariate homoscedasticity assumption is met.

The assumption of MIV for the data set of the five education levels of learners in this study was also checked. MIV in general terms refers to an undesirable situation where one independent variable is a linear function of other independent variables (Ibrahim et al., 2018). In other words, MIV relates to the correlation matrix, and it occurs when predicted variables are highly (0.9 and above) correlated (Ibrahim et al., 2018).

Revealed also is that the Tolerance and Variance Inflation Factor (VIF) has a maximum value of 1.289, which is within the acceptable range. Since all the VIF values are far below 10.0 and above 0.1, it, therefore, means that the multivariate MIV assumption is met for the data set used in this study (Ibrahim et al., 2018).

The assumption of the CD test of the data set of this study was also checked. It was discovered that the SPSS results for the CD were not produced, which means there is no case of CD in the data set used for this study.

Also depicted is that both maximum and minimum standardised residuals are within the required range of the residual. In general terms, the minimum standardised residual should not be less than -3.29 and the maximum standardised residual should not be greater than +3.29. That means there is no case of CD. Hence the multivariate CD assumption is met for the data set used in this study.

Lastly, the approximately normally distributed residual test was also conducted for the data set used in this study, using the histogram and the normal probability plot. It showed the superimposed normal curve as expected. Likewise, P-P plots showed that the dots were distributed approximately close along the 45-degree line of the plot. It means that the assumption for normal P-P plots of the data set used in this study has been met. This shows that the residual is approximately normally distributed. Hence all the assumption tests for the data set used in this study have been

182   A. MICHAEL AND A. BAYAGA

met and there is nothing to transform or fix in the data set. The next section expounds on the interpretation and discussion of the multivariate analysis.

9.7   Multivariate Analysis of Variance (MANOVA)

It was hypothesised that there is no significant relationship between learners' prior mathematical cognition compared to their performance at current educational levels. Thus, MANOVA was employed for the data analysis of this study. Generally, MANOVA is a procedure for comparing multivariate sample means and it is used when there are two or more dependent variables. MANOVA was therefore used in this study because data obtained comprises five categorical educational levels of learners (high school, certificate, diploma, degree, and postgraduate) that were independent variables and four dependent variables which were coded as (MAS, CUP, CUI, and IUP).

The MANOVA test showed a significant effect of the dependent variables (MAS, CUP, CUI, and IUP) on the educational levels of learners as it reported the Wilks' Lambda to be 0 646, $F = 2.537$, $P < 0.05$, partial eta squared =0 135, the power to detect the effects =0.942. In addition, the significance levels of all four (Pillai's Trace, Wilks' Lambda, Hotelling's Trace, and Roy's Roots) SPSS MANOVA tests are less than 0.05 indicating the significant effect of the MAS, CUP, CUI, and IUP on learners' educational levels collectively. Levene's test of equality of error variances was also computed at $p < 0.05$ only "MAS had significant mean values less than 0.05". This means that in this study, it is only the MAS category under which Levene's test for homogeneity is significant, which is also an indication that variances are not equal between the CUP, CUI, and IUP categories. However, Levene's test proved insignificant in the CUP, CUI, and IUP categories for the data set of this study, indicating that the assumption of equal variances across groups holds for the CUP, CUI, and IUP categories under which the learners' educational levels were predicated.

Between-Subjects Effects for MAS, CUP, CUI, and IUP, tests were carried out and the results of the multiple comparisons between the MAS, CUP, CUI, and IUP categories were also conducted to determine the level of significance and their effects on learners' educational levels. Since half (6 out of 12) of the multivariate test for the multiple comparisons among MAS, CUP, CUI, and IUP categories are less than 0.05, it revealed that the performance of learners in their various categories of educational levels is significantly dependent on their prior mathematical skills. Hence the null hypothesis for this study is rejected at a 5% significance level and concludes that there is a significant relationship between learners' prior mathematical cognition compared to their performance at current educational levels.

The mean of the educational levels was also compared in a pairwise format across all categories (high school, certificate, diploma, first degree, and postgraduate) to determine which mean differences are significant.

The mean performance difference was also checked since the multivariate test revealed that the performance of learners in their various categories of educational levels is significantly dependent on their prior mathematical skills. The means of high school, certificate, diploma, first degree, and postgraduate were compared in a pairwise format across MAS, CUP, CUI, and IUP categories to determine which mean differences are significant. In general terms, significant mean differences are always associated with significant values less than 0.05 and the confidence interval is either between 2 positive or negative real numbers. The mean differences in this research work were significant across all educational levels. However, the test largely reveals that differences occurred between the diploma and postgraduate levels.

10   DISCUSSION

This study investigates whether learners' prior mathematical cognition could be a predictor of their performance at current educational levels. The theoretical perspective for this study was the revised Bloom's taxonomy related to the cognitive process dimension and hinged upon Adams (2015) and Anderson et al. (2001). The results as presented in the foregoing sections of this chapter appear to support the claim that the performance of learners in their various categories of educational levels is significantly dependent on their prior mathematical skills.

The authors formulated four research questions (R1–4) which border on ease of learning (fluency) mathematics topics, relationships between learners' current underperformance in mathematics to their past years in primary school, learners' current underperformance, and how to improve learners' achievement by making mathematics topics visible in a real-world situation.

Again, the major gaps emanating from the reviewed literature were lack of study in the South African context on learners' prior mathematical cognition compared to their performance at current educational levels; a gap exists in understanding the negative effects of primary school learners' mathematical knowledge on their performance at high school. Learners' prior underperformance in mathematics compared to their performance at current educational levels needed further exploration as well as the practical application of mathematics topics with respect to learners' performance.

Findings in terms of the analysis of the formulated research questions (R1–4) for this study show as follows: for R1, learners at current certificate levels seem to agree that ease of learning (fluency) mathematics and acquisition of skills in any of the topics in mathematics and related programmes matter to them, unlike the learners at current high school, postgraduate, and diploma. According to R2, learners at current postgraduate levels agree and seem to associate their current underperformance with past underperformance in their early (primary school) years of mathematics instruction. However, learners with current certificates, diplomas, and high schools did not agree. R3: learners at current certificate levels agreed to associate their current underperformance to inappropriate preparatory practices and lack of understanding of the topics of concern. However, learners at current postgraduate, high school, and diploma did not. R4: learners at current certificate levels agreed that making topics visible and presenting them in the real-world situation by mathematics teachers would improve their achievement, while it did not matter for learners at current high school, diploma, and postgraduate levels.

From the multivariate test analysis of variance (MANOVA) and the results of the descriptive statistics, this study has found that ease of learning (fluency) mathematics and acquisition of skills in any of the topics in mathematics and related programmes matters to learners. This agrees with the existing research (Sasman, 2011) which reported that learners committed errors in solving mathematical problems because they had a poor understanding of the basics and foundational competencies in some of the topics taught to them in their earlier grades. Therefore, the authors speculate that practically oriented learning methods at the primary school level should be adopted to enable learners to have solid foundations and ease the acquisition of skills in any of the topics in mathematics.

The overall mean rating of learners associating their current underperformance to past underperformance in their early (primary school) years of mathematics instruction is satisfactory with the postgraduate level having the higher score. Such a finding means that most of the learners seem to manifest those learners need motivation and guidance from their teachers on how to solve mathematical problems using cognitive approaches in other levels for them to perform better in their higher grades/high schools. This finding supports the study of Zan and Martino (2008) that high school learners' attitude towards mathematics is key. Their likes and dislikes as well as their belief in mathematics can affect their learning.

Learners agreed to associate the current underperformance with inappropriate preparatory practices and lack of understanding. This agrees with Pournara et al. (2016) that learners' errors in simplifying algebraic expressions evince that they were not paying attention to signs and operations. These errors as discussed earlier often lead to underperformance and are also an indication that learners lack basic knowledge of some content areas in mathematics.

This study confirms that of Mji and Makgato (2006) that due to the lack of practical application of some mathematics topics by teachers, most learners could not effectively acquire the skills expected of them. Mji and Makgato (2006) rule that mathematics teachers in schools needed to adopt a practical approach to their teaching/learning process of the subject to enhance learners' understanding. Again, when teachers employ this method, learners would believe that practical learning is very useful for their learning activities and goals.

11 Advancing Mathematics and Cognition Education Conceptual Level

In this study, the authors have marched the course of mathematics and in particular cognition education at the conceptual level using multivariate analysis of variance to investigate learners' prior mathematical cognition as a predictor of performance at current educational levels in Kwa-Zulu province, South Africa. There exists a foundational connection in the mathematics content area at the primary school level with other HE levels. Conceptual knowledge is an integral aspect of mathematical ideas, thus learners with conceptual understanding have more ideas, facts, and methods of solving problems in the subject. Hence, learners with a conceptual understanding of mathematical cognition learn new ideas by connecting those ideas with what they already know previously (NRC, 2001).

12 Notions of Learner-Centred Pedagogy (LCP)

12.1 Global Context

Over the years, educators and researchers (Adams, 2015; Can & Yetkin Özdemir, 2020; DBE, 2014; Egodawatte, 2011) have propagated both locally and globally the notion of LCP as a "best practice". Notwithstanding its prominence in education policies, implementation has been thought-provoking, and changes to classroom practice are limited. This study has empirically extended the notion of LCP by its findings in which learners were viewed as active participants in their learning, with their education shaped by their interests, and prior knowledge. The study has extended the use of the revised Bloom's taxonomy as a theoretical framework in the mathematical context since South African classrooms are also within/part of the international community.

12.2 African Context

Despite sufficient evidence of application failure, LCP has been accepted in sub-Saharan African countries connatural with Rwanda. Most research (e.g., Hester et al., 2022) has examined its implementation at the primary and secondary levels in Rwanda. This Rwandan case uncovers that most primary and secondary teachers encouraged open and respectful classroom communications. In a similar study on LCP in Botswana Mungoo and Moorad (2015) found that to maximise learning in diverse-ability classrooms and to be effective, teachers need to employ a range of instructional approaches. It is therefore important that this study conducted in South Africa extended the notion of LCP in the African context.

13 Implications of This Study

Empirical studies in cognition education in high schools continue to extend the borders of prospects for theoretical frameworks, research designs, and interrogated questions. This study points to a new direction for education researchers in mathematics cognition. The authors of this chapter proposed the following as some of the new directions for research in mathematics cognition at the high school level: (1) adapting and extending the revised Bloom's taxonomy theory to answer new questions;

(2) evolving local theories that inform learning of mathematics in South African contexts; (3) exploring methods of knowledge shift in the South African classrooms for better understanding of LCP; (4) continuing to study new practices to learn mathematics in high school settings and sustain such practices.

14   Contribution of Chapter

This chapter contributes to the literature on knowledge building in mathematical cognition. The chapter has contributed to expanding the studies (Egodawatte, 2011; Sasman, 2011; DBE, 2014; Can & Yetkin Özdemir, 2020; Pournara et al., 2016) conducted in South Africa and internationally and concluded that learners' underperformance in mathematics at HE levels (above primary school) has a direct link with the primary school. First and foremost, the chapter offers verifiable data analysis that learners' prior mathematical cognition is a predictor of their performance at current educational levels. Second, the chapter has shown that learners need to be well-prepared in mathematics at their primary school for better performance in their future educational levels. Finally, the study has also revealed that much work is needed for the foundation phase of mathematics educators in South Africa and beyond.

15   Specific Lessons for Other African Countries

There is an urgent need to research LCP with a wider approach that aims at the role of the learner in an effective learning process to guarantee quality education. Educational institutions in Africa must encourage experts and new arrivals (e.g., doctoral students) to the world of expertise in international education with a focus on LCP in mathematics classrooms. The authors of this chapter have provided additional literature that could serve as a reference on LCP in schools. The study explored if learners' prior mathematical cognition could be a predictor of their performance at current educational levels. The findings from this study have provided a way forward for research into the LCP approach. Further investigations could focus on the challenges of using LCP in teacher education institutions, with a view of proposing several strategies of action with concrete steps to be taken by stakeholders in teacher education development in Africa.

16 Conclusion

This study investigated the relationship between learners' prior performance in their various categories of educational levels which is found to be significantly dependent on their prior mathematical skills and that the model used was generally sensitive and hence fit for the study.

Learners need to be well grounded at the lower level (primary schools) and be encouraged as well as motivated towards mathematical skills that have a direct bearing on their current educational levels. Mathematics teachers should also be encouraged and motivated to improve their quality of teaching and learning. In-service training, workshops, and conferences should be regularly organised for teachers, as this will help equip them with relevant learner-centred pedagogical skills that would keep them relevant in their profession.

This study was restricted to descriptive and inferential statistics, concerning the context. Further research is, therefore, encouraged to be performed using other methods and to cover other rural and urban settlements of schools in South Africa, as this will ensure the generalisability of the findings. Upcoming studies could also examine other factors—other than learners' prior mathematical cognition and their performance at current educational levels within several other contexts of South Africa.

References

Academic Impact. (2021). Impact Rankings 2021: Results announced by UK university tops third edition of global ranking measuring institutions' social and economic impact. Accessed Mar 20, 2023, from https://www.timeshighereducation.com/news/impact-rankings-2021-results-announced

Adams, N. E. (2015). Bloom's taxonomy of cognitive learning objectives. *Journal of the Medical Library Association: JMLA, 103*(3), 152–153. https://doi.org/10.3163/1536-5050.103.3.010

Agyedu, G. O., Donkor, F., & Obeng, S. (2011). *Teach yourself the research method, with 2010.* APA Update.

Alexopoulos, E. C. (2010). Introduction to multivariate regression analysis. *Hippokratia, 14*(Suppl 1), 23–28.

Anderson, L. W., Krathwohl, D. R., Airasian, P. W., Cruikshank, K. A., Mayer, R. E., Pintrich, P. R., Raths, J., & Wittrock, M. C. (2001). *A taxonomy for learning, teaching, and assessing: A revision of Bloom's taxonomy of educational objectives* (Complete ed.). Longman.

Boone, H. N., & Boone, D. A. (2012). Analysing Likert data. *The Journal of Extension, 50*, 1–5. Accessed Nov 18, 2022, from https://joe.org/joe/2012april/tt2.php

Can, D., & Yetkin Özdemir, I. E. (2020). An examination of fourth-grade elementary school Students' number sense in context-based and non-context-based problems. *International Journal of Science and Mathematics Education, 18*(7), 1333–1354.

Egodawatte, G. (2011). *Secondary school students' misconceptions in algebra. Teaching and learning Ontario Institute for Studies in.* Education University of Toronto.

Department of Education (DoE). (2001). *National Strategy for mathematics, science and technology education in general and further education and training.* Government Printer.

Department of Basic Education (DBE). (2014). *Report on the Annual National Assessment of 2014: Grades 1, 6 and 9.* DBE. Accessed July 27, 2022, from http://www.saqa.org.za/docs/rep_annual/2014/

Geary, D. C. (2011). Cognitive predictors of achievement growth in mathematics: A five year longitudinal study. *Developmental Psychology, 47*(6), 1539–1552.

Hester, V. K., Hulya, K. A., Joke, M. V., & Wenceslas, N. (2022). Recontextualization of learner-centred pedagogy in Rwanda: A comparative analysis of primary and secondary schools. *Compare: A Journal of Comparative and International Education, 52*(6), 966–983. https://doi.org/10.108 0/03057925.2020.1847044

Ibrahim, A., Mavis, A.-G., & Bawa, A. K. (2018). Factors affecting the adoption of ICT by administrators in the university for development studies tamale: Empirical evidence from the UTAUT model. *International Journal of Sustainability Management and Information Technologies, 4*(1), 1–9.

Linnenbrink, E. A. (2007). The role of affect in student learning: A multidimensional approach to consider the interaction of effect, motivation, and engagement. In *Emotion in education* (pp. 107–124). Academic Press.

Mji, A., & Makgato, M. (2006). Factors associated with high school learners' poor performance: A spotlight on mathematics and physical science. *South African Journal of Education, 26*(2), 253–266.

Mungoo, J., & Moorad, F. (2015). Learner-centred methods for whom? Lessons from Botswana Junior Secondary Schools. *African Educational Research Journal, 3*(3), 161–169.

NRC. (2001). Adding it up: Helping children learn mathematics. In J. Kilpatrick, J. Wafford, & B. Findell (Eds.), *Mathematics learning study committee; Center for Education; Division of Behavioral and social sciences and education.* The National Academies Press.

Opie, C. (2004). Research approaches. In C. Opie (Ed.), *Doing educational research* (pp. 73–94). SAGE.

Orakci, S., Durnali, M., & Aktan, O. (2019). Fostering critical thinking using instructional strategies in English classes. In S. P. Robinson & V. C. Knight (Eds.), *Handbook of research on critical thinking and teacher education pedagogy*. Hershey, PA.

Pekrun, R., Elliot, A. J., & Maier, M. A. (2009). Achievement goals and achievement emotions: Testing a model of their joint relations with academic performance. *Journal of Educational Psychology, 101*(1), 115.

Pournara, C., Hodgen, J., Sanders, Y., & Adler, J. (2016). Learners' errors in secondary algebra: Insights from tracking a cohort from grade 9 to grade 11 on a diagnostic algebra test. *Pythagoras, 37*(1), 1–10.

Sasman, M. (2011). Proceedings of the seventeenth National Congress of the Association for Mathematics Education of South Africa (AMESA). "Mathematics in a globalized world" 11-15 July 2011 the University of the Witwatersrand Johannesburg.

Sekaran, U., & Bougie, R. (2010). *Research methods for business: A skill building approach*. John Willey & Sons Ltd.

UNESCO. (2021). UNESCO Final Report 2021 Global Education Meeting From recovery to accelerating SDG 4 progress Ministerial Segment (online) – 13 July 2021 13:00–16:00 CEST. https://en.unesco.org/sites/default/files/global-education-meeting-2021-final-report-en.pdf

United Nations. (2015). Global Sustainable Development Report 2015 Edition Advance Unedited Version. Accessed Dec 20, 2022, from https://sdgs.un.org/sites/default/files/publications/1758GSDR%202015%20Advance%20Unedited%20Version.pdf

World Bank. (2020). World Development Report 2020 Chapters and Data. Accessed Jan 15, 2023, from https://www.worldbank.org/en/publication/wdr2020/brief/world-development-report-2020-data

Zan, R., & Martino, P. (2008). Attitude toward mathematics: Overcoming the positive/negative dichotomy, in Beliefs and Mathematics. In B. Sriraman (Ed.), *The Montana mathematics enthusiast: Monograph series in mathematics education* (pp. 197–214). Age Publishing & The Montana Council of Teachers of Mathematics.

PART II

Higher Education and Innovations

Scaling Education Innovations in Tanzania, Kenya, and Zambia: *Assessing the Design of School In-service Teacher Training*

Katherine Fulgence ⓘ

1 INTRODUCTION

Scaling of education innovations has received international recognition over the past few decades. Scaling is relevant for the attainment of Sustainable Development Goal 4 (SDG4) which focuses on ensuring inclusive and equitable quality education and promoting lifelong learning opportunities for all. With the advent of COVID-19, more than 1.7 billion children and youth especially those from difficult socio-economic situations have experienced learning loss (Vvob, 2021). Instead of reinventing the wheel of enabling learning for every student, education innovations that have proved to work having successfully passed the proof of concept and brought impact after piloting should be scaled up to benefit

---

K. Fulgence (✉)
Dar es Salaam University College of Education, University of Dar es Salaam, Dar es Salaam, Tanzania
e-mail: katherine.fulgence@udsm.ac.tz

© The Author(s), under exclusive license to Springer Nature Switzerland AG 2024
P. Neema-Abooki (ed.), *The Sustainability of Higher Education in Sub-Saharan Africa*, Sustainable Development Goals Series, https://doi.org/10.1007/978-3-031-46242-9_9

learners from different contexts (Vvob, 2021). Scaling refers to a systematic, principle-based science to optimize impacts and increase the likelihood that innovations will benefit society (Price-Kelly et al., 2020).

According to Cooley (2020), the scaling process embraces three successive stages: effectiveness (developing a solution which can be an innovation, a model, a design, or an intervention that works), efficiency (finding a way to deliver the solution at an affordable cost), and expansion (developing a way to provide the solution on a larger scale). For this realization, the approaches to scaling innovations should be scientific and should use rigorous research designs for the results to optimize impacts for the public good (Conn, 2017; McEwan, 2015; Vvob, 2021). In the understanding of Olson et al. (2021), successful scaling requires evidence to inform the contextualization of the innovation to be scaled up, with higher education institutions (HEIs) endowed to perform the research along the process in collaboration with other stakeholders. OECD (2009) defines innovation as a new or significantly improved product (good or service), a process, or a new marketing or organizational method in business practices, workplace, or external relations. In education, innovations are observed in curriculum, teaching methods, administrative practices, institutional structure, and funding mechanisms to mention a few (Thomas Jr. et al., 2004; Serdyukov, 2017). Thus, scaling of an education innovation implies a deliberate expansion of an externally developed design that successfully worked in one or a small number of school settings to many settings (Stringfield et al., 1998).

Research shows frameworks reflecting practical steps and tools as well as pathways for guiding the design of successful and sustainable education innovations for scaling up have been developed (Cooley, 2020, 2021; Price-Kelly et al., 2020). However, few education innovations at scale pass through the rigorous research process along the frameworks (Thomas Jr. et al., 2004; Kohl, 2021). According to Serdyukov (2017), conducting research and innovation has cost implications and requires time for efficient learning. Scaling in the education sector further involves changing the practices of teachers as well as creating institutional mechanisms relevant to support and sustain the practices of the scaled-up innovation, with these aspects not easily realized over a short period (Gallagher et al., 2016; Thomas Jr. et al., 2004). Indeed, factors synonymous with the involvement of multiple actors, each with different roles and a high number of decision-makers have continued to make the process of scaling education innovations complex and challenging (Dede et al., 2005; Cooley, 2020,

2021; Price-Kelly et al., 2020; Vvob, 2021). According to Kohl (2021), complexity in scaling is reflected in the problem being addressed, the system(s) in which that problem is embedded, and the innovation itself in terms of its features. Price-Kelly et al. (2020) share similar views. A consensus is now reached that systems are central to scaling (Kohl, 2021).

Following the growing trend towards national and local actors (governments, local communities, NGOs) to initiate and/or drive the scaling process, there is a need for local scaling activities particularly principles, guidance, and tools to inform and be informed by the international experience for counter learning (Kohl, 2021). Thomas Jr. et al. (2004) conclude that the scaling process is iterative and complex and requires the support of multiple actors with this likely to remain so for the foreseeable future.

Besides the demand for scientific evidence to inform the decision-making process before scaling of innovation, as well as framework and pathways to guide the scaling process, literature shows that decisions to undertake large-scale educational innovations are taken without rigorous scientific evidence including external validity of the innovation assumptions (Bates and Glennerster, 2017). Thomas Jr. et al. (2004) explored the experiences of 15 K–12 developers (kindergarten (K) and the 1st through the 12th grade (1–12) publicly supported schools before college) of different reform efforts to identify problems in scaling up reforms they had designed and implemented, the lessons learned, and the related solutions. The study found that the designers and developers of the education innovations had to learn through experience along the scaling process. Kohl (2021) explored crosscutting issues affecting scaling. The findings highlight the relevance of systems; change and complexity; risk, uncertainty, and de-risking; variance and impact of robustness; demand-driven scaling; costing and economies of scale and scope as well as digital technologies in scaling education innovations. Vvob (2021) presents the approaches to scaling education initiatives through government systems to increase the chances of the education initiatives reaching a significant scale.

Cooley (2020) categorized scaling into three pathways; expansion, replication, and collaboration and the role of intermediation under each pathway with the categorizations based on the degree to which the originator of the innovation continues to control its implementation as the model goes to scale. The expansion entails growth where the piloting organization increases the scope of operations by branching out into new locations or target groups; franchising the model to agents and spinning off where aspects or parts of the originating organization operate independently. Under this pathway, the role of intermediation is to enable the originating organizations to plan for growth. Replication involves scaling a particular aspect of the model me-to with the process, technology, or mode of service delivery by getting others, including the public sector and/or commercial providers, to take up and implement the model while maintaining a relationship between the originating and adopting organization. The collaboration includes the formation of formal partnerships and informal networks meant to disseminate innovations with the same achieving scaling indirectly by strengthening the capacity of, or support for, the whole field of the innovation actors. There also exist indirect pathways, termed as field building that blend replication and collaboration strategies thereby emphasizing the key players in the ecosystem, particularly policymakers, community groups, NGOs, advocacy groups, service delivery groups, think tanks, funders, investors, and beneficiaries towards the scaling process (Cooley, ibid.).

Price-Kelly et al. (2020) developed an education scalability checklist to be used by practitioners including governments and funders to guide the initial design of a large-scale education innovation as well as drive the scaling process. Cooley (2020) presents a three-step, ten-task management framework to guide the scaling of economic, social, and political outcomes. The framework guides designing interventions with scale in mind while assessing the scaling potential of the prototypes and pilot programmes. Task 1 of the planning step (Step 1), for example, clarifies the tacit elements of the model central to its effectiveness. Cooley (2016) further identified four key elements to be considered during the scaling process under the planning step to include: (1) "what" particularly what is being scaled up, reflecting the features of the innovation, intervention, or design to be scaled up; (2) "how of scaling up" with this placing emphasis on the scaling pathway particularly the types and methods of scaling up including the projects' theory of change; (3) "who of scaling up" with this involving identifying and determining the roles of the organizations best suited to implement and deliver at scale and to perform the intermediation functions; and (4) "where" of scaling up, particularly the expected scope and nature of the scaling-up effort and the dimension(s) along which scaling will occur.

Cooley (2021) further developed practitioners' scaling-up the toolkit, including tools for creating a scaling strategy, developing scaling plans, securing needed political support and financing, implementing the scaling process, and tracking progress towards sustainability. According to Cooley (2020), the beginning point of scaling is to have an innovation, an unscaled model, or a model component to be scaled up and the model has to be embedded in a project. The model should parade technical, process, and organizational components with the components being well-identified and repackaged to enable scaling.

This chapter applies the Cooley (2021) framework to detail how the innovation "School-based In-service Teacher Training (SITT)" currently being scaled up in Tanzania, Kenya, and Zambia managed Task 1 of the Planning Step 1 of the scaling process. The Cooley (2021) framework has been applied in 47 countries to more than 500 interventions across a range of disciplines. The SITT innovation is implemented under a 30-month project titled "Strengthening In-service Teacher Mentorship and Support" (SITMS). The design of the innovation did not systematically and explicitly follow any scaling framework. The chapter addresses the following questions: (a) What are the features of the SITT approach as implemented under the SITMS project? (b) How does the SITT approach fit into the scaling pathways? (c) Who are the implementers of the SITT approach? (d) What is the expected scope of scaling the SITT approach "The where"?

SITT is a practice-based teacher mentorship and support approach that involves training mentor teachers to coach and mentor fellow teachers through peer learning exchange, model lessons, and team teaching. SITT was first piloted in select districts in Tanzania in 2012 and has since gained momentum across Tanzania's primary schools. The SITMS project adapts and scales up the model to secondary schools in Tanzania, Kenya, and Zambia. The project's intended outcome is to support government efforts to implement well-functioning inclusive school-based in-service teacher training programmes that are effective in improving the quality of teaching, empowering students, and enhancing the quality of basic education.

In terms of organization, after the introductory section, this chapter first highlights the background of the SITMS project reflecting the context of mentorship and support across the countries. Scaling is further conceptualized reflecting on the related frameworks, with this section followed by the study methodology. Findings are further presented showing how the SITT innovation aligns with Task 1 of the Cooley (2021) framework. The chapter is further discussed and concluded by offering recommendations and areas for further research.

2 Project Background and Country Context

Teachers are the key determinants of quality education for sustainable development. However, many teachers in Sub-Saharan Africa (SSA) including the SITMS project countries have low competence to impact students' learning outcomes, especially at the secondary level (UNESCO, 2020). According to Serdyukov (2017), learning outcomes can be qualitative (improved knowledge, skills, competencies, character development, attitude, values, and dispositions) and quantitative (improved academic performance, increased enrolment, retention, attrition, graduation rate, number of students in class, cost, and time efficiency) to mention a few. Besides significant progress in school enrolment in SSA, there are limited remarkable efforts to support the quality of teaching with this reflected in the inadequate engagement of teachers in meaningful Continuous Professional Development (CPD) (Ajani, 2020; UNESCO, 2020). In Tanzania, for instance, many teachers struggle to teach larger classes with diverse learning needs due to increased student enrolment (Mwakabenga, 2018). Tanzanian local studies have reported teachers' limited capabilities in implementing learner-centred curricula (Bermeo et al., 2013). Associated with the curriculum challenges (Kafyulilo, 2014), identified several concerns related to teachers' inability in integrating media technology into teaching. Declining students' performance in examinations is also largely attributed to poor teaching quality (TIE, 2018).

In Kenya, studies have proved that many students are not learning due to teachers' inadequate pedagogical skills (UNESCO, 2020; KNEC, 2010). For instance, the Early Grade Mathematics Assessment (EGMA), the Southern and Eastern African Consortium for Monitoring Educational Quality (SACMEQ), and the National Assessment/Monitoring Learner Achievement (NASMLA) have shown that many teachers have inadequate mastery of subject content and teaching pedagogy (Uwezo, 2013). The teachers have also exhibited low competence in other teaching-related practices that included classroom-based gender matters. For example, the tendency of boys to outperform girls on every assessed mathematical skill as reported by NASMLA can be partly attributed to teachers' low competence in classroom-gendered practices. Zambia too is not spared from these problems associated with inadequate teachers' competence.

Generally, the shared challenges across the three countries tend to affect the quality of teaching and students' achievements. Even initial teacher programmes in Africa do not adequately prepare teachers to undertake their roles (UNESCO, 2020; Akyeampong et al., 2011). Therefore, investment in effective CPD and particularly secondary school teachers' mentorship and support is indispensable if teachers' competence and better student learning outcomes are to be improved. Teacher mentorship and support programmes enable teachers to engage in ongoing professional learning and develop the required competence. Teacher mentorship and support is a form of CPD, in which experienced teachers provide guidance and expertise to novice or beginning teachers in schools (Alabi, 2017). In CPD, mentorship can be provided among experienced teachers based on their levels of expertise to help develop innovations in teaching. There are usually formal and informal mentoring and support programmes. Informal mentorship might be happening in schools, but experiences have shown that there is no sense of responsibility among mentors. In formal mentorship, both the mentor and mentee set objectives and are accountable for them.

Initiatives to improve teacher mentorship and support in Tanzania, Kenya, and Zambia recognize the role of mentorship and support in CPD to realize SDG4. Tanzania established the Teacher Resource Centres (TRCs) in the 1970s where teachers obtained mentorship and support through local facilitators or master subject teachers, with TRCs currently restored. In line with the national initiatives, teachers have been participating in various CPD programmes organized by individual educators, institutions, and non-governmental organizations, with most of these programmes, particularly, Education Quality Improvement Programme in Tanzania (EQUIP-T), Tusome Pamoja (Lets' Read), Right to Play, and School-based In-service Teacher Training (SITT) being implemented at the primary education level (TIE, 2018). There is, however, evidence of EQUIP-T as done at the primary level being adapted, tested, and scaled up to the secondary level. These include Secondary Education Quality Improvement Project in Tanzania (SEQUIP) and Kenya (SEQIP). Recently, the Tanzania Institute of Education (TIE) developed a national framework to guide the implementation of all sorts of CPD (MoEST, 2019).

Similarly, Kenya institutionalized a Teacher Professional Development Framework to meet existing teaching challenges. For instance, the Education Sector Plan (2018–2022) and the Teachers' Service Commission anticipate a framework to explore global perspectives in teacher CPD to establish a coordinated and structured programme for teachers to acquire the requisite competencies and expectations of the twenty-first-century learning outcomes. In the past, CPD has taken various forms, ranging from self-sponsored grading of qualifications to those supported by government and donors being TUSOME, Kenya Primary Education Development (PRIEDE) project, and SEQIP.

In Zambia, anecdotal evidence prompts that there is limited information on CPD for teachers within schools. One would maintain that CPD which focuses on how teaching and learning could be improved between and among secondary school teachers is hardly known and rarely promoted. For instance, there are no thematic issues that guide how secondary school teachers participate in teacher professional development programmes. This makes the whole CPD setting unclear and too general as well as lacking some focus in the process.

In reviewing the initiatives in Tanzania, Kenya, and Zambia there are limited large-scale CPD programmes for secondary education teachers compared to primary schools. For example, SEQUIP-T which started in 2017 is only operational in 9 out of 30 regions of Tanzania (30% of the regions). In Kenya, SEQIP is implemented in 2146 out of 9966 secondary schools nationally, representing only 21.5% of schools, which leaves out 78.5% of the schools unattended (MoEST, 2019). Likewise, most of the existing secondary school teachers' support programmes are implemented as general means of CPD with little mentorship. The programmes are largely ad-hoc and lecture-based with little follow-up on the classroom practical teaching (Hardman et al., 2015).

The SITMS project was thus conceived to address the challenge of limited and meaningful Continuous Professional Development (CPD) supported by mentorship among in-service teachers. The general objective of the project is to strengthen the capacity of teachers in terms of content mastery and pedagogical skills by generating and strengthening the use of knowledge on effective teacher mentorship and support models. The SITMS project is managed by four partner institutions: (1) Dar es Salaam University College of Education, Tanzania as an intermediary organization, (2) HELVETAS[1] Tanzania, an international NGO, (3) University of Kibabii (Kenya), and (4) University of Zambia (Zambia). The project aims to: generate lessons on effective teacher mentorship and support models for continuous professional development, build the capacity of teacher mentors and teachers to improve teaching and learning outcomes at the classroom level, and mobilize policy uptake of teacher mentorship and support models for improved secondary school teacher capacity. Besides

---

[1] https://www.helvetas.org/en/tanzania.

improving mentorship and support to in-service teachers in the member countries, competent school-based mentors will be trained and they will continue working with teachers and help nearby schools with fewer cost to improve the teaching and learning process.

3   Literature Review

3.1   Conceptualizing Scaling

MacGregor et al. (2018) conceptualizes scaling as the diffusion, dissemination, and implementation of innovative and effective public intervention. Hartmann and Linn (2008) define scaling up as the process of expanding, adapting, and sustaining successful policies, programmes, or projects in geographic space and over time to reach a greater number of people. Cooley (2020 p. 10) identified five trajectories along which the scaling of products or services can occur as follows: geographic coverage (extending to new locations), breadth of coverage (extending to more people in currently served categories and localities), depth of services (extending additional services to current clients), client type (extending to new categories of clients), and problem definition (extending current methods to new problems). Accordingly, and as commented by Cooley (2016) and Kohl (2021), successful scaling up of a pilot intervention to the national application requires multiple projects over 15 years demanding the need to educate education practitioners including funders about the realities of scaling up.

Price-Kelly et al. (2020 p. 5) introduce the concept of "scaling science" as a component of research for development with the view that scaling requires a strong culture of research and development in education. While research for development intends to achieve impacts that promote development through discovery and applied sciences, scaling complements these approaches to research as it moves beyond outputs (papers, books, chapters, data sets) and outcomes (policy influence, behavioural change, programme improvement, product development) to mention a few towards impact which can be measured along better health, economic returns, social stability, environmental protections, and the general sustainable development.

Price-Kelly et al. (2020) further comment on the concept of knowledge translation, meaning moving research-generated knowledge into action and scaling. According to Cooley (2021), scaling up an intervention requires "drivers" (as champions, incentives, market, and/or community) to push the scaling process, "enabling conditions" (fiscal, institutional, and political) and partnerships whose presence or absence can significantly spearhead and/or stifle the scaling process. In this regard, Price-Kelly (ibid.) came up with a scaling system comprising initiators, enablers, competitors, and the impacted. Accordingly, initiators comprising of innovators or researchers, funders or investors, the know-how, a willing community, land and cultural acceptance bring innovations making it possible to begin a change in scale. Enablers are the actors who facilitate the scaling process and may include service providers, laws, policymakers, culture, markets, communities, and government. Competitors offer better alternatives to scaling the innovation and may include commercially competing companies or products, substitute ideas, social or cultural norms, deep-rooted habits, and traditions. The impacted are the actors that experience the scaling results in an either a positive or negative way. Scaling efforts further need to come up with explicit strategies for integrating innovations into commercial markets and government policy (Cooley, 2020). While commercial market scaling mechanisms may not be cost-efficient for all, especially the marginalized, government institutions can be able to fund and deliver either directly or indirectly most innovations sustainably at scale, also facilitating access to the population at risk.

The decision of the government to take up innovation to large-scale implementation, however, is dependent upon the innovation's ability to respond to an urgent local problem. The innovation should prove effective at different levels of scale and should be both cost-effective and politically appealing. For effective integration of innovation in the government system, the innovation and scaling strategy must be aligned with the decision-maker's needs including politicians, scarce financial and human resources, and the government's agenda. In this regard, the decision of a government to scale an innovation is the interaction between the technical and financial features of innovation, personal relationships, political incentives, and competing innovations (Cooley, 2020). Overall, planning for scale goes way beyond proof of concept and should be integrated from the initial design of a pilot project (Cooley, 2020, p. 3). The design should also incorporate; doing a baseline survey; documenting the model; developing a method for monitoring, measuring, evaluating, and publicizing results; practising adaptive management and design to minimize complexity and unit cost; and incorporating mechanisms to gain buy-in from policymakers and potential adopting organizations.

4 Scaling Process in the Education Context

Scaling in the education setting is viewed as an iterative process of learning by doing where different actors interact and engage in an attempt to bring an in-depth and sustainable change in teacher practices across many sites and over time (Thomas Jr. et al., 2004). Thomas Jr. et al. (2004) see scaling to comprise the developers or the designers of the innovation on the one side and the role of adopters, for instance, institutions, or entities supporting the implementation of the new practices on the other side. There also exists intermediation between the two with the roles of the intermediary organization including strategic planning, systems strengthening, coordinating stakeholders, fundraising, evaluation, advocacy, as well as the management of the change and the process to mention a few (Cooley, 2020). According to Thomas Jr. et al. (2004), the developers provide implementation support and ensure the quality of its delivery; teachers and schools on the other hand as the impacted need to reciprocate by attending the training sessions provided and adapting the design constructs in their context. Education leaders might also need to change incentive structures to motivate teachers towards adapting the intervention constructs by providing supportive and coherent infrastructure to sustain the new practice within the education system. The supporting resources and infrastructure may include: (a) curriculum, instructional practices, and related resources (textbooks, libraries, and technology); (b) the human-resource infrastructure, including professional development; policies for hiring, assigning, and retaining teachers; and performance incentives; (c) governance systems, including realigning decision authority, especially among the teacher, school, and district; (d) resource allocations; and (e) systems of data collection and accountability (2004, p. 69). There is also a need for having a coherent set of practices for teaching, learning, and assessment and the associated infrastructure to sustain the new change.

The scaling process in the education contexts further needs to evolve systematically and significantly by changing the practice of teachers in classrooms through interactions among developers, teachers, schools, districts, and other major actors (Thomas Jr. et al., 2004). In this regard, the process of scaling up needs to be: (a) interactive, involving the developers, teachers, schools, and districts in relationships that continue over time; (b) adaptive, involving reciprocal relationships among the actors and reactions to unfolding situations; (c) iterative, with continuous re-examination and learning over time; and (d) non-linear, with the sequence of activities depending, to some extent, on the need for adjustments as the actors adapt to unfolding circumstances.

To formally support the implementation of educational interventions or a design in a school setting through scaling, multiple approaches and tools are needed to enable multiple actors to implement the design at specific sites simultaneously (Thomas Jr. et al., 2004). According to Thomas (ibid. p. 655), the approaches and tools include: (a) planning that is based on data to inform the designs and development of interventions to address needs; (b) development and/or provision of specific curriculum and instruction modules, guidelines; (c) provision of professional development particularly training, seminars, and workshop opportunities for teachers and other education leaders, with the training modality in terms of frequency and duration on the content to be covered; the capacity building pieces of training can be supported by mentorship visits to provide on-site technical assistance to teachers; (d) instituting structures, particularly schedules of classes, schedules to ensure planning time for teachers, student assignment methods, or structures as a way to ensure that schools and teachers create the infrastructure needed to support new curriculum and instruction; (e) facilitators and coaches where schools provide facilitators to support implementation or provide such facilitators themselves to provide on-site support; (f) funding, to support the implementation; and (g) forums for exchanging information about new practices, with this done by developers where they host annual forums, and dissemination to provide opportunities for teachers to exchange lessons on best practices or find solutions to problem they share.

For successful scaling up of education interventions, attention should be placed on capacity building for education leaders including teachers, the use of structured pedagogy, collaboration, and systemic policies as drivers to improve learning outcomes (Fullan, 2016; Snilstveit et al., 2016). Institutional infrastructure to support those approaches is also important to support scaling (Fleisch et al., 2016). According to Fleisch (2017), institutional infrastructure entails a variety of high-quality learning materials, implemented in combination with support to teachers (in the form of ongoing coaching) to enable them to utilize the same ensuring professional instructional accountability.

Cooley (2020) insists that the context particularly social, political, cultural, and economic actors (which are specific to localities, states, and countries) matter when it comes to the effective adoption and scaling of education innovation to a large scale by the government. Indeed, embedding the intervention in government systems has proved to be a valid prerequisite for successful large-scale implementation (Piper et al., 2018). In this regard, achieving scale in education requires innovation that can flexibly adapt to effective use in a wide variety of settings and contexts across a spectrum of learners and teachers (Dede et al., 2005). Dede, Honan, and Peters (ibid.) assert that the setting particularly the teacher's content preparation, students' self-efficacy, and prior academic achievement play a major influence in shaping the desirability, practicality, and effectiveness of educational interventions. An element of trust needs also to be exercised between the government, NGOs, universities, and the private sector along the scaling process.

5   Scaling Frameworks

Various frameworks have been developed to offer relevant guidance during the design of a scaling project as well as support the scaling process (Crouch & DeStefano, 2017; Cooley, 2020, 2021; Price-Kelly et al., 2020; Kohl, 2021). Crouch and DeStefano (2017) identified three core functions in terms of capacities that education systems should focus on to support large-scale instructional or educational changes. The core functions include: (1) setting and communicating expectations, (2) monitoring and guaranteeing accountability for meeting those expectations, and (3) intervening to ensure the support needed to assist students and schools that are struggling.

Price-Kelly et al. (2020) developed four guiding principles for scaling an impact which include: justification, optimal scale, coordination, and dynamic evaluation. The justification principle stresses the need for scientific evidence before scaling innovation. The findings on the one side enable the adopting organization to understand the feasibility of taking the innovation to scale and on the other hand, the values of those impacted further information if it is relevant to scale the innovation. According to Vvob (2021 p.3), scaling of quality education innovations calls for evidence along: (a) a strong instructional core, comprising three interdependent components: teachers' knowledge and skill, students' engagement in their own learning, and academically challenging content that demonstrably improves learning; (b) effective teacher and school leader professional development delivered at scale; (c) widely available high-quality low-cost teaching and learning materials to accompany professional development;

and (d) context-sensitive, long-term capacity development support to government institutions at different levels of the education system. Building on the evidence and values obtained during the justification process, the developers identify key stakeholders to engage during the scaling process, with similar stakeholders endorsing the innovation.

The second principle is optimal scale. Set by the diversity in viewpoints among different stakeholders during the design and implementation of a scaled-up intervention, determining an optimal scale requires, first, an understanding of the concept to be scaled up with the same realized through data collection along the four aspects; magnitude, sustainability, equity, and variety; and second, ongoing considerations and trade-offs between the aspects. According to Price-Kelly et al. (2020), magnitude refers to how much impact will the intervention create, which may include the average size or quality of impacts; how many people will benefit or be harmed; and the importance, value, or merit of such impacts as judged by stakeholders; variety means the range and different types of impacts the research will create as health, economic, environmental; sustainability reflects how long will impacts last, and what factors might affect this; and equity reflects the benefits and/or harm the innovation will bring to different sub-groups (based on gender, religion, or class) and if they will be impacted differently.

The third principle is coordination. Since scaling occurs in complex systems across multiple actors, there is a need to coordinate them along the scaling process. Coordination involves the need to plan for the scaling journey where many actors are involved in bringing impact to scale. Coordination highlights the need of the adopters or developers to consider the wider range of the four categories of the scaling system, particularly the initiators, enablers, competitors, and the impacted as earlier communicated. Accordingly, the impacted may affect, or be affected, by scaling in ways that alter the intended impacts demanding a strong understanding of the system and ongoing monitoring. In this regard, the scaling efforts should ensure gender equality and social inclusion to address unequal power relations experienced by people on the grounds of gender, wealth, ability, location, caste/ethnicity, language, and agency or a combination of these dimensions (Bagale, 2016).

The fourth principle is dynamic evaluation which aims to measure the collection of impacts of scaling as an intervention. Since scaling is a dynamic change, there is a need for a dynamic evaluation of the scaling process before, during, and after scaling to measure the related impact using various tools, also recommended by Price-Kelly et al. (2020).

Cooley's (2021) framework proposes three steps to scaling supported with ten manageable tasks, with each task having relevant tools (see Table 1). The first step with four tasks aims to develop a plan for taking an intervention to scale (the "Planning Step"). The second step with three tasks aims to create the necessary preconditions for scaling (the "Political Step") and the final step with three tasks focuses on the actual management of the scaling process (the "Operational Step").

Regarding the challenges of scaling education innovations, Vvob (2021) presents three hurdles as follows: (a) a high number of decision-makers are involved in adoption at scale, with teachers deciding mostly individually and behind closed doors; (b) pilot initiatives are designed and executed and supported in ways that make it difficult for government systems to adopt them and make them fit within existing resources, structures, and incentives; and (c) government systems are not that receptive to

Table 1 Tools and Guides

Scaling Up Steps	Tasks	Associated Tools
Crafting an Overall Scaling Strategy		Tool 1: Scaling Task Model
		Tool 2: Scaling Plan Template
		Tool 3: Real-time Scaling Lab
Step 1: Developing a Scaling Up Plan	Create a Vision	Tool 4: Second Theory of Change
	Assess Scalability	
	Fill Information Gaps	Tool 5: Intervention Profile
	Prepare a Scaling Plan	Tool 6: Scalability Assessment Checklist
Step 2: Establishing the Preconditions for Scaling	Legitimize Change	Tool 7: Drivers of Change Analysis
	Build a Constituency	
	Realign and Mobilize Resources	Tool 8: Stakeholder Analysis
		Tool 9: Advocacy Strategy Profile
		Tool 10: Scale Costing Protocol
Step 3: Managing the Scaling Process	Modify Organizational Structures	Tool 11: Guidelines of Evidence Generation and Use
	Coordinate Action	Tool 12: Adaptive Management Protocol
	Adapt Strategy and Maintain Momentum	Tool 13: Institutionalization Tracker

Adopted from Cooley (Cooley, 2021 p. 2)

change, especially when additional resources are called for to implement the scaled-up innovation. Other challenges to achieve impact are project-specific ranging from the interventions being supply versus demand led, overly dependence on technological solutions to instructional programmes, higher expectations from teachers to invest more time and effort than they were willing to provide on the intervention as well as the implementation of the initiatives outside government systems (Piper et al., 2018). In that respect, successful scaling that involves sustaining and institutionalizing education initiatives heavily depends on the decision and system capacity of governments. Thus, building the institutional capacity of an education system, where front-line implementers like teachers and school leaders are enabled and supported to integrate the new approaches is the key to realizing a scaling impact.

6   METHODOLOGY

It was important to conduct a documentary review of the SITT innovation as well as the SITMS project design proposal. To facilitate the planning and monitoring of the project objectives, a situational analysis and baseline study of the project was conducted to inform the design of the secondary SITT as well as the capacity-strengthening components of the project beneficiaries (the impacted). The findings also aimed to form the basis for monitoring the project's progress, refining project objectives, and measuring the impact of the project outcomes.

Across the countries, the baseline study adopted mixed methods where both qualitative and quantitative data were collected. The selection of the study regions in Tanzania focused on stakeholders with whom SITT primarily worked. The selection of Teacher Colleges in Kenya and Zambia was based on the offering of education courses at the diploma level for secondary school teachers. Both open-ended interviews and focus group discussions were used to collect data from the key education stakeholders particularly ministry officials, education agencies, teacher education institutions, district education officers, school heads, teachers, and students. A total of 2962 (43% being female) education stakeholders participated in the study; where 1186 (45% female) were from Tanzania, 987 (41% female) from Kenya, and 789 (43% female) from Zambia.

7 Findings

The findings summarize how the SITT approach as the SITMS project innovation went through Task 1 of Step 1 of the Cooley (2021) scaling framework. The findings first describe the elements of the innovation—the "What" of the SITT model. This is followed by the SITT methodology—the "How" of scaling up an innovation. The subsequent part describes the partner institutions and their roles in the scaling up of the SITT innovation—the "Who". The final part describes the expected scope of the scaling-up effort and the scaling dimension(s)—the "Where" of the SITT innovation.

8 The SITT Model "The What"

The SITT model can be traced back to a private initiative in 2012 which aimed at enhancing the quality of primary education in Tanzania. The goal of the SITT model in Tanzania was to have a well-functioning inclusive school-based in-service teacher training leading to more effective teaching, empowerment of pupils, and quality of basic education for the children. The project was a collaboration between HELVETAS Swiss Inter-cooperation and the Tanzania Teachers' Union (TTU).

SITT has seven components aligning to the competence-based curriculum as advocated by the curriculum agencies across the countries:

(i) Practice-oriented teaching and learning. This means teaching and learning that link the concepts with practical examples inspired by everyday life.

(ii) Active participation of learners. In this component, pupils are encouraged to participate actively in and outside the classroom.

(iii) Use of local materials as teaching aids. The model encourages the use of affordable and locally available materials developed by pupils and teachers to illustrate concepts and provide hands-on learning.

(iv) Emphasis on model lessons to promote the transfer of learning. In this component, trained teachers invite fellow teachers to observe how they employ SITT techniques.

(v) Emphasis on teacher peer learning groups to encourage teamwork. In this element, trained teachers organize peer learning groups to exchange and share experiences among themselves.

(vi) Emphasis on team teaching to promote cooperation among teachers and manage large classrooms. In this component, trained teachers invite other teachers to collaboratively deliver lessons in classes.

(vii) Addressing public health issues among pupils and improvement of the school environment. Promoting best health practices like hand washing practices, menstrual hygiene, and general body cleanliness and making the learning environment green and friendly have formed a part of the SITT model.

At the school level, trained teachers train, coach, and mentor fellow teachers through peer learning exchange, model lessons, and team teaching. Moreover, under the HELVETAS initiative, schools create school SITT core groups which are composed of the school head, ward education coordinator, school quality assurance officer, district education officer, teacher college representative and the Ministry of Education, Science and Technology (MoEST) representative whose roles are to train, coach, mentor, monitor, and assess teachers to embed SITT elements in their teaching. In each district, the SITT core group team offers support visits to schools and teacher colleges (TCs). SITT reached teachers and schools through two different approaches: (1) intensive approach, where schools are reached through training organized by national facilitators and (2) extensive approach where teachers/schools are reached through TCs. Overall, a total of 532 schools from 19 districts in 7 regions of Tanzania have so far benefited from the primary SITT model, with 507 schools (HELVETAS 378 and district/school 129) applying the intensive SITT approach and 25 schools applying the extensive approach as done by TCs.

SITT contributed to improved competencies and confidence of teachers in Mathematics, English, and Education for Sustainability (ESD) subjects, better knowledge and skills in teaching methods, and more joyful learning of pupils. The SITT approach helped to instil the teamwork spirit among teachers through the application of peer learning and team teaching as articulated by the project beneficiaries.

Comparatively, the SITT model has been unique when compared to other CPD and other mentorship and support programmes in various ways. Firstly, while other CPDs aim to build capacity for teachers, the SITT model comprises school-based learning practices with peer-to-peer sharing that build competencies for teachers and pupils by learning from each other. Secondly, while other CPDs focus on teachers only, the SITT model engages government authorities and other education actors to support coaching, mentoring, and monitoring. Thirdly, whereas other CPDs are teacher-centred in character, the SITT model is learner-centred, as learners are actively engaged and collaboratively prepare resources with their teachers. Lastly, while other CPDs focus on a cascade approach, the SITT model promotes participation, peer-to-peer learning as well as exchange learning and mentorship with neighbouring schools.

The SITT model further aligns with the Tanzania CPD framework and has been institutionalized along the Tanzania education system and structures. HELVETAS has a working relationship with various stakeholders from the national level to the ward level. The stakeholders at the national level include the Ministry of Education, Science and Technology, the President's Office–Regional Administration and Local Government (PO-RALG), and the Tanzania Institute of Education which develops curriculum. At the zonal, district, and ward levels, HELVETAS works with quality assurers, district education officers, and ward education officers, respectively. While implementing SITT, HELVETAS has also been working with the Mathematics Education Primary Programme (MEPP) from South Africa to particularly strengthen teaching in the area of Mathematics as well as build the capacity of local facilitators. Both the weaknesses and strengths observed in the implementation of the model call for the need to adapt, test, and scale the SITT model to another level.

9   Demand for Scaling Primary SITT to Secondary SITT

In order to improve teacher performance and students' learning outcomes, the SITMS project sought to adapt and pilot the primary SITT model to gain knowledge exchange on how teacher mentorship and support can be strengthened at the secondary education level in Tanzania, Kenya, and Zambia. Indeed, the SITT model achievements have continued to kindle stakeholders to inquire regarding the modalities on how the SITT approach can be adapted in secondary schools, inclined by comparatively poor performances among students, especially in Mathematics.

10   Scaling of the SITT Approach "The How"

The SITMS project has been using Participatory Action Research (PAR) that is embedded in the theory of change (ToC) to guide the project intervention during planning, implementation, and evaluation. PAR facilitates a systematic collection of information in the course of intervention activities intending to improve the project practices, methods, and approaches (Reason and Bradbury, 2008). The PAR approach increases ownership and sustainability of the interventions beyond the life span of the project since the findings are evaluated in terms of local applicability. The project's ToC underlines the view that "if mentoring and support of in-service teachers are done through the SITT model, the teachers' competence skills will be improved, effective teaching will be guaranteed, and student learning outcomes will be improved". Ultimately, improved teaching will result in improved quality of education in Tanzania, Kenya, Zambia, and Africa at large.

11   SITMS Theory of Change

To practice PAR along with implementing the project activities, a baseline study was conducted across the project countries to contextualize the innovation. According to Fraser and Galinsky (2010), the creation of an intervention content demands contextualized knowledge. The baseline study findings identified the existing mentorship and support programmes, mostly training with limited mentorship and support initiatives. The findings also established competence profile gaps in Mathematics teachers emphasizing the topics that were identified as difficult by teachers and students. The findings informed the design of the secondary school SITT manual and a geometry mathematics module since the topic was perceived as difficult by both teachers and students. Therefore, throughout the project, the researchers will continue to collaborate with the intended intervention agents particularly schools, TC, and government authorities to implement the designed project interventions.

11.1   The "Who"

The project adopted replication and collaboration as a direct pathway to scale SITT as an education innovation. The SITMS project is thus a collaboration between four partner institutions at different capacities; an

NGO which is the originator of the innovation and three universities acting as intermediaries. HELVETAS Tanzania is the originator of the innovation and has been willing to share the knowledge and understanding of the innovation with the partner institutions. Dar es Salaam University College of Education (DUCE) is an intermediary institution and the project's lead overseeing the overall implementation of the project across the countries. From the design of the scaling innovation, DUCE searched for the funder particularly the International Development Research Centre (IDRC) and responded to an open call on the scaling education innovation. During the application process, DUCE searched for the scaling idea, particularly the innovation that aligns with the call requirements, in this case, the SITT innovation, which was already operated at the primary school level for over ten years by HELVETAS Tanzania. DUCE also searched the project partners, particularly Kibabii University of Kenya and the University of Zambia. During the project design, it was important to involve teacher colleges to build the capacity of secondary school teachers as the primary beneficiaries of the innovation. In terms of specialization, the three universities and the teacher colleges (TCs) are offering public education courses for pre-service teachers at a graduate and diploma level, respectively. The selection of the government project partners particularly TCs and the beneficiaries was intentional to enable ownership and the sustainability of the innovation within the education system even beyond the project lifespan.

Along with implementing their roles, the development of the SITMS project interventions, particularly the secondary SITT manual and the geometry module, happened centrally in Tanzania with expertise from HELVETAS Tanzania and Mathematics experts. The Kibabii University and the University of Zambia are responsible to introduce the SITT innovation in their contexts. DUCE in addition has been working with HELVETAS Tanzania to capacitate tutors from TCs across the countries enabling them to later train secondary school teachers and provide mentorship and support. Other activities coordinated by DUCE and HELVETAS Tanzania include planning meetings by tutors at the TCs as well as mentorship and monitoring visits at the schools by the education leaders and the TC tutors. The project design also integrated baseline, midline, and end-line evaluation to enable close monitoring along with implementing the project activities throughout the scaling process. While the baseline study was done by all the project implementers, midline evaluation will be done by DUCE, and the end-line evaluation will be done by an external consultant along with the project objectives.

11.2 *"The Where"*

"The Where" of the SITT innovation is discussed along the five trajectories of scaling as identified by Cooley (2020). Regarding geographic coverage, SITT innovation was originally implemented in Tanzania and is now scaled up to Kenya and Zambia. The innovation is thus extended to new locations, with the partner countries engaging in the contextualization of the intervention from the design of the project, towards its implementation and along the scaling process. Regarding the breadth of coverage, the original SITT has been operating at the primary level and it is now extended to the secondary level in all countries including Tanzania thus serving new categories of beneficiaries across the countries. New resources particularly the secondary SITT manual and a geometry module have been designed to facilitate capacity building at the secondary level. Regarding the depth of services, the primary SITT capacitated tutors of TCs to build the capacity of primary school teachers on the geometry module using the SITT approach. The secondary SITT, on the other hand, continues using TC tutors, and thus tutors have been capacitated to deliver a secondary geometry module designed based on the curriculum from the three countries. The tutors will build the capacity of secondary school teachers on the designed module using secondary SITT. Education leaders are also capacitated to mentor secondary school teachers. Regarding the types of clients, the primary SITT benefitted primary school teachers and pupils, while the secondary school SITT will benefit secondary school teachers and students as new categories of clients. Regarding problem definition, the SITT primarily built the capacity of teachers at the primary level in all subjects. While the secondary SITT innovation is building the capacity of Mathematics teachers only, the assumption is that the trained teachers will share knowledge with other subject teachers at the school level.

12 Discussion

Research on education innovations (Serdyukov, 2017) and the scaling of education innovations have been growing (Cooley, 2020, 2021; Kohl, 2021). By and large, the scaling of education has been perceived as complex and challenging due to the multiplicity of actors playing different roles including decision-making along the scaling process (Price-Kelly et al., 2020; Vvob, 2021). It also takes time to scale up an education innovation (Cooley, 2020, 2021). According to Cooley (2020) based on experience and theory, successful scaling occurs where there is an intermediary organization whose role is to strengthen the capacity of other institutions as well as design and form innovative partnerships among others. In this study, four partner institutions (three public universities and one international NGO) are involved to implement the SITT innovation, with DUCE, a constituent college of education acting as a Lead and an intermediary institution. According to Kohl (2021), good scaling practices also termed balanced scaling include crosscutting partnerships, collaboration, and participation, among the key stakeholders (the innovators, funders, and local actors) from the design of the innovation towards and along the scaling process.

In Tanzania, for example, particularly the two ministries MoEST and PO-RALG are responsible for making education decisions, also involving education agencies especially the Tanzania Institute of Education, the national overseer of curriculum design and review at the primary, secondary, and teacher education. The project also works with 21 teacher colleges, each mentoring 5 schools making a total of 105 secondary schools. Each of these schools is attended by education leaders at the district and ward levels with all involved in making decisions along with project interventions. The baseline study findings received representation from all these categories across the countries enabling the initial introduction of the project for familiarization and continual support along the scaling process. The intermediary institution, in this regard DUCE, coordinates the scaling of the SITT intervention in collaboration with the partner institutions. The project partners formed working teams in the areas of advocacy and gender, with the teams developing related strategies. The design of the SITT by itself is inclusive; thus it addresses the crosscutting issues as advocated by Kohl (2021). The distribution of roles has made it easier for scaling activities to advance as planned.

Scaling breakthroughs have been realized as evidenced by the development of frameworks to guide the design of scaling interventions (Crouch & DeStefano, 2017; Cooley, 2020, 2021; Price-Kelly et al., 2020; Kohl, 2021). However, not all the scaled-up interventions apply the frameworks systematically, which was also the case with the SITT innovation. This chapter applied the Cooley (2021) framework to assess the innovation of SITT which is currently under pilot since 2021. According to Cooley

(2020), innovations can comprise pilot projects (new technical products, services, processes, and partnerships), second-stage pilots, designed to test whether the factors responsible for success in one context are transferable to other settings and demonstration projects, which creates awareness about an existing model to make it more public and widely accepted by decision-makers and potential users.

Building on the findings, SITT falls under the categories of second-stage pilots as well as demonstration projects. SITT as currently implemented is not only scaled up to the secondary level across the countries but also transferred to Kenya and Zambia as new contexts. To contextualize the innovation pertaining to its relevance in scaling up education innovations, a situation analysis and a baseline study were conducted across the countries. The findings formed the basis for designing the secondary school SITT manual and guided the capacity-strengthening facilitation. In this regard, a facilitation guide for geometry has been developed in collaboration with Mathematics experts, since the topic was perceived as difficult by teachers and students across the countries.

The project designs, particularly the secondary SITT and the geometry facilitation guide, will continue to be improved through further research along with piloting the scaled-up SITT innovation. Besides research, the wisdom and knowledge of the project implementers; the originating organization, the researchers, teacher education institutions and the impacted, particularly education leaders, teachers, and students will continue to inform the pilot of the scaling process, views also supported by Thomas et al. (2004). On the demonstration aspect, an advocacy strategy has been created with one component being awareness creation about the model across the countries, to make the innovation more appealing to the government as decision-makers for policy uptake and to the potential users for further ownership. According to Cooley (2020), for an innovation to be adopted, the scaling strategy has to align with the government's agenda. In this regard, plans are underway to create a scaling strategy to further facilitate policy up-taking of the model for its sustainability after the end of the project.

Regarding the model elements, the SITT innovation is categorized under evolutionary innovations since the elements enable changes in the way students learn, with the same resulting in enhanced student learning outcomes. In this regard, the SITT innovation aligns with a structured pedagogy where effective teaching methods are applied to conventional courses to improve the existing instructional approaches, with the practice resulting in better learning, views also shared by Serdyukov (2017). Indeed, SITT applies competence-based curricula where locally available resources are used to prepare teaching aids. The SITT approach enables learners to relate mathematics to real-life and beautifying of the environments (mathematics garden) using geometric shapes. Such approaches make learning more authentic and practical to learners by aligning with the competence-based curriculum as advocated by the respective countries.

As per the application of SITT model in HE, there is limited integration of evidence-based innovative approaches in teacher education that adopt a competency-based curriculum. While curriculum review considers the views of diverse stakeholders, the views of specific innovations considering their uniqueness are not integrated into a wholesome one. Integrating the SITT approach in HE curriculum for teachers can prepare competent teachers along the application of a competence-based curriculum regardless of their specialization for improved students' academic performance and linkage of the subject with real-life environment. Further research regarding evidence-based education innovations and how they can be adopted in HE curricula should be carried out to inform policy and practice.

13   Conclusion, Recommendations, and Future Research

The study aimed to assess SITT as an education innovation currently being scaled up and adopted in selected secondary schools in Kenya, Tanzania, and Zambia using the Cooley (2021) framework. Besides the existence of frameworks to guide the design for scale, theory, and experience show that adopters of education innovations do not systematically use evidence along designing for scaling (Price-Kelly et al., 2020; Vvob, 2021). The study was empirical and it combined documentary review and mixed research methods to address the study questions. Following the Cooley (2021) Task 1 framework, the study first described SITT as an innovation reflecting its elements. The SITT approach is a pedagogy that supports the implementation of the competence-based curriculum as it enables peer learning and practice-based learning encouraging learners to apply the learned directly to the real-life environment.

To scale up the innovation, Vvob (2021) advocates for the need to consider the core (particularly the views of the teachers and students), professional development of leaders, low-cost teaching, and learning materials that are of quality and are context-specific as well as provide long-term capacity development support. In this regard, a baseline study was conducted to obtain the views of all project stakeholders emphasizing the direct beneficiaries (teachers and students), education leaders from the school, ward and district levels as well as policymakers across the countries. The findings enabled the design of the project piloting interventions, particularly the SITT secondary manual and a geometry module with the same used to build the capacity of education leaders and teachers as the direct beneficiaries. Likewise, the design of SITT along with its elements integrates low-cost teaching and learning resources reflecting individual contexts.

Overall, the design of SITT as an innovation for scaling, though it did not fully apply the framework during its design, aligns with the Cooley (2021) Step 1 Task 1 aspects. On the "What", the innovation has elements that are well elaborated, and these innovations stand out differently compared to other innovations in place, with these being attributed to the mentorship and support components. On the "How", the piloting of the innovation is participatory, where all the actors after being identified have been assigned different roles and are well coordinated by the intermediary organization. On the "Who", the piloting partners adopted the replication and collaboration scaling pathways while at the same time maintaining a working relationship with the originating organization. On the "Where", the study fits all the related attributes since the innovation is scaled up from primary to secondary and is transferred to other contexts, particularly Kenya and Zambia. Plans are there to design a scaling strategy to enable the sustainability of the innovation at hand, while at the same time making it more publicly appealing to the government and other policymakers.

Delineated by the relevance of scaling, there is a need to explore more education innovations for scale considering diverse contexts including developing context and a combination of different actors including the HE sector, views also recommended by Olson et al. (2021) who further provides key lessons, particularly the role of researchers and practitioners along the scaling process, with most of these aspects put forward in this study.

REFERENCES

Ajani, O. A. (2020). Investigating the quality and nature of teachers' professional development in South Africa and Nigeria. *Gender Behaviour, 18*(2), 15813–15823.

Akyeampong, K., Pryor, J., Westbrook, J., & Lussier, K. (2011). *Teacher Preparation and Continuing Professional Development in Africa; Learning to teach early reading and mathematics.* Centre for International Education, University of Sussex.

Alabi, A. O. (2017). Mentoring new teachers and introducing them to administrative skills. *Journal of Public Administration and Governance, 7*(3), 65–74.

Bagale, S. (2016). Gender equality and social inclusion in technical and vocation education and training. *Journal of Training and Development, 2,* 25–32.

Bates, M. A., & Glennerster, R. (2017). The Generalizability Puzzle. *Stanford Social Innovation Review Summer,* 1–6. Retrieved Mar 24, 2022, from https://ssir.org/articles/entry/the_generalizability_puzzle

Bermeo, J. M., Kaunda, Z., & Ngarina, D. (2013). Learning to teach in Tanzania: Teacher perceptions and experiences. In F. Vavrus and L. Bartlett (Eds.), *Teaching in tension: International pedagogies, national policies, and teachers' practices in Tanzania* (pp. 39–59). Boston: Sense Publishers.

Conn, K. M. (2017). Identifying effective education interventions in sub-Saharan Africa: A meta-analysis of impact evaluations. *Review of Educational Research, 87*(5), 863–898. https://doi.org/10.3102/0034654317712025

Cooley, J. (2020). *Scaling up—From vision to large-scale change: A management framework for practitioners.* MSI.

Cooley, J. (2021). *Scaling up–From vision to large-scale change: Tools for practitioners.* MSI.

Cooley, L. (2016). *Scaling up from vision to large scale change: A management framework for practitioners* (3rd ed.). Washington D.C.: Management Systems International, a Tetra Tech Company.

Crouch, L., & DeStefano, J. (2017). *Doing reform differently: Combining rigour and practicality in implementation and evaluation of system reforms.* International development group working paper no. 2017-01, RTI International, Research Triangle Park, NC. Retrieved June 18, 2018, from https://www.rti.org/publication/doing-reform-diferently-combining-rigor-and-practicality-implementation-and-evaluation.

Dede, C., Honan, J., & Peters, L. (Eds.). (2005). *Scaling up success: Lessons learned from technology-based educational improvement.* Jossey-Bass.

Fleisch, B. (2017). Teachers, the politics of the governed and educational development: Insights from South Africa. In C. Day (Ed.), *The Routledge international handbook of teacher and school development* (pp. 185–193). Routledge.

Fleisch, B., Schöer, V., Roberts, G., & Thornton, A. (2016). System-wide improvement of early-grade mathematics: New evidence from the Gauteng primary language and mathematics strategy. *International Journal of Educational Development, 49*, 157–174. https://doi.org/10.1016/j.ijedudev.2016.02.006

Fraser, M. W., & Galinsky, M. J. (2010). Steps in intervention research: Designing and developing social programs. *Research on Social Work Practice, 20*(5), 459–466. https://doi.org/10.1177/1049731509358424

Fullan, M. (2016). The elusive nature of whole system improvement in education. *Journal of Educational Change, 17*, 539–544. https://doi.org/10.1007/s10833-016-9289-1

Gallagher, M. J., Malloy, J., & Ryerson, R. (2016). Achieving excellence: Bringing effective literacy pedagogy to scale in Ontario's publicly-funded education system. *Journal of Educational Change, 17*(4), 477–504. https://doi.org/10.1007/s10833-016-9284-6

Hardman, F., et al. (2015). Implementing school-based teacher development in Tanzania. *Professional Development Education, 41*(4), 602–623.

Hartmann & Linn. (2008). *Scaling up: A framework and lessons for development effectiveness from literature and practice.* Wolfensohn Center Working Paper No. 5, Brookings.

Kafyulilo, A. C. (2014). Access, use and perceptions of teachers and students towards mobile phones as a tool for teaching and learning in Tanzania. *Educational Information and Technology, 19*(1), 115–127.

Kenya National Examination Council (KNEC). (2010). *Monitoring learner Achievement Study for Class Three in Literacy and Numeracy.* (NASMLA) Class 3 Study. Nairobi.

Kohl, R. (2021). *Crosscutting issues affecting scaling: A review and appraisal of scaling in international development.* Global community of practice on scaling development outcomes. Strategy and Scale LLC.

MacGregor, H., McKenzie, A., Jacobs, T., & Ullauri, A. (2018). Scaling up ART adherence clubs in the public sector health system in the Western Cape, South Africa: a study of the institutionalisation of a pilot innovation. *Globalization and Health, 14*(40). https://doi.org/10.1186/s12992-018-0351-z

McEwan, P. J. (2015). Improving learning in primary schools of developing countries: A meta-analysis of randomized experiments. *Review of Educational Research, 85*(3), 353–394. https://doi.org/10.3102/0034654314553127

Ministry of Education Science and Technology (MoEST). (2019). *National framework for teachers' continuous professional development.* MoEST.

Mwakabenga, J. R. (2018). Developing teacher-led professional learning in a Tanzanian secondary school. Unpublished doctoral dissertation, Massey University, Palmerston North, New Zealand.

OECD. (2009). *Introduction Innovation in Firms: A microeconomic perspective.* Accessed July, 2022, from https://www.oecd.org/berlin/44120491.pdf

Olson, B., Hannahan, P., & Arcia, G. (2021). *How Do Government Decisionmakers Identify and Adopt Innovations for Scale?* Brookings Institution.

Piper, B., Destefano, J., Kinyanjui, E. M., & Ong'ele, S. (2018). Scaling up successfully: Lessons from Kenya's Tusome national literacy program. *Journal of Educational Change, 19,* 293–321.

Price-Kelly, H., van Haren, L., & McLean, R. (2020). *The scaling playbook: A practical guide for researchers.* International Development Research Centre.

Reason, P., & Bradbury, H. (2008). *The SAGE Handbook of Action Research: Participative Inquiry and Practice* (2nd ed.). London: SAGE Publication.

Serdyukov, P. (2017). Innovation in education: What works, what doesn't, and what to do about it? *Journal of Research in Innovative Teaching & Learning, 10*(1), 4–33.

Snilstveit, B., Stevenson, J., Menon, R., Philips, D., Gallagher, E., Geleen, M., et al. (2016). *The impact of education programmes on learning and school participation in low- and middle-income countries* (Systematic review summary) (Vol. 7). 3ie. Retrieved April 5, 2018, from http://www.3ieimpact.org/media/fler_public/2016/09/20/srs7-education-report.pdf

Stringfield, S., Datnow, A., Ross, A., & Snively, F. (1998). Scaling up school restructuring in multicultural, multilingual contexts: Early observations from Sunland County. *Education and Urban Society, 30*(3), 326–357.

Thomas, K. G., Jr., Bodilly, S. J., Galegher, J. R., & Kerri, A. K. (2004). *Expanding the reach of education reforms perspectives from leaders in the scale-up of educational interventions.* Research Brief, RAND Corporation. Retrieved from https://www.rand.org/pubs/monographs/MG248.html on 3rd May, 2022.

TIE. (2018). *Inception report for testing school-based continuos professional development in-service training modules.* Dar es Salaam.

UNESCO. (2020). Global Parternership for education knowledge and innovation exchange (KIX) Africa 19 HUB, UNESCO-IICBA.

Uwezo. (2013). *Kenya annual report are our children learning?.* Uwezo Kenya report 2012. Uwezo.

Vvob. (2021). *Putting SDG4 into practice: Moving education innovations from pilot to scale.* Technical Brief No. 6. Retrieved August 25, 2022, from 202102_vvob_tech_brief_p2s_web_spreads.pdf

Mechanisms for Enhancing Employability Skills Among Students Within Vocational Education Training Institutions in Tanzania

Mwaka Omar Makame ⓘ *and Katherine Fulgence* ⓘ

1 INTRODUCTION

Research shows that VET appears as the major counter-balancing force in producing a competent workforce to operate independently and create self-employment (Woyo, 2013). According to UNESCO (2020), the number of VET pupils dramatically increased all over the continents. For example, from 2008 to 2018 vocational pupils increased from 56.2 million to 62.5 million. Research done by Rojewski (1997) and Saleh (2017) shows that a workforce that receives vocational training has a higher likelihood of employment than that without vocational training. VET is

---

M. O. Makame (✉)
Mboga Secondary School, Chalinze, Pwani Region, Tanzania
e-mail: mwakanazmni@gmail.com

K. Fulgence
Dar es Salaam University College of Education, University of Dar es Salaam, Dar es Salaam, Tanzania
e-mail: katherine.fulgence@udsm.ac.tz

© The Author(s), under exclusive license to Springer Nature Switzerland AG 2024
P. Neema-Abooki (ed.), *The Sustainability of Higher Education in Sub-Saharan Africa*, Sustainable Development Goals Series, https://doi.org/10.1007/978-3-031-46242-9_10

thus seen as a key to reducing graduates' unemployment rate. In this regard, the promotion of VET should be emphasised. For. the skills acquired are expected to enhance the production of human capital that can sustain the global economy and the labour market demands.

VET implements a Competency-Based Curriculum (CBC) which places emphasis on the acquisition of self-employability skills. Although the conceptualisation of skills is not homogeneous among different VET actors particularly the policymakers, management, teachers and students, this study defines self-employability skills as the key skills and personal attributes needed to enter, operate and thrive in the world of work (Lindsay & McQuaid, 2005). Employability skills include communication skills, teamwork skills, problem-solving skills, initiative and enterprise skills, planning and organising skills, self-management skills, learning skills and technology skills (Cornford, 2006; Curtis & McKenzie, 2002; Nagarajan & Edwards, 2014).

In VET, employability skills are enhanced through extra and core curriculum activities, work-integrated learning and practical training (Bennett et al., 2016). Other mechanisms include the use of practical and interactive teaching methods (Mbugua et al., 2012), national and institutional policy stipulations (Fulgence, 2016) and institutional leadership to spearhead the implementation of the policies and directives (Nkirina, 2010). Among the policies spearheading VET in Tanzania *(the focus of this study)* include the Educational and Training Policies (ETP) of 1995 now 2014, the Technical Education Policy of 1998, the Education Sector Development Programme (ESDP) and the National Strategy for Growth and Reduction of Poverty (NSGRP) (ILO, 2020). According to Martín-Garin et al. (2021), interactive pedagogies attuned to problem-based learning, project-based learning, practice-based assessments and research-based learning if used on sustainable basis by teachers enable students to develop employability skills. It is the VET institution's responsibility to identify and implement appropriate mechanisms in enhancing employability skills. Possessing employability skills is key to employability, which is defined by one's ability to sustainably create and secure employment and compete in the world of work (Yorke & Knight, 2004; Fulgence, 2016). With these efforts, the country expects to improve technical and vocational education and prepare more competent graduates with employability skills who can be able to compete in regional and global employment markets.

Vocational education is meant to empower the youth to be self-employed and to have employability skills in boosting the economy of the country. In Tanzania, VET aims at preparing vocational students with the skills and knowledge for employability in both formal and informal establishments (VETA, 2019). The provision of competence-based education is influenced by Tanzania's development vision 2025 which emphasises the need for having a well-educated and learning society (URT, 1999). This includes the production of competent educated people, sufficiently equipped with the requisite knowledge to solve society's problems and meet the development challenges at the regional and global levels while at the same time "seeking to ensure inclusive and equitable quality education and promote lifelong learning opportunities for all" (SDG 4). According to International Labour Organisation (ILO) (2015), developing employability skills can help achieve SDGs 4 (quality education) and 8 (decent work and economic growth) by improving access to quality education and increasing the number of decent work opportunities.

Despite the investment efforts in VET, the stress on how to improve and increase the number of self-employment graduates through VET is still a problem in developing countries (AU, 2007; Allais et al., 2020). VET remains weak and has several challenges across the African context, including Tanzania (Bashir et al., 2016). A study by Mbugua et al. (2012) in Kenya noted lecture to be the most popular teaching method used in VET, which is majorly theory-based and teacher-centred, which is contrary to the VET competence-based curriculum that requires 70% practical skills and less than 30% theories (Ismail et al., 2018). A similar challenge is experienced in Tanzania. Particularly, the adaptation of Competency-Based Education and Training (CBET) inadequately led to the provision of quality VET due to the lack of competent instructors in the area of competency-based training (URT, 2013). Teachers without industrial or hands-on experience are more likely to encounter difficulties in the classroom, especially where they are required to handle specific equipment and hand tools.

On the teaching and learning resources, Munishi (2016) established an inadequate supply of teaching and learning resources in VET centres and infrastructure like classrooms, workshops, well-equipped libraries and computer laboratories as well as internet and projector facilities. VET institutions also experience poor facilitation means for career guidance and counselling; this contributes to a lack of employable skills among graduates especially on career direction, aspiration and career planning options. Career guidance and counselling are comprehensive, developmental programmes that aid in the planning and implementation of educational and career choices (UNICEF, 2017). There has also been a poor perception and recognition of VET by the public with the view that it caters for the less academically qualified individuals against the strategy of training skilled workers for employability prospects (Li, 2012; Ismail & Abiddin, 2014).

Besides the efforts made by the government to emphasise the provision of quality vocational education, it is questionable if the current VET can produce graduates with the employable skills to meet the aspiration of Tanzania Development Vision 2025 considering the existing challenges. By addressing the challenges, Tanzania can create employment for a large group of young people, many of whom come from the poorest societies through skills acquisition in VET (Andreoni, 2018). This study focuses on the institutional mechanisms that are aimed at enhancing the employability skills of VET students. The study will address the following questions: How do VET actors particularly institutional leaders (principals) and teachers perceive employability skills? How do students rate themselves on their employability skills? What strategies and mechanisms are applied by VET institutions to enhance the employability skills of the students?

This study provides useful knowledge on the strategies used in enhancing the employability skills of students. Study findings will also assist policymakers in shaping the fields in the establishment of reforms, putting in place the appropriate institutional mechanisms, and modifying or introducing the scheme to support VET on skills development.

2 STUDY CONTEXT AND THE CONCEPTUALISING VOCATIONAL EDUCATION AND TRAINING

Countries worldwide including Tanzania recognise the role of VET in reducing youth unemployment and socio-economic inequalities (Tripney & Hombrados, 2013; URT, 2013; Eicker et al., 2016). In Tanzania, this is supported by the highlighted policies that support skills development including the acquisition of appropriate knowledge and values to youth through Technical Vocational Education and Training (TVET). TVET institutions have also been re-orientating themselves to promote lifelong learning and produce graduates who can create employment (Nkirina, 2010). Even in its modern world, few youths possess the appropriate skills, knowledge and attitudes relevant to find decent employment in the current labour market (Eicker et al., 2016).

VET is work-based learning, continuing training and professional development for a wider range of occupational fields, including production, service and livelihood that leads to work qualification (UNECSO, 2015). VET is thus a part of lifelong learning that can be taken at secondary, post-secondary and tertiary levels that offers craft courses in tailoring, masonry and bricklaying, painting, carpentry, electrical installation, motor vehicle mechanics, plumbing and pipe fitting, etc. The courses can last up to three years and participants are awarded National Vocational Training Award (NVTA) certificate, with the qualifications enabling students for the labour market and further education levels. In Tanzania, VET is offered through Folk Development Colleges (FDC) and Vocational Training Centers (VTCs). URT (2018) accords the number of students enrolled in VET is more than 225,683 in institutions owned by VETA and other providers.

Standard seven is the minimum entry level for National Vocational Training Awards (NVTA) level 1 which leads up to NVTA level 3. We also have National Technical Awards (NTA) levels 4 up to 10 for Technical Education and Training (TET). Through the national qualification, VET graduates can join HE and meet the standard qualification across each appropriate level. For this study, the researcher focused only on NVTA levels 1–3 because it concerns VET.

In terms of regulation, two bodies regulate the provision of quality vocational education in Tanzania, the National Council for Technical Education (NACTE) and the Vocational Education and Training Authority (VETA) (NACTE, 2012; URT, 2014) currently coordinated under the National Council for Technical Vocational Education and Training (NACTVET). NACTVET was established because of efficiency increase, rehabilitation and quality control of technical and vocational training provided in the country (NACTVET, 2022). NACTVET falls within the Department of Technical Vocational Education (DTVET) under the Ministry of Education, Science and Technology (MoEST).

3   Literature Review

3.1   The Study Theoretical Framework and the Conceptualisation of Employability Skills

This study is guided by human capital theory as proposed by Becker (1962). Human capital refers to the collection of individual attributes; for example, knowledge, skills, experience, ability, talent, intelligence, judgement and training in production activities and increasing income. The theory holds that investment in education is no waste, but rather a means to gain job performance, emotional intelligence and cognitive ability, which contributes to both individual and organisational adaptability, which consequently leads to economic development. The human capital theory assumes that education helps in the development of work skills and improves the capacity of the worker to be more productive and able to contribute to industries and the community (Sweetland, 1999).

The human capital theory is useful in this study as it assists in understanding human attributes that fit the world of work. According to Ma et al. (2020), schooling and training are the most useful components and observable sources of human capital development. Therefore, reforming education system is a way of improving human capital. Therefore, for workers to be more productive, it depends on the investment of students' skills through education. Indeed, low-quality human capital investment in training leads to the failure of the production of human capital (Hung & Ramsden, 2021). The human capital theory emphasises the need for policymakers to allocate significant resources to training. The resources needed in training are human resources, financial resources, physical resources and information resources. The lack of these resources makes it difficult for educational institutions, regarding VET in the study context to accomplish their goals. High investment in training produces competent students with employability skills.

Based on this belief, employability skills are defined as a combination of communication skills, technical skills, ICT skills, enterprise skills, team working, planning and organising and problem-solving, lifelong learning and initiative and enterprise (Cornford, 2006; Curtis & McKenzie, 2002; Nagarajan & Edwards, 2014). The combination of these skills produces perfect human capital. Employability skills are the basic skills that every worker must have to adapt to the workplace. The following are attributes one must possess to be an effective human capital. Other employability skills attributes include loyalty, commitment, honesty, integrity, enthusiasm, reliability, self-management, learning, common sense, positive self-esteem and a sense of humour (NCVER, 2003; Yorke & Knight, 2004). Moreover, there are employability attributes like career self-management, cultural competence, self-efficacy, career resilience, sociability, entrepreneurial orientation, proactiveness, project management, business with personal skills and emotional literacy (Coetzee & Potgieter, 2013; Al-esmail et al., 2015).

The study thus categorises employability skills into eight categories; communication skills which include listening, speaking, writing, negotiating, empathising, using numerals, establishing networks and sharing information (Chapman & Young, 2010; Fahimirad et al., 2019). Teamwork skills include one's ability to work across different ages, gender, races, religions or political persuasion (Brinkley et al., 2010; Curtis & McKenzie, 2002). Problem-solving skills involve creating solutions and applying problem-solving strategies (Ananiadou & Claro, 2009; Akhyar et al., 2020). Initiative and enterprise skills include adapting to a new situation, being goal orientated, identifying opportunities, changing ideas to actions and having options and innovation (Curtis & McKenzie, 2002; NCVER, 2003). Planning and organising skills involve managing time, being resourceful, establishing clear goals, predicting risks and organising information (Cornford, 2006). Self-management skills include personal goals, own evaluation, confidence and taking responsibility (Chapman & Young, 2010; Fahimirad et al., 2019). Technology skills involve having basic Information Technology (IT) skills, using Information Technology to organise data and being willing to learn new knowledge (Curtis & McKenzie, 2002). Technical skills focus on knowledge related to the job and a general understanding of the subject matter (Medina, 2010; Tan & French-Arnold, 2012). For example, in the engineering field, graduates should know the design development and the usage of modern tools to the environment.

3.2   Related Work

Studies linking vocational education and training and the state of skills acquisition have focused on the promotion of youth employment through information and communication technology (Kissaka et al., 2020), challenges towards the development of employability skills (ILO, 2019) and factors contributing to the lack of employability among TVET graduates. ILO (2019) explored the state of the skills in Tanzania and found out the challenge in the development of employability skills is the poor implementation of the policies designed. Also, there are leadership challenges in Technical and Vocational Education and Training (TVET) in Tanzania. It seems that VETA, NACTE and TCU operate separately which affects skills acquisition. Kissaka et al. (2020) in their study regarding the role of

ICT in promoting youth employment in Tanzania found out, the main barriers to VET graduates entering the world of work are mismatches between skills acquired in college and skills needed in the workplace. A study by Munishi (2016) established the factors contributing to the lack of employability of TVET graduates which include the lack of foundation training at the primary and secondary level, poor organisation of the TVET courses, lack of English language competence for students, poor infrastructure and poor policy formation in TVET. Tambwe (2017) conducted a study to establish the challenges faced by TVET in the implementation of CBET along with enhancing employability skills. The study reveals that the preparation and recruitment of teachers in both pedagogic and industrial practical skills are poor in VET. The shortage of VET teachers, lack of institutional support and large class size makes it difficult for teachers to interact with students. Based on the reviewed literature, little attention is placed onto the institution's mechanisms in place towards developing employability skills among VET students. The studies came to a conclusion regarding the need to emphasise the mechanisms for developing employability skills through VET.

3.3 Mechanisms for Enhancing Employability Skills Among VET Students

Developing student skills is an ongoing process. Students enter VET with the expectation that they will acquire knowledge, skills and ability to enter the job market or advance their careers. VET enhances students' employability through extracurricular activities, which support character development. For example, participating in student government in the councils, giving service, sports, music, arts and clubs can build teamwork, problem-solving and leadership skills. A field study programme can play a great role in helping one make the right choice. This encourages the student to immerse in the study environment, exposes students to campus life and looks out of the class environment to the world of work (Behrendt & Franklin, 2014). The establishment of the conference and seminars also creates students' confidence and develops communication skills, time management and motivation (Lice, 2019). Conferences and seminars are extremely beneficial to students since they enable them to evaluate, stay up to date on current issues and build their network.

The use of computer and libraries create Information Communication and Technology (ICT) skills. According to Pirzada (2013), ICT skills are associated with employability and increased productivity, leading students to succeed at their professional level and also enhancing lifelong learning (Chen & Ni, 2016). In the recent employment market, ICT skills become very important due to advancements in technology. The preparation of students with employability skills in today's advancement of technology requires teachers to apply ICT in delivering the curriculum to equip students with digital literacy. This involves the effective use of digital devices for communication, collaboration and advocacy (Lynch, 2018).

The use of internship, placement and work-based learning provides the student with experience and skills, which help students to get to know their first opportunity in the world of work. Also, it can help to link with employers and have the chance to sustain their ability, and create motivation, technical skills and career goals (Elliot et al., 2011). Effective use of the workshops and laboratories helps in strengthening technical skills. Workshops and laboratory practice give the basic working knowledge required for production. They explain the construction functions, use and application of the working tools, and pieces of equipment, as well as manufacturing the products from raw materials (Pangestu & Sukardi., 2019).

Career guidance and counselling are related to skills development programmes that aid in the planning and implementation of educational and career choices (UNICEF, 2017). It is an important mechanism that helps students to have career goals and motivation. VET institutions should collaborate with industries that help in designing teaching and learning contexts and modification of the curriculum based on the skills gap demand (Setiadi et al., 2019).

Enhancing employability skills also depends on effective and efficient teaching and learning strategies (Obeta et al., 2013). However, effective teaching and learning depend on the ability of teachers or facilitators in motivating and bringing out the interest of students through different instructional strategies. Different strategies and methods should be used to develop employability skills (Nwazor & Onokpanu, 2016). The foregoing observation is also espoused by Ekwue et al. (2019) who further hold that good teaching and learning strategies are needed in VET in enhancing practical skills, entrepreneurship knowledge and the establishment of a network. The most important teaching methods in VET which influence the acquisition of employability skills are: simulation, computer-based instruction, problem-based learning, context-based learning, demonstration, field trip and other student-centred methods, discussion, and project engagements (Anindo, 2016; Audu et al., 2014; Mbugua et al., 2012; and Martín-Garin et al., 2021).

4 Methodology

This study used a mixed research approach to understand the study phenomena. A mixed approach allows triangulation and each approach complements the other during data analysis, as Airasian et al., 2012 and Creswell and Plano (2018) testify. The study was conducted in four VETs in the Kibaha District, Tanzania. The district has a diversity of VTCs in terms of ownership; some are owned by the central government, Faith Based Organisations (FBOs), private institutions and VETA-owned centres (*VETA.*, 2022). Study participants were the principals, teachers and final-year students purposively selected based on their roles and positions. Accordingly, purposive sampling is based on choosing respondents according to their knowledge of the information desired (Airasian et al., 2012).

Survey and interviews were used to collect data from the study participants along the study questions. The survey questions were adapted from Anindo (2016) and Dasmani (2011) but with modifications. For example, students were asked to self-rate their employability skills along with the study eight attributes in a four-point scale: 4 = Expert—high level of competence, extensive experience in the application of the skill; 3 = Advanced—moderate to high level of competence; 2 = Intermediate—the average level of competence; 1 = Beginners—low level of competence. Students opined on a four-point scale how often they engage in the activities meant to enhance their employability skills. The questionnaire for teachers aimed to establish the teaching and learning methods they use in with their teaching roles and how frequently they use them. For the management, particularly VET principals, the interview questions aimed to explore their perception regarding employability skills and the strategies they use to develop employability skills among VET students. The management questions were also established from the different categories of teachers and students through interviews.

The qualitative data were recorded, transcribed, coded and analysed with the aid of MAXQDA 2018 software, a software for analysing qualitative data. Thematic coding dominated the analysis, whereby the mechanisms aimed to enhance employability skills formed the themes and sub-themes for reporting the study findings. Research clearance was sought from the University of Dar es Salaam research committee.

MECHANISMS FOR ENHANCING EMPLOYABILITY SKILLS... 233

Participants' consent was sought before data collection, including permission to record the interviews. Likewise, data were used for the study purpose, as Cohen et al. (2011) do recommend.

In terms of academic qualifications, 50% of the VET principals had a diploma and the remaining 50% had a bachelor's degree in vocational education. Of the teachers, 58% had a certificate qualification and the rest (42%) had a bachelor's degree in non-vocational education. In terms of working experience, 75% of the principals had over 10 years of working experience, contrary to the teachers of whom 58% had less than 5 years of working experience. Additional descriptive statistics show that 79.8% of the vocational students end their training in level 2 with the remaining few 20.2% progressing to level 3. Regarding the age of the students, 67.7% were aged between 21 and 25, 27.4% were below 20 years and only 4.8% were aged 26 years and above. Overall, VET students study 10 courses. It was of interest to note that 36.3% of the VET students study electronic installation. Further findings show that there are more formal and self-employment prospects attached to studying this course. Another course more studied is motor vehicle mechanics where 16.9% attested to have studied this course.

5 FINDINGS

5.1 *Awareness of the Concept of Employability Skills*

The first objective of the study was to find out if principals and teachers within VET institutions are aware of employability skills and their attributes. On this objective, two themes emerged where employability skills were first related to entrepreneurship skills and second as competencies (knowledge, skills and attitudes) developed through the vocational academic qualifications. Each is further discussed and supported with quotations.

5.1.1 *Employability Skills as Entrepreneurial Skills*
Some study participants conceptualise employability skills as entrepreneurship skills. They use the terms synonymously. As narrated:

> Employability skills entails the possession of entrepreneurial attributes as commitments, finishing work within the agreed time, knowing your

competitors, and looking and evaluating the market. (**Interview, Teacher 1 VETCD, 18.07.2022**)

Another narration supports this observation:

> Yes, employability skills are like being creative…, and having entrepreneurial characteristics which include self-reliance, also having self-discipline and respect. (**Interview, Teacher 2 VETCA, 18.07.2022**)

5.1.2 Employability as Technical Competence in the Field of Qualification

Employability skills were associated with the core competence acquired in one's field of vocational qualification. As narrated:

> In general, a person is regarded to possess employability skills, if he/she is competent in what he or she is doing or intends to do. For example, in this electrical field, you must know what electricity is. (**Interview, T3 VETC C, 17.07.2022**)

Another similar response under this aspect:

> Employability involves skills that students get from the core courses. You know this is not the secondary school…, here we train technical and indulge in practices on the core subject. The technical expertise enables students to be employed or employ themselves. (**Interview, Principal 4 VETC D, 18.07.2022**)

Indeed, about 44% of principals mentioned technical skills, particularly the skills acquired from the core subjects to be the best skills to be possessed by VET students. This was supported by teachers in VETC D who maintains that:

> I think if they are good in their core courses, they can be employed or employ themselves. (**Interview, Teacher 4 VETC D, 18.07.2022**)

5.1.3 Students' Self-rating About Their Employability Skills

The study also aimed to assess the level of employability skills among students. Descriptive statistics were conducted from students' questionnaires to calculate the mean as the measure of central tendency. The findings revealed that the level of employability skills among VET students at

NVTA Levels 2 and 3 is intermediate. None of the students attested to the advanced and expertise level along the study employability skills attributes. Apart from that, students apparently uphold ICT skills at the beginner's level. Impliedly, ICT skills are less likely to be developed among the VET students.

5.2 Mechanisms Used by VET Institutions to Enhance the Employability Skills of Students

The study's third objective was to identify institutional mechanisms and strategies used to enhance the employability skills of VET students. Under this objective, five sub-themes emerged, reflecting the views of VET principals and teachers under Table 1. Each is further discussed.

5.2.1 Practice-Based Assessment to Monitor Progress
Monitoring helps to recognise the pupil's academic abilities. During the interview, most teachers and principals mentioned regular monitoring and assessment as the effective mechanism for acquisition of the employability skills. As narrated;

> We do live projects and assess students in every lesson. The assessment is meant to measure progress..., to establish if students can demonstrate what I have taught... (Interview, Teacher 1 VETC C, 17.07.2022).

The other teacher in VETC B postulated thus:

Table 1 Institutional mechanisms for enhancing employability skills among VET students

SN	Topic	F	Percentage
1	Practice-based assessment	10	25%
2	Using workshops	7	17.5%
3	Professional development of teachers	6	15%
4	Fieldwork studies	5	12.5%
5	Career guidance and counselling	5	12.5%
6	Seminar presentations	4	10%
7	Extra curriculum	3	7.5%
	Total	**40**	**100%**

Source: Field Data 2022

I help my students during classroom sessions, assess them…, make some follow-ups during the practical time and keep encouraging them… (Interview, Teacher 1 VETC B, 18.07.2022).

5.2.2 Using Workshops

This sub-theme emerged in almost all centres visited. Practical classes are more needed in VET institutions and is through workshop the practical classes are conducted.

As the teacher in VETC A says:

Even though there are insufficient facilities…, we use workshops…, workshops provide a way to create an intensive experience for our students over a short time…, students also understand well. **(Interview, Teacher 2 VETC A, 16.07.2022)**

Meanwhile, the principal of the same VETC A said:

Workshops are very useful…, we have to practice more through workshops…, we need to use modern tools, and modern facilities…, vocational training needs more practice for students to be competent. **(Interview, Principal 1 VETC A, 16.07.2022)**

5.2.3 Professional Development

Professional development of teachers was mentioned as the source of the acquisition of employability skills for students. As narrated:

*About professional development, this institution provides different training, for example, training on entrepreneurship, we train among ourselves and sometimes we invite expert guests. We encourage teachers to increase their education level, especially for those who have certificates only. **(Interview, Principal 1 VETC A, and 16.07.2022)***

A similar observation was also shared by the teachers:

I have been trained by VETA twice. First, it was about entrepreneurship and the other was about teaching techniques in VET. The training helped me a lot, especially the training concerning entrepreneurship. I use this for my students. **(Interview, Teacher 1 VETC D, 18.07.2022)**

5.2.4 Fieldwork Studies

The results reveal that students improve their employability skills through field study as they get a chance to substantiate their abilities and competencies. Students can get actual practical skills. This is supported by the Principal in VETC A who said:

> We send students for fieldwork in order to attach themselves to the industries. We select big companies; as they return we discover a lot of changes in our students, many of them become experts in technical skills and full of confidence. **(Interview, Principal 1 VETC A, 16.07.2022)**

The Principal from VETC B had these to say on the fieldwork and how working relationships are established:

> Look, I send my students to TANESCO and most of them return to work there after graduation, this is because they establish connections during the field studies. **(Interview, Principal 2 VETC B, 17.07.2022)**

5.2.5 Career Guidance and Counselling

This is another mechanism mentioned by teachers and principals about the strategies used on enhancing employability skills. As narrated:

> We have career guidance and counselling in our centre, sometimes a parent brings students and asks you, what suitable course do you have for my child? So you have to counsel them, regarding students' dreams and educational background. **(Interview, P4 VETC D, 18.07.2022)**

It was however revealed that the provision of career guidance services is not sufficient in the VET institutions. Most guidance and counselling services available in the education system place focus on personal and social support particularly on relationship related aspects, family problems, health problems, decision-making and illegal drugs. For example, in one of the interviews, the principal in VETC C mentioned:

> We are conducting counselling and guidance to our students; here we have internal counselling programmes and sometimes we invite neighbouring doctors to counsel our students. **(Interview, Principal 2 VETC C, 17.07.2022)**

5.2.6 Seminar Presentations

Through seminars and other presentations, students acquire various skills, in particular, communication skills, problem-solving skills, build confidence and teamwork to mention a few. As narrated:

> I prefer to give them group work to make them analyse the issues then, they conduct a presentation. Presentations help them to understand the lesson and build their confidence. (Interview, Teacher 1 VETC A, 16.07.2022)

Another teacher in VETC C narrated:

> We use participatory methods, I do seminars and presentations…, These presentations help students a lot in their career development. (Interview, Teacher 1 VETC C, 18.07.2022)

5.2.7 Engaging in Extra Curriculum

Through extra-curriculum activities, one can get skills like teamwork, communication skills, problem-solving and others. This was elaborated by a teacher from VETC C:

> We encourage our students to do part-time work or volunteer as extra curriculum from their neighbouring place. It is a great way to showcase several skills including time management, responsibilities, and teamwork. (Interview, Teacher 2 VETC C, 17.07.2022)

Regarding extra-curriculum engagement from VETC B another teacher quipped:

> As for some students engage in football and netball, they compete inside the college and outside. Outside they interact with others from different VETCs. But this activity depends on the student's interest, some of them are not interested and you cannot force them. (Interview, Teacher 1 VETC B, 17.07.2022)

5.3 Students' Responses to Institution's Mechanisms for Enhancing Employability Skills

The study further sought information from students regarding the mechanisms VET institutions use to enhance the acquisition of employability skills and how often they engaged in them—Table 2 presents the findings.

The findings show that the strategies that are often used by VET institutions to enhance students' employability skills are workshops followed by e-libraries and extra-curriculum activities. The findings can be viewed that, workshops could assist graduates to acquire practical skills which help them take off smoothly in the world of work, especially in the area of self-employment.

Teaching and Learning Methods Used in Enhancing the employability Skills of Students

Students were further asked about teaching and learning methods frequently used by teachers where the average mean was computed as presented in Table 3.

The above findings show that majority of the teachers used to lecture and discussions during seminar presentations in the process of teaching and learning, with this largely attributed to a large class size and shortage of equipment. The use of the lecture method could have also been attributed to the few teachers with the same handling of many subjects. As narrated:

> Teachers have to reduce the number of subjects, to make it possible to do practical for a long time…, VET seems to be like normal secondary education…, which in practice need more practical. (**Interview, Teacher 2 VETC C, 17.07.2022**)

Another teacher exclaimed:

Table 2 Students' responses to institutions mechanisms for enhancing employability skills

Strategies and mechanisms	Not at all	Sometimes	Often	Very often
Workshops	14	42	32	36
Computers and e-libraries	67	22	17	18
Extra curriculum	25	52	32	15
Career guidance and counselling	61	26	23	14
Participation in conferences and seminars	66	30	16	12
Participation in professionals competitions	52	32	28	12
Collaborating with industry practitioners	50	38	24	12
Meeting with employers and graduates	62	35	17	10
Learning from other VET institutions	54	38	24	8

Source: Field data (2022)

240   M. O. MAKAME AND K. FULGENCE

Table 3 Teaching and learning methods used in enhancing the employability skills of students

Teaching method	Never	Rarely	Often	Always	Mean	Decision
Lecture	0.0	0.0	33.3	66.9	3.6	Often
Demonstration	0.0	25.8	67.7	6.5	2.8	Rarely
Work-based learning	56.5	26.6	16.9	0.0	1.6	Never
Simulation	63.7	27.4	8.9	0.0	1.4	Never
Field trip	39.5	30.6	29.0	0.8	2.0	Rarely
Context-based learning	54.8	25.8	8.9	10.5	1.7	Never
Discussion/seminar Presentations	3.2	11.3	59.7	25.8	3.0	Often
Project work	19.4	62.9	16.9	0.8	1.9	Never

Sources: Field Data (2022)

The number of subjects is high…, which causes students to miss a lot of practice time. (**Interview, Teacher 2 VETC A, 17.07.2022**)

Overall, the methods that are meant to develop employability skills are the ones reflected to be never used in the teaching roles by teachers. For VET institutions to develop employability skills, a new orientation regarding the sustainable teaching methods might need revisiting, especially in the professional development of the VET teachers about these methods.

6   DISCUSSION

6.1   The Conceptualisation of Employability Skills

Study findings show that the majority of the participants perceive employability skills to mean entrepreneurship skills. According to Atistsogbe et al. (2019), employability skills attribute have a positive relationship with entrepreneurship knowledge and skills. According to the findings, the extent to which graduates perceive themselves as employable has a high level of entrepreneurial skills. Pallawi et al. (2022) ascertain that there is a similarity between entrepreneurship and employability skills since they both involve soft transferable skills, interpersonal skills, communication skills, self-management skills, willingness to learn, optimism, resilience, ability to work under pressure, adaptability, positive attitude, digital literacy and commercial awareness. Few participants especially the principals view employability skills as the core competence in the field of qualification, which means being an expert in the core field. According to Ahmad et al. (2019), graduates are not only required to have the right knowledge or technical expertise but also the capacity to make ethical decisions and resilience to deal with emerging global changes. Therefore, based on these findings, not all principals and teachers understand well the concept of employability skills demanding a broader orientation in this direction. Employability skills involve communication skills, team working, problem-solving skills, ICT skills, creativity and enterprise and should integrate core fields in the world of work (Cornford, 2006; Curtis & McKenzie, 2002; Nagarajan & Edwards, 2014). Indeed, technical knowledge and skills are necessary for success in the workplace also supported by Tokarcikova et al. (2020) and Fulgence (2016).

7 EMPLOYABILITY SKILLS AND THE SUSTAINABLE DEVELOPMENT GOALS

This study shows employability skills to be an essential prerequisite for attaining sustainable development goals (SDGs), particularly SDG 4 and SDG 8. Employability skills are a constituently key element of the 2030 Agenda for sustainable development. SDG 8 on its part aims to promote economic growth and decent work for all. This involves sustaining good jobs to boost the economy and eradicate poverty (Comyn, 2018; ILO, 2015). In the realisation of the SDGs, VET and employability skills got a greater policy priority in the current years globally. The priority was due to the higher levels of youth unemployment, growing skills mismatch, changing the nature of employment, increasing skills migration and new skills demand arising from continued globalisation and introduction of new technology (ILO, 2015). The findings further show that students choose VET courses with the expectation of possessing the skills that enable them to employ themselves or to be easily employed, the aspect that aligns with the acquisition of quality education and decent job, especially for youths. Indeed, VET aims to prepare graduates with knowledge and skills for employability in the both formal and informal establishments (VETA, 2019). VET can also contribute to facilitating the transition from an informal to a formal economy.

8 Level of Employability Among VET Students

The study aimed also to assess the employability of the VET students using a self-rated measure. The findings show that out of the study's eight attributes of employability skills, they are at an intermediate level (average level of competence) across seven of them, with ICT skills demonstrating a low level. A study by Dasmani (2011) revealed that most VET graduates lack employability skills because of theory-oriented training and little practice. The study recommended more practical training that helps students to possess technical skills. On ICT skills, Basnet et al. (2009) mentioned that the use of ICT is still in the primary stage in many developing countries. The application of ICT in VET is low in poor countries, including the use of the internet, phone lines and email to link centres, classrooms, instructors and learners. The cost of recruiting qualified teachers in ICT as well as running programmes, development, and maintenance is high. These factors have continued to lead to the production of graduates with low ICT skills (Chinien, 2003; Inyiagu, 2014; Kissaka et al., 2020), also reflected in this study.

9 Mechanisms for Enhancing Employability Skills Among VET Students

This study spells out six institutional mechanisms for enhancing employability skills among VET students; practice-based assessments, workshops, teacher professional development, fieldwork studies, career guidance and counselling, students' engagement in extra-curriculum activities and seminar presentations. While career guidance and counselling are vital for enhancing employability skills development in VET institutions, it is less practised due to lack of officers or offices responsible for career guidance. These findings correlate with Munishi (2016) who reveals that in all ten visited technical institutions there was no office or officer for implementing career guidance and counselling. Regarding other mechanisms, Foshay and Silber (2009) maintain that seminar presentations have a direct impact on students' employability skills. Students can communicate and engage in logical reasoning and decision-making. Pangestu and Sukardi. (2019) reiterate that workshops in vocational schools are the first place where students acquire understanding and vocational skills.

Regarding students' level of employability skills, the study found computer use, e-library, attending conferences and seminars, and meeting with employers and graduates to be done on a small scale in the VETCs. This is the same as what was mentioned by Bello et al. (2013) and Kissaka et al. (2020) with this attributed to inadequate ICT infrastructure and unreliable internet network. This study further reveals that VET teachers use lecture as a teaching method. This finding agrees with what Mbugua et al. (2012) noted in Kenya VET institutions that the lecture method which is a theory-based method is the most popular teaching method. Asked to conjecture the best methods for enhancing employability skills, the majority of the students opted for discussion and practices related methods, to illustrate, project work and work-based learning. Accordingly, a large number of students and high workload among teachers might have attributed to the application of the lecture method. Therefore, from the above findings, it seemed that there is a need for VET institutions to establish different mechanisms that can help students to gain the expected skills, through sustainable pedagogies, views also supported by Martín-Garin et al. (2021).

10   Conclusion, Recommendations and Future Research

The study aimed to establish the mechanism used by VET institutions in enhancing the employability skills of students and was conducted in Kibaha District, Tanzania where four selected VETCs were involved. It employed mixed methods and was guided by three specific objectives: to evaluate the understanding of the employability skills within VET institutions, to assess VET students' employability skills, and to explore the mechanisms used by VET institutions including teaching and learning strategies to enhance the employability skills of graduates.

The study findings show that VET's principals and teachers conceptualise employability skills as entrepreneurship skills and some refer employability skills to technical skills, with the view that possession of the technical qualification is sufficient to corroborate employability skills. The latter assertion can hold if employability skills are well integrated into the vocational subjects and programmes through interactive pedagogies, the practice that is not valid as rated by students. The final-year students rated their employability skills to be at an intermediate level for the study's seven employability skills attributes. For the ICT skills, students rated their level of employability to be at the beginners' level. None of the students mentioned being at an advanced level or an expert on the study employability skills attributes, meaning the vocation education courses and programmes are sufficient to develop students' ICT skills.

It was of interest to note that students who completed NVTA level 3 authenticated higher levels of employability skills than those of NVTA level 2. While the management view practice-based assessment as the major mechanism used by VET institutions to enhance students' employability, students mentioned the use of workshops which are practice-based as the best way to acquire employability skills. It was also noted that teaching is dominated by the lecture methods, which are less likely to comprehensively develop employability skills. There is therefore a need for VET institutions to use other interactive teaching methods in the form of projects, project-based learning, demonstrations and field trips at NVTA Level 2 and Level 3.

The high number of subjects led students to consider VET to be no different from general secondary education since more time is spent on classroom activities versus practising in workshops. VET in Tanzania has little attachment with industries causing students to complete VET without special orientation. In this regard, study findings have shown VET in Tanzania to be more theoretical than practical-oriented. Consequently, graduates who are produced in VET institutions find themselves lacking employability skills.

Based on the study findings, recommendations for action and further studies are offered. The recommendations for actions are meant to enable VET students to be well-prepared for the world of work. In this regard, students should be independent to have self-motivation in studying hard to gain enough skills for self-employment and employability skills attributes. Students should also engage in volunteering activities to enable them to build a working spirit, and confidence and establish relevant networks and working collaborations with industry practitioners. It might also be important for students to advance to NTVA level 3 and higher levels along the national qualification framework since more employability skills are developed as one advances the framework levels.

At the institutional level, the recommendations for action are meant to improve teaching methods and more emphasis should be granted to the practical-based approaches. Field trips, engaging in minor and major projects and interacting with other VET institutions, should be emphasised. Moreover, the professional development for teachers has to be emphasised to enable them to learn new advancements, especially the ones arising from technological advancements. In further research, the study recommends the need to evaluate the effectiveness of the VET CBC towards the development of employability skills and the challenges VET teachers encounter along implementing the CBT curriculum. On the lessons, other VET institutions in the region can adopt the study mechanisms to enhance the employability of their students.

REFERENCES

Ahmad, N., Idris, A., & Kenayathulla, H. (2019). The gap between competencies and importance of employability skills: Evidence from Malaysia. *Emerald Insight, 13*(2), 97–112. https://doi.org/10.1108/HEED-08-2019-0039

Airasian, P., Gay, I. R., & Mill, G. (2012). *Education research competencies for analysis and application.* Pearson Education, Inc.

Akhyar, M., Budiyomo, Fajaryati, N., & Wiranto, T. (2020). The employability skills needed to face the demands of work in the future: Systematic literature reviews. *DE GRUYTER, 10*(1), 595–603. https://doi.org/10.1515/eng-2020-0072

Al-esmail, R., Eldabi, T., Hindi, N., Irani, Z., Kapoor, K., Osmani, M., & Weerakkody, V. (2015). Identifying the trends and impact of graduate attributes on employability: A literature review. *Tertiary Education and Management, 21*(4), 367–379. https://doi.org/10.1080/1358388 3.2015.1114139

Allais, S., Lotz-Sisitka, H., McGrath, S., Monk, D., Openjuru, G., Ramsarup, P., Russon, A., Wedekind, V., & Zeelen, J. (2020). Vocational education and training for Africa development: A literature review. *Journal of Vocational Education and Training, 72*(4), 465–487. https://doi.org/10.1080/1363682 0.20191679969

Ananiadou, K., & Claro, M. (2009). *21st-century skills and competencies for new millennium learners in OECD countries.* (No. 41). OECD Publishing. https://doi.org/10.1787/218525261154

Andreoni, A. (2018). Skilling Tanzania: Improving financing, governance and outputs of the skills development sector (issue October).

Anindo, J. (2016). *Institutional factors influencing acquisition of employable skills by students in public technical and vocational education and training institutions in Nairobi county.* The University of Nairobi.

Atistsogbe, K., Mama, N., Pari, P., Rossier, J., & Sovet, L. (2019). Perceived employability and entrepreneurial intention across university students and job seekers in Togo: The effect of career adaptability and self-efficacy. *Frontiers in Psychology, 10.* https://doi.org/10.3389/fpsyg.2019.00180

AU. (2007). Strategy to revitalize technical and vocational education and training (TVET) in Africa COMEDAFII+.

Audu, R., Kamin, Y., & Musta'amal, A. (2014). Assessment of the teaching methods that influence the acquisition of employability skills of mechanical engineering trade students at the technical college level. *International Journal of Physical and Social Science, 4*(9), 363–377.

Bashir, S., Tan, H and Tanaka, N. (2016). Skill use, skill deficits, and firm performance in formal sector enterprises evidence from the Tanzania enterprise skills survey, 2015. WB. https://doi.org/10.1596/1813-9450-7672.

Basnet, K., Eun, T., & Kim, J. (2009). Issues and challenges of technical education training TVET in Nepal. *Journal of The Korean Institute of Industrial Educators, 34*(2), 379–395.

Becker, G. (1962). Investment in human capital: A theoretical analysis. *Journal of Political Economy, 5*, 9–49.

Behrendt, M., & Franklin, T. (2014). A review of research on school field trips and their value in education. *International Journal of Environment and Science Education, 9*(1), 235–245. https://doi.org/10.12973/ijese.2014.213a

Bello, H., Bin Saud, M., Buntat, Y., & Shuaib, B. (2013). ICT skills for technical and vocational education graduates' employability. *World Applied Sciences Journal, 23*(2), 204–kke207. https://doi.org/10.5829/idos.wasj.2013.23.02.588

Bennett, D., MacKinnon, P., & Richardson, S. (2016). *Enacting strategies for graduate employability: How a university can best support students to develop generic skills part A.* Australian Government, Office for learning and teaching, Department of Educational and Training.

Brinkley, I., Clayton, N., & Wright, J. (2010). *Employability and skills in the UK: Redefining the debate.* The Work Foundation.

Chapman, E., & Young, J. (2010). Generic competency framework: A brief historical overview. *Education research and perspective.* Accessed Mar 20, 2022, from https://eric.ed.gov/?id=EJ945700.

Chen, Y.-C., & Ni, A. (2016). A conceptual model of information technology competence for public managers: Designing relevant MPA curricula for effective public service. *Journal of Public Affairs Education, 22*(2), 193–212.

Chinien, C. (2003). *The use of ICTs in technical and vocational education and training.* UNESCO.

Coetzee, M., & Potgieter, I. (2013). Employability attributes and personality preferences of post graduates' business management students. *The Journal of Individual Psychology, 39*(1), 1064–1074.

Comyn, P. (2018). Skills, employability and lifelong learning in the sustainable development goals and the 2030 labour market. *International Journal of Training Research, 16*(3), 200–217. https://doi.org/10.1080/14480220.2018.1576311

Cornford, R. (2006). Making the generic skills more than a man traits in vocational education policy. *Australian Association for Research in education conference.*

Creswell, J., & Plano, V. (2018). *Mixed methods research* (3rd ed.). SAGE Publication.

Curtis, D., & McKenzie, P. (2002). *Employability skills for Australian industry: Literature review and framework development report to Business Council of Australia, Australian Chamber of Commerce and Industry.* Accessed May 3, 2022, from http://hdl.voced.edu.au/10707/40939

Dasmani, A. (2011). Challenges facing technical institute graduates in practical skills acquisition in the upper east region of Ghana. *Asia Pacific Journal of Cooperation, 12*(2), 67–77.

Eicker, F., Haseloff, G., & Lennartz, B. (2016). *Vocational education and training in sub-Saharan Africa: Current situation and development.* W. Bertelsmann Verlag GmbH & KG.

Ekwue, C., Ojuro, I., & Udemba, F. (2019). Strategies for improving employability skills acquisition of business education students. *Nigerian Journal of Business Education (NIGJBED), 6*(1), 94–106.

Elliot, D., Hall, S., Lewin, J., & Lowden, K. (2011). *Employer's perception of the employability skills of new graduates.* Edge Foundation.

Fahimirad, M., Feng, J., Kotamjani, S., Kumar, N. P., & Mahdinezhad, M. (2019). Integration and development of employability skills into Malaysian higher education context: A review of the literature. *International Journal of Higher Education, 8*(6), 26–35. https://doi.org/10.5430/ijhe.v8n6p26

Foshay, R., & Silber, H. (2009). *Handbook of improving the performance in the workplace place, instruction design and training delivery.* John Wiley & Sons.

Fulgence, K. (2016). *Employability of higher education institution graduates: Exploring the influence of entrepreneurship education and employability skills development programmes activities in Tanzania.* Doctoral Dissertation. University of Siegen.

Hung, J., & Ramsden, M. (2021). The application of human capital theory and educational signalling theory to explain parental influence on the Chinese population's social mobility opportunities. *Social Science, 10*(362), 1–7. https://doi.org/10.3390/socsci

ILO. (2015). *Sustainable development goals.* ILO.

ILO. (2019). *State of skills Tanzania.* ILO.

ILO. (2020). *Global employment trends for youth 2020 global employment trends for youth 2020.* ILO.

Inyiagu, E. (2014). Challenges facing technical education in Nigeria. *Journal of Education Policy and Entrepreneurial Research, 1*(1), 40–45.

Ismail, A., & Abiddin, N. Z. (2014). Issues and challenges of technical and vocational education and training in Malaysia towards human capital development. *Middle-East Journal of Scientific Research (Innovation Challenges in Multidisciplinary Research & Practice), 19*, 07–11.

Ismail, K., Mohd, R., & Mohd, Z. (2018). Challenges faced by vocational teachers in public skill training institutions: A reality in Malaysia. *Journal of Technical Education and Training, 10*(2), 13–27. https://doi.org/10.30880/jtet.2018.10.02.002

Kissaka, M. M., Mtebe, J., Raphael, C., & Stephen, J. K. (2020). Promoting youth employment through information and communication technologies in vocational education in Tanzania. *Journal of Learning for Development, 7*(1), 90–107.

Li, W. (2012). *Education and training for rural transformation: Skills, jobs, food and green future to combat poverty.* UNESCO International Research and Training Centre for Rural Education (UNESCO INRULED).

Lice. (2019). *Managing facilitation of employability of Vocational education graduates in Latvia.* University of Latvia. https://doi.org/10.13140/RG.2.2.26362.52163

Lindsay, C., & McQuaid, R. (2005). The concept of employability. *Urban Studies, 42*(2), 197–219. https://doi.org/10.1080/0042098042000316100

Lynch, M. (2018). How digital media literacy impacts today's classroom. *The Tech Edvocate.*

Ma, X., Nakab, A., & Vidart, D. (2020). *Human capital investment and development: The role of On-the-job Training.* Accessed Oct 20, 2021, from acsweb.ucsd.edu

Martín-Garin, A., Millán-García, J. A., Leon, I., Oregi, X., Estevez, J., & Marieta, C. (2021). Pedagogical approaches for sustainable development in building in higher education. *Sustainability, 13*, 10203. https://doi.org/10.3390/su131810203

Mbugua, Z. K., Muthaa, G. M., & Sang, A. K. (2012). Challenges facing technical training in Kenya. *Creative Education, 3*(1), 109–113. https://doi.org/10.4236/ce.2012.31018

Medina, R. (2010). Upgrading your self-technical and non-technical competencies. *IEEE Potentials, 29*, 10.

Munishi, E. (2016). Factors contributing to lack of employable skills among technical and vocational education (TVET) graduates in Tanzania. *Business Education Journal, I*(2), 1–19. Accessed Mar 5, 2022, from http://www.cbe.Ac.tz/bej FACTORS

NACTE. (2012). *NACTE and the quality in technical education: A handbook for morning the quality in the technical institutions in Tanzania.* MSM.

NACTVET. (2022). Striving the world-class excellence in technical and vocational education and training. *NACTVET Newsletter, 1*–50.

Nagarajan, S., & Edwards, J. (2014). The relevance of university studies to professional skills requirements of IT workplaces: Australian IT graduates' work experiences. *Journal of Perspectives in Applied Academic Practice, 2*(3), 48–61.

NCVER. (2003). *Defining generic skills.* National Centre for Vocational Education Research.

Nkirina, S. P. (2010). The challenges of integrating entrepreneurship education in the Vocational training system. An insight from Tanzania's Vocational Education Training Authority. *Journal of European Industrial Training, 34*(2), 153–166.

Nwazor, C., & Onokpanu, O. (2016). Strategies considered effective for transforming business education programs to the needs of the 21st-century workplace. Delta state Nigeria. *Africa Journal of Education and Practice, 1*(2), 74–82.

Obeta, C., Onoh, C., & Rufai, A. (2013). Human capital development in technical vocational education (TVE) for Sustainable National Development. *Journal of Education and Practice, 4*(7), 100–106.

Pallawi, Kumar, M., & Singh, S. (2022). Employability and entrepreneurial skills in the digital era: a critical review. *Academy of Marketing Studies Journal, 26*(4), 1–11.

Pangestu, F., & Sukardi. (2019). Evaluation of the implementation of workshops and laboratory management in vocational high schools. *Journal Pendelikon Vokasi, 9*(2), 172–184. https://doi.org/10.21831/jpv.v9i2.25991

Pirzada, K. (2013). Measuring the relationship between digital skills and employability. *European Journal of Business and Management, 5*(24), 124–133.

Rojewski, J. (1997). Effects of economically disadvantaged status and secondary vocational education on adolescent work experiences and post-secondary aspirations. *Journal of Vocational Education Research, 14*(1).

Saleh, I. (2017). The role of vocational training in reducing the unemployment rate in the outlying states of the United States of America. *TVE @ Asia, 9*(1), 1–14. Accessed June 30, 2021, from http://www.tvet-online.asia/issue9/saleh_tvet9.pdf

Setiadi, R., Sumbodo, W., & Yudiono, H. (2019). The role of industry partners to improving student competency in vocational high school. *Journal of Physics: Conference, 1387*(1), 1–7. https://doi.org/10.1088/1742-6596/1387/1/012031

Sweetland, S. (1999). Human capital theory: Foundations of the field of inquiry. *Review of Educational Research, 66*(3), 341–359. https://doi.org/10.2307/1170527

Tambwe, M. A. (2017). Challenges facing the implementation of Competency-Based Education and Training (CBET) system in Tanzanian Technical Institutions. *Education Research Journal, 7*(11), 277–283. Accessed Apr 11, 2022, from https://www.researchgate.net/publication/331070860/Challenges

Tan, L. C., & French-Arnold, E. (2012). *Employability of graduates in Asia: An overview of Case studies.* Bangkok Asia and Pacific Regional Bureau for Education.

Tokarcikova, E., Kucharcikova, A., & Malichova, E. (2020). Importance of technical and business skills for future IT professionals. *Amphitheatre Economic, 22*(54), 567–578.

Tripney, S., & Hombrados, J. (2013). Technical and vocational education and training (TVET) for young people in low- and middle-income countries: a systematic review and meta-analysis. *Journal of Empirical Research in Vocational Education and Training, 5*(3), 1–14. https://doi.org/10.1186/1877-6345-5-3

UNECSO. (2015). Proposal for the revision of the 2001 revised recommendation concerning Technical and vocational education. *UNESCO, General Conference, 38th, 2015.*

UNESCO. (2020). *Secondary education, vocational pupils.* UNESCO.

UNICEF. (2017). *Education equality and quality in Tanzania.* UNICEF.

United Republic of Tanzania (URT). (1999). *The Tanzania development vision 2025.* Ministry of State and Vice Chairman.

URT. (2013). *Tanzania Technical and Vocational Education and Training Development Programme (TVETDP) 2013/2014–2017/2018.* Ministry of Education and Vocational Education.

URT. (2014). *Education and training policy.* Ministry of education and Culture.

URT. (2018). *Education sector development plan (2016/17–2020/21) TANZANIA mainland.* Ministry of Education, Science and Technology.

VETA. (2019). *Tracer study report for 2010–2015 vocational education and training graduates.* VETA Head Office.

VETA. (2022). *VETA registered centers.* VETA. Accessed Feb 1, 2022, from https://www.veta.go.tz/vetcat/center/listCenter

Woyo, E. (2013). Challenges facing technical and vocational education and training institutions in producing competent graduates in Zimbabwe. *Open Journal of Education, 1*(7), 182–189. https://doi.org/10.12966/oje.11.03.2013

Yorke, M., & Knight, P. (2004). *Learning, curriculum, and employability in higher education.* Routledge Falmer.

Higher Education: *Towards a Model for Successful University-Industry Collaboration in Africa*

Ngepathimo Kadhila ⓘ, *Kyashane Stephen Malatji* ⓘ, *and Makwalete Johanna Malatji* ⓘ

1 INTRODUCTION

World Bank (2012) opines that most governments throughout the world stressed the importance of strengthening the industry and research partnerships, as well as concentrating on industrial attachments that may enable fresh graduates to meet the industry's changing needs. In addition,

---

N. Kadhila (✉)
University of Namibia, Windhoek, Namibia
e-mail: nkadhila@unam.na

K. S. Malatji
School of Interdisciplinary Research and Graduate Studies, University of South Africa, Pretoria, South Africa
e-mail: emalatks@unisa.ac.za

© The Author(s), under exclusive license to Springer Nature Switzerland AG 2024
P. Neema-Abooki (ed.), *The Sustainability of Higher Education in Sub-Saharan Africa*, Sustainable Development Goals Series,
https://doi.org/10.1007/978-3-031-46242-9_11

academic courses will be more market-oriented to help students move into the workforce and increase their overall employability. To improve the efficiency of structured industrial attachment programmes, HEIs have created a common policy framework in conjunction with industry stakeholders. Most governments are increasingly focusing on fostering science-industry collaboration and creating high-tech industries.

Prigge and Torraco (2006) encourage universities to form strategic partnerships with international research institutes and foreign universities to improve research and development (R&D) activities, particularly those involving new and emerging technologies. Such partnership broadens the scope of innovations in very collaborative ways which allows institutions to benchmark with one another. Malatji et al. (2018) concur that collaborations between stakeholders create space for the learner from one another and come up with innovative strategies. Salleha and Omar (2013) propose three strategies for achieving the objectives. The first is to introduce more industry attachment programs, which will allow academics to share their experience and ideas, therefore improving the quality of their research. The second is to enhance the governance of research activities by increasing the management of intellectual property created in institutions. The third one is that industry partnership in R&D activities at universities is expanding the function of centres of excellence to promote and expedite the commercialization of discoveries and new technologies. Salleha and Omar (2013) perceive that active collaboration between universities and industry should be encouraged to boost any country's economy. Academia, industry, and government in Africa should collaborate to improve people's knowledge and abilities so that they can contribute positively to the nation's development. In this context, the Triple Helix Model of Innovation is proposed in this chapter to illustrate the dynamics that arise from the interplay of the three principal institutional realms in an economy: academia, industry, and government.

Collaboration between universities and industries is essential for societal concerns like job creation and economic growth to be solved. Yet, there are several obstacles to effective university-industry collaboration in

---

M. J. Malatji
Department of Curriculum and Instructional Studies, College of Education, University of South Africa, Pretoria, South Africa
e-mail: makwalete.malatji@up.ac.za

Africa, including a lack of trust, insufficient money, poor communication, and cultural differences. Successful university-industry partnerships in Africa can aid in economic growth, job creation, and the achievement of Sustainable Development Goals (SDGs) claims a report by the Team et al. (2013). The research also emphasizes the necessity of creating clear goals, promoting communication, and safeguarding intellectual property rights in university-industry partnerships.

Team et al. (2013) pledge by creating clear objectives, promoting communication, fostering trust, making sure there is sufficient financing, developing capacity, protecting intellectual property rights, and managing cultural differences, also acknowledging the significance of these elements for effective university-industry partnership in Africa in 2021 in the African Development Bank.

Building trust, promoting communication, defining clear goals, having enough money, growing capacity, protecting intellectual property rights, and managing cultural differences are all necessary for a successful university-industry partnership in Africa. An enabling environment for successful university-industry cooperation in Africa can be created through cooperation between institutions, industrial partners, governments, and international organizations.

2   THEORETICAL PERSPECTIVE

This chapter is based on Etzkowitz et al. (2007) theory of social capital and Ansell and Alison (2007) collaborative governance theory. The theory of social capital asserts that social ties are resources that can lead to human capital development and accumulation (Machalek & Martin, 2015). It also advocates for stronger inter-organizational ties to obtain and update external information constantly, as well as use it for long-term competitive advantage. The Triple Helix Model of Innovation bears this out. Three types of Triple Helix structures have been researched by organizations. The state or government dominates the university and industrial structures in Triple Helix 1. The university and industry structures are governed by the state. The Triple Helix 2 depicts the three organizations as separate and distant from one another, with relatively few interactions across strong borders (Etzkowitz et al., 2007). This structure illustrates the three actors' limited relationships. Each institutional sphere in Triple Helix 3 retains its traits while simultaneously taking on the roles of the others. The Triple Helix system's evolutionary process is depicted graphically below.

The process of building, guiding, facilitating, managing, and monitoring cross-sectoral organizational structures to address public policy problems that cannot be easily addressed by a single organization or the public sector alone is known as collaborative governance. Joint efforts, reciprocal expectations, and voluntary participation among officially autonomous entities from two or more sectors—public, for-profit, and non-profit—are characterized by these arrangements to exploit the distinctive traits and resources of each. While education is indispensable for human capital development, and equal access to higher education (HE), job-skills mismatch, and delivery of high-quality education continue to be major concerns among parents, government, industry, and policymakers, the collaboration between academia, industry, and the government is paramount, and hence the focus of this study. The interplay between the academe, industry, and government (Triple Helix Model of an Innovation) is depicted by the Triple Helix Model in Fig. 1.

Figure 1 depicts the Triple Helix as apt to the dynamics arising from the interactions between the three principal institutional spheres in an economy—academe, industry, and government. Academe, industry, and

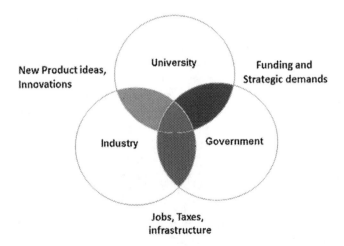

Fig. 1 Triple Helix Model showing the interplay between academia, industry, and government (Triple Helix Model of an innovation). Source: Hermosura (2019: 801)

government have equal roles in a knowledge-based society in stimulating innovation. A stable regulatory framework is necessary, but it is not sufficient. A university's transformation from teaching to research to an entrepreneurial institution is unparalleled. Changes in the regulatory environment, tax incentives, and the allocation of public venture capital are all things that the government may do to help foster discoveries. In developing training and research, the industry assumes the function of the university. If knowledge-based industries don't yet exist, university-government relations can help kick-start their development; if they are, they can help them expand. In this connection, the academe's primary responsibility is to ensure that the knowledge it generates is beneficial enough to be widely shared and implemented, resulting in community and national growth. The industry would be interested not just in the use of knowledge, but also in knowledge production and sharing as a means of generating income. The government's role would be to facilitate university-industry relations by establishing proper policy frameworks for identifying, among other things, research and development priorities in light of current socio-economic conditions and allocating money among these objectives (Saad & Girma, 2011).

3   THE INTERNATIONAL CONTEXT

University-industry collaboration has become a defining aspect of the twenty-first-century knowledge economy, particularly in developed nations where it plays a large role in the innovation process (Perkmann & Walsh, 2007). Similarly, Tumuti et al. (2013) counter that collaborations between universities and industry are becoming increasingly widespread around the world. The importance of symbiotic relationships between universities and industry in fostering a country's economic prosperity cannot be overemphasized. Since there is a growing competition as a result of globalization and continuous technical improvements, colleges must engage with the industry to improve knowledge dissemination, promote research and development, patent innovations, and build the nation's organizational ability. As a result, it has become increasingly evident that tight university-industry collaboration is required to ensure national economic growth (Ekaterina et al., 2020). According to Tumuti et al. (2013), university-industry partnerships have risen to the top of the agenda in HE policymaking, as well as in the national and institutional economic environments. The strategic role of HE in national and continental economic development is imperative in regard to the context of knowledge-intensive economies, specified by their ability to upgrade labour force skills and knowledge, as well as contribute to the production and processing of innovation through technology transfer.

Tumuti et al. (2013) submit that universities have grown to become catalysts in a nation's progress as a result of these connections, as their functions are not limited to human capital development but also include technology transfer, research and development, and innovation. Universities have three main responsibilities to play in this collaborative process. Firstly, they engage in a broad scientific research process, which has a long-term impact on the industry's technological frontier. Secondly, they contribute to the development of knowledge that is directly useful to industrial production (prototypes, new processes, etc.). Thirdly, universities give significant human capital inputs to industrial innovation processes, either through the education of graduates who go on to become industry researchers or through people's mobility from universities to businesses (Hermosura, 2019). As a result, collaborations and partnerships are becoming more common at universities for lifelong learning. Internationally, universities are increasingly acknowledged as major drivers of economic development, and the benefits of university-industry partnerships can be observed in the world's factor and efficiency economies (Hadidi & Kirby, 2017).

University-industry partnerships are unsubstantiable for boosting universities' ability to conduct high-quality, relevant research and enhancing the industry's ability to compete worldwide. Collaboration is commonly viewed as a means of achieving some of these goals and fostering a greater degree of competitiveness (Salleha & Omar, 2013). Meanwhile, Hadidi and Kirby (2017) consider that with the expansion in university-industry collaboration in recent years, it has become clear that there are various types of collaboration and that they often address knowledge commercialization (patenting, licensing, spin-off ventures, incubators, etc.); and academic engagement (research collaboration, contract research, consulting, etc.) between academics and industry.

It is also acknowledged that companies in different industrial sectors use different types of technological and market information and value access to university-developed knowledge differently (Perkmann & Walsh, 2007). Similarly, Salleha and Omar (2013) note that in countries with little R&D commitment, there is no motivation for businesses to work with universities, and those who do are those with innovation plans.

4   THE AFRICAN CONTEXT

When universities collaborate with industry and government, they can provide powerful and unique inputs. However, to do so, African universities must become autonomous institutions capable of designing and implementing strategic initiatives to collaborate with industry (AU, 2013). They must be able to make strategic decisions about which new domains of knowledge to develop and act towards establishing an interdisciplinary workplace. They must be capable of establishing high-quality support facilities for university-industry-government relationships and better incentivizing individual academics to collaborate with the industry.

In light of the current studies, university research productivity appears to be a key factor in predicting the degree of industry collaboration. The organization culture of universities has been highlighted as an important dimension in their technology transfer success when it comes to elements that allow university-industry relationships (AU, 2013; The World Bank 2012). Others have looked into the role of geographical proximity in the formation of university-industry partnerships. Although proximity to universities can be motivating, the quality of HEIs is the most important determinant factor for industries to engage with universities in their region (Ruwoko, 2021).

Even when regional and country differences are taken into consideration, university research capability in Africa appears to be severely constrained. Ruwoko (2021) defines research capacity as "the institutional and legal frameworks, infrastructure, investment, and adequately competent personnel to do and publish research." Research capacity varies widely across the continent. Indeed, research by the RAND Corporation found that African countries, except for South Africa, Egypt, Mauritius, and Benin, were among the scientific laggards (AU, 2013; The World Bank 2012). According to a 2007 report, African HE lacks ability not only at the system and institutional levels but also at the level of individual academics (Ogunwale, 2021).

The poor research capability of universities in developing countries, notably on the African continent, has been frequently described in the literature (The World Bank 2012). Universities in Africa have long struggled to support themselves due to a lack of public resources. Building research programmes in relevant domains of science and technology that would be of interest to industry is a challenge for universities. A more persistent research function is hampered by a general lack of research capacity and limited R&D funding (Ruwoko, 2021). Universities are unable to train a bigger number of scientists or retain productive researchers due to structural constraints. Universities are often unable to begin and sustain research projects due to a lack of financing and support for their research goal (AU, 2013; The World Bank, 2012).

Individual African academics appear to have been particularly badly struck by difficult working conditions, even though these issues are not specific to Africa (Ogunwale, 2021). To begin with, academic staff in African universities have very minimal resources, which results in a lack of desire. They lack the resources essential to serve as mentors, train graduate students, and contribute to knowledge development and dissemination. They are afflicted by low pay and overcrowding in their institutions. Since African academics are reliant on academic research conducted in rich nations, their location on the "periphery" of knowledge creation (Ruwoko, 2021) makes it difficult for them to contribute significantly to innovation. As a result, African university professors face numerous hurdles in contributing to innovation.

The majority of current studies on university partnerships in Africa are based on grey literature from various national, regional, and international institutions (AAU, 2012a, 2012b; The World Bank, 2012, 2013). Even in the extant literature, university-industry collaborations are discussed from the standpoint of universities. For example, some studies analyse university institutional capacity to handle business linkages (AAU, 2012a, 2012b; The World Bank, 2012), while others investigate similar collaborations at the national level or compare university-industry partnerships across countries. This dearth in the literature on university-industry collaboration is especially troubling as many international donors recognize the value and contribution of HE to economic development (Ogunwale, 2021).

Research on the status of university-industry partnerships in Africa (AAU, 2012a, 2012b) highlighted important facts that serve as a worrisome alert including the low percentage of academic staff with PhD training and qualifications, as well as the brain drain of trained scientists, which limits university research output; several African universities have attempted to create linkages with corporations by creating offices and staff roles in charge of such activities. However, such offices lack the necessary resources and expertise to effectively manage industrial relationships and technology transfer; also, academic institutions have a limited number of research parks and technological incubators. Only a small minority of the universities polled said they were involved in science park management and technology transfer. Support for the establishment and management of business incubators and research parks, according to studies, would respond to the needs and goals of African universities (Lubbe et al., 2021).

The nature of Africa's industry has an impact on the possibilities for university collaboration. In general, the informal sector dominates African economies, employing between 50% and 75% of the workforce (AU, 2013). It even has a majority share in industries including manufacturing, commerce, and mining (Ogunwale, 2021). In terms of employment, the remaining formal and structured sector accounts for only a small portion of the entire economy. African industries are primarily extractive and based on natural resources. The majority of these industries are populated by subsidiaries of multinational businesses with headquarters in more developed nations, where they do research and development. As a result, multinational companies in Africa use foreign technology. As a result of this situation, a possible area of collaboration between local universities and businesses that may be able to form collaborations with HEIs has been eliminated. This also contributes to the continent's overreliance on foreign technology, hindering the development of indigenous inventions (Lubbe et al., 2021).

Another study concentrating on science parks and business incubators gives a background for the organizational structures in place that may facilitate university-industry links, albeit not all types of partnerships would rely on such structures, as shown above. According to the study, scientific parks and business incubators are a new phenomenon in Africa, with a few exceptions. The one-way transfer of knowledge from university to the industry is frequently emphasized in the literature on these partnerships in Africa.

5   The Perspective of the Industry's Engagement with African Universities

In the last decade, countries and governments around Africa have engaged in strengthening the relationship between universities and industries with the view to boosting the economies of different African countries. Dill and Van Vught (2010: 43) are of the view that "stimulating technical advance in industries is a necessity to promote economic growth." Therefore, policymakers, business owners, and government officials in Africa should stimulate universities to integrate entrepreneurial knowledge within their curriculum and engage more actively and productively. As universities become more involved in promoting economic development, there is a push from various governments for more relevant research and training to strengthen partnership with industries (Welch, 2005). One way to address this call for relevance is by encouraging more linkages between HEIs and the business sector (Martin, 2000). Such linkages are particularly relevant in Africa as the majority of universities were created with a mission to contribute to nation-building. Initially focused on training the workforce for the newly independent countries, HEIs are now asked to contribute to national economic development. This chapter aims to establish the importance of university linkages with the economic sector in Africa. In this regard, the literature on university-industry partnerships remains concentrated on advanced industrial economies, while very little appears to be known about university-industry linkages in Africa.

Looking at the literature, several studies are conducted on university-industry partnerships. However, most of these studies are focusing on Western countries which leaves a lot of knowledge gaps in the context of African countries. Geiger and Sá (2009) focus their university-industry focus on e-Learning or technological innovations. In most cases, universities contribute to industries by conducting research in the area/field of technology relevant to the industry to provide community engagement projects by assisting local firms, in-service technological training of professionals, and supporting faculty to engage in consulting and commercialization activities.

In the corporate sector, there is a trend in high technology industries towards more and closer linkages with university research. Firms' readiness to seek out multiple sources of knowledge is viewed as a non-such for their success in fiercely competitive markets (Chesbrough, 2003). This drives large companies to establish more partnerships with research institutions. Typologies have been developed to categorize the types of partnerships existing between universities and industries. Cohen et al. (2002) recount that partnership should include several types of activities through which academic research interacts with industry. Such activities should be used to describe partnerships in more advanced industrial economies, which should be followed by a memorandum of understanding. The focus of this chapter articulates the nature of collaboration and partnership of the university-industry in the African context. Reflecting on the literature engaged in this chapter, universities' research productivity appears to be indispensable in shaping the extent of partnerships with the industries.

HIGHER EDUCATION: *TOWARDS A MODEL FOR SUCCESSFUL...* 261

When it comes to ensuring an effective university-industry partnership, Archer (1995), in Social Realist theory, talks of the organizational structure of universities as an important dimension in their technology transfer performance. Bercovitz et al. (2001) assessed the role played by geographical proximity in the development of partnerships between universities and industry. Although geographical proximity can have a motivating factor, the literature broaches that the quality of HEIs is the most important determinant factor for industries to engage with universities in their region (Vedovello, 1997; Laursen et al., 2011). In the African context, university research capacity appears to be very limited, even considering regional and country variations. Research capacity, defined by Volmink (2005), as "comprising the institutional and regulatory frameworks, infrastructure, investment, and sufficiently skilled people to conduct and publish research," varies greatly across the continent. In the deed, a study by the RAND Corporation revealed that, except for South Africa, Egypt, Mauritius, and Benin, African countries were part of a group of scientific laggards (Rand Corporation, 2001). The 2007 report recognized that African HE lacks capacity not only at the system and institutional levels but also at the level of individual academics (Jones et al., 2007). The literature has also often described the limited research capacity of universities in developing countries, including in the African continent (Altbach, 2006). At the institutional level, universities in Africa have long been facing funding difficulties due to limited state resources. Universities face constraints in building research programmes in relevant fields of science and technology that would be of interest to industry.

Generally, weak research capacity and insufficient research development funding inhibit a more sustained research role (Atuahene, 2011). These structural issues prevent universities from training a larger number of scientists and retaining productive researchers. With limited funding and support for their research mission, universities are usually struggling to initiate and sustain programmes of research (Munyoki et al., 2011; Mwiria, 1995). Individual academics in Africa appear to have been particularly hard hit even by constraining working conditions, although these challenges are not unique to Africa (e.g., Enders & Teichler, 1997; Welch, 2005). First, academic staff in African universities have very limited resources, leading to lower motivation. They do not have the necessary resources to play the role of mentors, train graduate students, and contribute to knowledge production and dissemination to benefit communities and industries. They are faced with low remuneration and overcrowded institutions. Moreover, the position of African academics at the "periphery" (Altbach, 2006) of knowledge production makes it difficult for them to substantially contribute to innovation because they are reliant on academic research performed in developed countries. Hence, university faculty in Africa face many challenges in contributing to innovation. Current studies on university partnerships in Africa mostly consist of grey literature in the form of reports and conference publications of various national, regional, and international organizations (AAU, 2012a, 2012b; Massaquoi, 2002; The World Bank, 2009, 2010). Even in the existing literature, university-industry partnerships are addressed from the perspective of HEIs. For instance, some studies assess the existence of institutional capacity in universities to handle business linkages (Martin, 2000; AAU & AUCC, 2012), and others explore these partnerships at the national level (Jansen, 2002; Adeoji, 2009) or make a cross-country comparison of university-industry partnerships (Mwiria, 1995). This gap in the literature on university linkages with industry is particularly concerning as many international donors have also acknowledged the importance and contribution of HE economic development (Koehn, 2012; Yusuf et al., 2009). A recent report on the state of university-industry linkages in Africa revealed relevant findings that serve as a cautionary warning (AAU & AUCC, 2012):

- University research output is limited by the low percentage of academic staff with PhD training and qualifications and the brain drain of qualified scientists.
- Many African universities have attempted to foster linkages with firms through the creation of offices and staff positions in charge of such affairs. However, such offices lack the material resources and expertise to handle industry partnerships and technology transfer effectively.
- There is a low number of science parks and technology incubators in academic institutions. Only a small percentage of universities surveyed reported being involved in managing science parks and engaging in technology transfer.
- The study exhorts that support for establishing and managing business incubators and science parks would respond to the needs and priorities of African universities. The nature of the industry in Africa also relates to the potential for university partnerships.

Generally, African economies are dominated by the informal sector, which employs between 50% and 75% of the workforce (African Union, 2008). It even represents the dominant share in sectors as manufacturing, commerce, and mining (Sparks & Barnett, 2010). The remaining formal and structured sector only represents a fraction of the overall economy in terms of employment. Moreover, African industries are mainly extractive and natural resources-based. Most of these industries are populated by branches of multinational corporations with headquarters in more economically advanced countries, where they perform their research and development activities. Therefore, the technology used by multinational companies in Africa is imported. This situation has the consequence of removing a potential area of collaboration between local universities and industry that may have the capacity to engage in partnerships with HEIs. Also, this contributes to the overreliance of the continent on imported technology, preventing the development of local innovations. Another recent study focusing on science parks and business incubators provides a context of the organizational structures in place that may support university-industry linkages, although as seen above, not all forms of partnerships would depend on those structures. This study showed that science parks and business incubators are an emergent phenomenon in Africa, concentrated in a few regions (AAU, 2012a, 2012b). Key findings include:

- There is evidence of a growing interest in the creation and support of university-related science parks and business incubators in recent years.
- Multiple organizational models are being experimented with. Universities have established their units, usually in partnership with other stakeholders, and have also collaborated with ventures created in their regions by other governments and industries.
- University researchers and students are an important audience for science parks and business incubators, some of which mostly serve these groups.
- A focus on technology-based firms was common among the parks and incubators identified, particularly in ICT, biomedical sciences, and engineering.
- Most units identified were small and had fewer than ten tenants. However, most science parks were operating under capacity, and as relatively new ventures, had room to grow.

- Consistent with previous studies, multiple barriers remain to facilitating university-industry research development collaborations. The most important issues identified include ambivalent academic culture, lack of funding/financial incentives for research development partnerships, lack of industry interest in university partnerships, lack of industry capacity, and mismatch between university research, strengths, and regional industry sectors.
- Reflecting the somewhat incipient nature of many of the parks and incubators identified, most experienced difficulties related to the overall environment for business development, as well as with sustaining their budgets.

Despite these difficulties, the units displayed goals and sought to provide services similar to their peers internationally. This plugs that the availability of appropriate resources, such activities could be expanded. The literature on these partnerships in Africa often stresses the one-way transfer of knowledge from university to industry. The following quote illustrates the view that the benefits of these sorts of linkages are overwhelmingly on the side of the industry: "The ideas of technology transfer and university-industry linkage are related in the sense that the former deals with the transfer of ideas and skills between those who have them and those who need them, while the latter addresses the issue of the bond between generators of ideas and users of the ideas" (Munyoki et al., 2011). Rarely is the conversation about the contribution of industry to universities. This chapter has attempted to balance that point of view by looking at different angles on the issue.

6 Key Drivers of University-Industry Partnerships

The assumption that collaborative research efforts can lead to innovation has sparked interest in university-industry partnerships. For example, industry funding for a variety of programmes can have a significant impact on the overall success of these universities. Partnering can also open up new avenues for universities to rethink how they fund, produce, advertise, deliver, and support their education. However, a successful industry-university partnership must support each partner's missions and motivations. The improvement of teaching, access to finance, reputation promotion, and access to empirical data from industry are all common reasons for colleges to engage with industry (Prigge & Torraco, 2006).

The industry may collaborate with universities for a variety of reasons, including access to complementary technological knowledge (including patents and tacit knowledge), tapping into a pool of skilled workers, providing training to current or future employees, gaining access to the university's facilities and equipment, and gaining access to public funding and incentives; firms may also seek to reduce risks by sharing R&D costs, as well as to influence the overall telecommunications landscape (The World Bank, 2013).

According to World Bank (2013), the goals, scopes, and institutional arrangements of the various forms of university-industry partnerships vary. Collaboration might be intense or light, and it can focus on instruction or research. Formal equity partnerships, contracts, research initiatives, patent licensing, and other forms of collaboration range from human capital mobility, to publications, and encounters in conferences and expert groups, among others. It is also necessary to differentiate between short-term and long-term partnerships. Short-term collaborations are typically defined by on-demand issue solving with set outcomes and are typically expressed through contract research, consulting, and licensing. Long-term collaborations are often associated with joint projects and public-private partnerships in the name of privately funded university institutes or chairs, joint university-industry research centres, and research consortia), which allow businesses to contract for a core set of services and re-contract for specific deliverables on a flexible basis. Longer-term collaborations are more strategic and open-ended, giving a diverse platform on which businesses can create a stronger inventive capacity over time by leveraging university talents, processes, and resources (Prigge & Torraco, 2006). The World Bank (2013) identifies a typology of university-industry links, from higher to lower intensity as summarized in Table 1.

In line with Table 1, universities and industries in Africa could both gain from collaboration. These partnerships provide financial support for universities' teaching, research, and service purposes; broaden students' and faculty's experience; find major, fascinating, and relevant challenges; boost areas of economic growth; and increase student job prospects. Industry benefits from such partnerships because they gain access to the knowledge they didn't have before; they help with technology renewal and expansion; they increase access to students as possible employees; they expand pre-competitive research; and they use internal research capabilities (Prigge, 2005). While that human capital is a key determinant of firm performance, in the aggregate, human capital is also a key determinant of

Table 1 A typology of university-industry links, from higher to lower intensity

High (relationships)	Research partnerships	Inter-organizational arrangements for pursuing collaborative R&D, including research consortia and joint projects.
	Research services	Research-related activities commissioned to universities by industrial clients, including contract research, consulting, quality control, testing, certification, and prototype development.
	Shared infrastructure	Use of university labs and equipment by firms, business incubators, and technology parks located within universities.
Medium (mobility)	Academic entrepreneurship	Development and commercial exploitation of technologies pursued by academic inventors through a company they (partly) own (spin-off companies).
	Human resource training and transfer	Training of industry employees, internship programmes, postgraduate training in the industry, secondments to the industry of university faculty and research staff, and adjunct faculty of industry participants.
Low (transfer)	Commercialization of intellectual property	Transfer of university-generated IP (e.g., as patents) to firms (e.g., via licensing).
	Scientific publications	Use of codified scientific knowledge within the industry.
	Informal interaction	Formation of social relationships (e.g., conferences, meetings, social networks).

Source: World Bank (2013: 2)

a nation's productivity and economic prosperity (Aguinis & Vaschetto, 2011). Governments promote university-industry knowledge transfer initiatives because the commercialization of research has the potential to stimulate economic growth and ensure the relevance and accessibility of academic research to industry (Robertson et al., 2019). Commercialization is a prime example of generating academic influence on an institutional level since it represents immediate and measurable market acceptability for academic research outputs (Robertson et al., 2019). The availability of complementary resources (funding, human resources, knowledge, etc.) and relational drivers (trust, commitment, shared goals, and balancing of differing expectations) have been identified as the primary drivers of the firm–university relationships (Ekaterina et al., 2020).

University-industry knowledge transfer has been identified as one of the key drivers for university-industry partnerships. Prigge and Torraco (2006) offer the benefits of university-industry partnerships as follows:

- Financial support for the university's education, research, and service mission
- Broadened experiences for university students and academics
- Enhancement of regional economic development
- Increased employment opportunities for students
- Identification of significant, interesting, and relevant problems
- Access for the industry to the expertise they do not possess
- Aid in the renewal and expansion of industry technology
- Access for the industry to trained labour pool (students)
- Expansion of pre-competitive industrial research
- Leveraging of internal corporate research capabilities

These benefits to both the universities and the industry are deemed essential in maintaining a long-term relationship (Prigge & Torraco, 2006). Particularly, as public funding for HE continues to decline, universities will be forced to aggressively seek different sources of private funding through partnerships with industry to survive. Meanwhile, Robertson et al. (2019) advance that in recent years, researchers and business executives have paid more attention to knowledge transfer within university-industry relationships. Examples of successful university-industry knowledge transfer exist in strategic high-tech industries like pharmaceuticals and biotechnology (e.g., the Boyer-Cohen "gene-sequencing" rDNA technique), where university-based knowledge derived from research in specific areas has been transferred to industry for commercial exploitation and use. Licensing technology, which is an essential and expanding stage of the innovation process, is one way to achieve this type of transfer (McCarthy & Ruckman, 2017).

McCarthy and Ruckman (2017) tangle that due to the recognition that collaboration between industry and universities is becoming beneficial to both parties, on the one hand, the industry is increasingly implementing open innovation policies to improve access to and integration of external sources of information, resulting in a greater interest in collaborating with universities. On the other hand, since the 1990s, universities' strategic mission has shifted away from the traditional teaching and research missions and towards a "third mission" focused on better-addressing industry needs and contributing directly to economic growth and development (The World Bank, 2013).

7 Barriers to University-Industry Collaboration

Historically, industry and universities historically have a long tradition of collaborating on education, research, and innovation (Ekaterina et al., 2020). One of the well-known kinds of corporate is collaboration on innovation with universities (Perkmann & Walsh, 2007). These partnerships, on the other hand, are not without risk. Conflicts of interest between the university and industrial researchers, the withholding of information from colleagues, and the "undermining of academic norms" are all genuine possibilities that must be managed carefully in such collaborations (Prigge, 2005). University partnerships are complicated by the vast differences in organizational structures and cultures between academia and business, which generate specific barriers and may hinder the efficiency and efficacy of such collaborations (Ekaterina et al., 2020).

Barriers enlisted by Ekaterina et al. (2020) include a lack of understanding of external organization skills, a lack of relationships, and difficulties in locating the proper partner. Interaction of knowledge exchange and internal organizational learning processes produces innovative capabilities, which are vital for strategic partnerships. The other set of barriers is a lack of resources on both sides, for instance, financing, human resources, knowledge, and so on (Saad & Girma, 2011). Differences in organizational cultures visible in various levels of motivation, styles of communication, language (academic vs. business), and time horizons, and bureaucracy also hamper collaboration between universities and industry. Other barriers involve disparities in internal organizational characteristics—controversies over intellectual property rights and constraints on companies' ability to access university-developed information (Prigge & Torraco, 2006). The World Bank (2013) states that despite the growing strength of these motivations, many barriers to university-industry collaboration persist, including the following:

- There is an inherent mismatch between industry and university research orientations, with corporations placing an undue emphasis on speedy commercial results and universities emphasizing basic research. Collaboration is expensive, and the benefits only appear in the medium to long term, yet businesses want immediate results and tangible contributions to their present business lines.

- In terms of outputs, industries are typically concerned with how quickly they may obtain new patents or products, and they prefer to postpone publication to prevent releasing knowledge. University researchers, on the other hand, are usually motivated to publish their findings as soon as feasible.
- The industry is concerned about secrecy and mismatch of expectations when it comes to monetizing intellectual property (IP) rights. As a result, agreements must be made in a commercially expedient manner to assure the ability to commercialize with adequate returns.
- Lack of knowledge, difficulty identifying contact persons, and transaction expenses of finding the proper partner are just a few of the challenges that come up when negotiating a collaboration.

Robertson et al. (2019) highlighted that barrier to knowledge transfer are knowledge differences and differences in goals resulting from different institutional cultures. These barriers result in ambiguity, problems with knowledge absorption and difficulties in applying the knowledge. Facilitators of knowledge transfer are trust, communication, the use of intermediaries and experience, which all assist in resolving the identified barriers. The reasons for this were believed to be (Hadidi & Kirby, 2017):

- a lack of collaboration among the different initiatives,
- a shortage of technology transfer offices,
- a lack of support from senior university management,
- a lack of commercial and professional awareness,
- a lack of support for inventions that solve national problems, and
- a lack of formal courses on technology transfer and commercialization.

For the obstacles to university-industry collaboration, the World Bank (2013) offers the following suggestions and guiding principles for university-industry endeavours:

- Successful university-industry partnerships should advance each partner's objective. Any endeavour that is incompatible with either partner's purpose will ultimately fail.

- Institutional policies and national resources should be geared towards developing proper long-term collaborations between universities and industry.
- Universities and industry should concentrate on the benefits that collaborations will provide to each party by streamlining negotiations to ensure timely research and development of research findings.

This chapter contents that these propositions are equally relevant for all countries in Africa. Therefore, universities and industries in Africa need to find solutions to these barriers to strengthen collaboration to enhance human capital and innovation. This is more so that the realities of the twenty-first century including globalization, hyper-competitiveness, and challenges in HE require innovative and creative win-win solutions for both businesses and universities (Aguinis & Vaschetto, 2011). A key for universities is to proactively manage university-industry partnerships by putting processes in place to minimize the risks to the greatest extent possible while maximizing the benefits (Prigge & Torraco, 2006).

8   A TRIPLE HELIX MODEL FOR SUCCESSFUL UNIVERSITY-INDUSTRY-GOVERNMENT COLLABORATION IN AFRICA

Based on the model developed by Ansell and Gash (2007), this chapter insists that governments, universities, and industries in Africa should be linked to each other through three interrelated elements in order to enhance the level of learning and innovation in African countries. The Triple Helix Model is based on institutional agreements between universities, governments, and industries, in which universities and industries are fundamentally part of the state and have direct relationships with each other. Fig. 2 represents the proposed Triple Helix Model of university-industry-government linkage for Africa.

According to this model, the tripartite linkage consists of:

1. The connection between the activities (technical, administrative, commercial, and other activities of an organization)
2. The link to the resource (availability and accessibility of resources have a significant impact on the quality of the relationship)

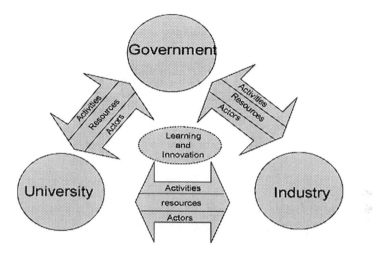

Fig. 2 The Triple Helix Model of university-industry-government linkage for Africa. Source: Ansell and Alison (2007)

3. The link between the actors (relationships, attitudes, and behaviours). Greater trust and synergy will be built inside the partnership as a result of these exchanges.

The Triple Helix university-industry-government model's third level is regarded as a requirement for its success (Aguinis & Vaschetto, 2011). Triple Helix's major goal is to create an environment conducive to bi- and tri-lateral relationships so that knowledge can be shared among industry, government laboratories, and university research groups. A key feature of a statist approach is that the government takes the lead in "driving" academia and industry while also controlling and regulating them to promote innovation. Meanwhile, the industry is viewed as the national champion, with the university's function limited to teaching and research (Etzkowitz, 2003). However, with this model, government or industry will be unable to take advantage of the potential knowledge-generation activities within universities, as both teaching and research are often disconnected from industry needs, and universities have little incentive to engage in research commercialization (Etzkowitz, 2003). One of the Triple Helix Model's primary points is that each actor is connected to the others and aids in the structure of interfaces between them. For example, the sector will adopt some of the university's ideals towards sharing and conserving knowledge. Firms will work together with the government and universities to achieve long-term strategic objectives (Ansell & Alison, 2007). Triple Helix's stakeholders, in addition to executing their conventional functions, also take on the roles of others (Aguinis & Vaschetto, 2011). The relevance of universities in the commercialization of knowledge is another major argument of the Triple Helix concept. Universities are redefining their goal, creating new links with businesses, and becoming more entrepreneurial in this dynamic new environment, according to Prigge (2005). All three spheres will be able to actively engage and collaborate under this approach, promoting robust innovation activities. They will all get value from one another, which will aid them in achieving long-term strategic goals. According to Etzkowitz (2003), universities must be the primary drivers of growth towards the Triple Helix paradigm.

9 The Role of University-Industry-Government Collaboration in Addressing Sustainable Development Goals

According to Makoni (2017), Africa's capacity for research is intimately linked to its ability to achieve the Sustainable Development Goals (SDGs), which places a special duty on universities on the continent to improve collaboration and engagement with government and industry. To achieve the SDGs, policy decisions must be supported by evidence that has been co-designed and co-produced with the appropriate stakeholders, taking political and local circumstances into account. Universities are strategically located to drive the 2030 agenda's cross-sector implementation of the SDGs (El-Jardali et al., 2018).

According to Neary and Osborne (2018), partnership working that ensures that public and private organizations collaborate for the common good is essential for the SDG agenda. The university is one partner that is frequently underrepresented in SDG discussions. Universities can connect their agenda to real-world sustainability issues due to their role in social and technological innovation and their so-called third mission, where they are increasingly asked to contribute to broader society beyond the research and teaching responsibilities. Makoni (2017) declared that African universities can help achieve the SDGs by increasing capacity, working collaboratively, and engaging with the public and private sectors. Nonetheless, some African universities in the post-independence era continued to train undergraduates; research does not form a component of their curricula, and they continued to offer courses for the public sector.

Certain African governments have established independent research institutes and centres in recognition of the value of research, but these are typically founded and funded independently of universities, contributing to the perception that research takes place elsewhere. Several other issues limit the involvement of African universities, which are seen in the poor quality of their research outputs. As a result, the production of research on the continent is today subpar, with only 1% of it coming from Africa and the majority of it coming from South Africa, where it is concentrated in a certain university (Makoni, 2017).

African universities might support the SDGs through collaborating, building capacity, and engaging with the public and private sectors more. In light of this, African universities may aid in accomplishing the SDGs through capacity building, teamwork, and active collaboration with both business and government. This chapter, therefore, advocates for increased intellectual cooperation between African universities and researchers. According to Makoni (2017), African researchers currently have the challenge of entering international collaborations as subordinates and staying there. Thus, it is decisive to promote intra-African collaborations and the sharing of research and instructional facilities. African universities cannot afford to have every university in Africa own the same equipment and perform at the same level due to the continent's limited resources.

The partnerships formed can focus on a specific issue or have a broad capacity for cooperation. Neary and Osborne (2018) skirmish that the key to a successful collaboration is combining various resources, interests, and points of view to comprehend the issue at hand and find a workable solution. There are some obstacles to be overcome to ensure that universities in the Global South are better integrated into the larger socio-economic system and to become better able to utilize knowledge exchange to interested parties to meet these global challenges and the SDGs.

Apart from collaboration in research, universities can integrate the SDGs into curricula. Students who are future leaders in sustainable development need to be trained and shaped, and this is the responsibility of universities. The SDGs can be incorporated into curricula to give students the knowledge and abilities they need to address them. Moreover, universities can design educational courses that focus on interdisciplinary learning and promote a multidisciplinary, systems approach to solving the increasingly complex challenges facing society today. For instance, achieving health-related goals involves experts skilful in creating and evaluating cross-cutting interventions within resource-constrained settings proposing new solutions and pushing for collaborations (El-Jardali et al., 2018).

10   LESSONS FOR AFRICA

There is a substantial amount of research on university-industry interactions in the developed world, particularly in Western countries where Africa could learn from the best practices. Universities can contribute to technological innovation in a variety of ways, including conducting research in industry-relevant technological fields, providing technical assistance to local firms, educating well-trained professionals, and assisting faculty in consulting and commercialization activities (Ogunwale, 2021). In the industry, there is a trend towards more and closer partnerships with academic research in high-tech industries. In today's intensely competitive markets, companies' willingness to seek out different sources of knowledge is seen as vital to their success (Ruwoko, 2021). This encourages huge corporations to form greater collaborations with research universities. University-industry collaboration also plays a role in increasing student employability (Lubbe et al., 2021). In this respect, collaboration can empower students by getting them "work ready" and allowing them to gain and retain employment (Lubbe et al., 2021).

According to Ruwoko (2021), Africa's policymakers, universities, and industry need to acknowledge that collaboration is important and cannot be postponed to encourage scientific and technological progress and create jobs. However, reliable data is required to optimize collaboration and data on the skills that the industry demands from graduates, available financing, university capacities, resources, and expertise, and data on trends that advise institutions about research, innovation, and education on offer. Policymakers across the continent are the main facilitators in promoting university-industry cooperation. African governments should promote collaboration between industry and universities by reviewing university funding models, particularly those related to research and development, and offering incentives to industries to sponsor research programmes. Collaboration between universities and industry is requisite for skill development, both in terms of graduate education and training, as well as the generation, acquisition, and adoption of new ideas and information transfer. Collaboration between universities and industry could boost R&D spending and allow both parties to benefit from synergies and complementarities in scientific and technological breakthroughs (Ogunwale, 2021).

11 Conclusion

All over the world, universities are well-known for their contributions to knowledge development, creativity, and technological progress. They are being positioned as strategic assets in terms of creativity and economic competitiveness, as well as problem-solvers for socio-economic challenges that affect their countries, all over the world (AU, 2013; The World Bank 2012). Collaboration between universities and industry can help secure and leverage additional resources for universities, promote innovation and technology transfer, and ensure that graduates have the skills and knowledge needed to contribute effectively to the workforce. In addition, governments have a role to play in championing policy initiatives that promote university-industry partnerships. Increased intermediary involvement, improved rewards for inventors, better government funding of near-market products, increased financial resources, and the availability of competent technology transfer office employees are all examples of improvements in technology transfer processes. However, African governments face a challenging task in determining which policy instruments best serve national needs in conjunction with important stakeholders. Therefore, with limited budgets, African governments, industries, and universities must make difficult decisions about whether to collaborate in education or research, whether to collaborate with established firms (marching grants, consortia), or new firms (spin-offs, incubators), and whether to provide grants or develop science parks, among other things. Governments must take responsibility for architecting a national innovation system with suitable structures and rules that control and incentivize university-industry interactions, while industries must be brought on board as active participants from the start. Finally, developing a suitable, enabling environment for facilitating university-industry relations necessitates a multifaceted approach that includes interventions. The initiatives should be synchronized with efforts to strengthen the government-university-industry partnership.

REFERENCES

Adeoji, J. (2009). *University–firm interaction in Nigeria: An analysis of the constraints and emerging opportunities.* Unpublished report. Nigerian Institute for Social and Economic Research.

African Union (AU). (2008). *Study on the informal sector in Africa. Sixth ordinary session of the labor and social Affairs Commission of the African Union.* African Union (AU).

African Union (AU). (2013). *Agenda 2063: Africa we want.* African Union. Accessed May 12, 2022, from https://au.int/en/agenda2063/overview

Aguinis, H., & Vaschetto, S. J. (2011). A new model for business-university collaboration to enhance human capital in emerging markets. *The Business Journal of Hispanic Research, 5*(1), 31–38.

Altbach, P. (2006). Peripheries and centres: Research universities in developing countries. *Higher Education Management and Policy, 19*(2), 111–134.

Ansell, C., & Alison, G. (2007). Collaborative governance in theory and practice. *Journal of Public Administration Research and Theory, 8*, 543–571.

Archer, M. S. (1995). *Realist social theory: The morphogenetic approach.* Cambridge University Press.

Association of African Universities (AAU). (2012a). *Perspective of industry's engagement with African Universities.* Accessed May 10, 2022, from https://www.heart-resources.org/wp-content/uploads/2015/09/Report-on-University-Industry-Linkages.pdf

Association of African Universities (AAU). (2012b). *Strengthening university-industry linkages in Africa: A study on institutional capacities and gaps.* Association of African Universities (AAU).

Association of African Universities (AAU) & Association of Universities and Colleges of Canada (AUCC). (2012). *Strengthening university-industry linkages in Africa: A study on institutional capacities and gaps.* Accessed May 10, 2022, from https://www.researchgate.net/publication/328749075_Strengthening_University-Industry_Linkages_in_Africa_A_Study_on_Institutional_Capacities_and_Gaps

Atuahene, F. (2011). Re-thinking the missing Mission of higher education: An anatomy of the research challenge of African universities. *Journal of Asian and African Studies, 46*(4), 321–341.

Bercovitz, J., Feldman, M., Feller, I., & Burton, R. (2001). Organizational structure as a determinant of academic patent and licensing behaviour: An exploratory study of Duke, Johns Hopkins, and Pennsylvania state universities. *The Journal of Technology Transfer, 26*(1), 21–35.

Chesbrough, H. (2003). *Open innovation: The new imperative for creating and profiting from technology.* Harvard Business School Press.

Cohen, Nelson, & Walsh. (2002). Links and impacts: The influence of public research on industrial R&D. *Management Science, 48*(1), 1–23.

Dill, D. & Vught, F. (2010). *National innovation and the Academic Research Enterprise.* Accessed May 12, 2022, from https://www.amazon.com/National-Innovation-Academic-Research-Enterprise/dp/0801893747

Ekaterina, A., Marcel, B., & Daria, P. (2020). Companies' human capital for university partnerships: A micro-foundational perspective. *Technological Forecasting and Social Change, 157,* 120085. Accessed April 20, 2022, from https://www.sciencedirect.com/science/article/pii/S0040162520309112

El-Jardali, F., Ataya, N., & Fadlallah, R. (2018). Changing roles of universities in the era of SDGs: Rising to the global challenge through institutionalising partnerships with governments and communities. *Health Research Policy and Systems, 16,* 38. https://doi.org/10.1186/s12961-018-0318-9

Enders, J., & Teichler, U. (1997). A victim of their success? Employment and working conditions of academic staff in comparative perspective. *Higher Education, 34,* 347–372.

Etzkowitz, H. (2003). Innovation in innovation: The triple helix of university-industry-government relation. *Social Science Information, 42*(3), 293–338.

Etzkowitz, H., Dzisah, J., Ranga, M., & Zhou, C. (2007). *The triple helix model of innovation.* Accessed April 20, 2022, from https://pdfs.semanticscholar.org/4bcc/884ed691ff919ae18c974e15b6baeba08e7f.pdf

Geiger, R., & Sá, C. (2009). *Tapping the riches of science: Universities and the promise of economic growth.* Harvard University Press.

Hadidi, H. H. E., & Kirby, D. A. (2017). University–industry collaboration in a factor-driven economy: The perspective of Egyptian industry. *Industry and Higher Education, 31*(3), 195–203.

Hermosura, J. B. (2019). *Fostering human capital development through the Triple helix model of innovation: Cases from selected local colleges and universities (LCUs) in Metro Manila.* Accessed April 20, 2022, from https://journal.iapa.or.id/proceedings/article/download/264/176

Jansen, J. D. (2002). Mode 2 knowledge and institutional life: Taking gibbons on a walk through a South African university. *Higher Education, 43,* 507–521.

Jones, N., Bailey, M., & Lyytikäinen, M. (2007). *Research capacity strengthening in Africa: Trends, gaps and opportunities.* DFID & IFORD. Accessed April 20, 2022, from http://www.odi.org.uk/sites/odi.org.uk/files/odi-assets/publications-opinion-files/2774.pdf

Koehn, P. H. (2012). Donors and higher education partners: A critical assessment of US and Canadian support for transnational research and sustainable development. *Compare, 42*(3), 485–507.

Laursen, K., Reichstein, T., & Salter, A. (2011). Exploring the effect of geographical proximity and university quality on university–industry collaboration in the United Kingdom. *Regional Studies, 45*(4), 507–523.

278 N. KADHILA ET AL.

Lubbe, B.A., Ali, A. & Jarmo, R. (2021). *Increasing student employability through university/industry collaboration: A study in South Africa, the UK and Finland.* Accessed May 01, 2022, from https://scholarworks.umass.edu/ttra/2021/research_papers/42

Machalek, R. & Martin, M. (2015). *Sociobiology and sociology: A new synthesis.* Accessed April 20, 2022, from https://www.researchgate.net/publication/304193830_Sociobiology_and_SociologyA_New_Synthesis

Makoni, M. (2017). *University research collaboration is key to meeting SDGs.* Accessed February 23, 2023, from https://www.universityworldnews.com/post.php?story=20170206184302712

Malatji, M. J., Mavuso, P. M., & Malatji, K. S. (2018). The role of school-community partnership in promoting inclusive and quality education in schools. *Journal of Educational Studies, 17*(2), 72–86.

Martin, M. (2000). *Managing university-industry relations: A study of institutional practices from 12 different countries* (A working document in the series "Improving the managerial effectiveness of higher education institutions"). International Institute for Educational Planning/UNESCO.

Massaquoi, J. G. M. (2002). *University-industry Partnership for Cooperative Technology Development in Africa: Opportunities, challenges, strategies and policy issues.* Report on the meeting of university-industry partnerships in Africa. Meeting of university-industry partnerships in Africa, Harare, Zimbabwe, may 21–24, 2001. United Nations Educational, Scientific and Cultural Organization (UNESCO).

McCarthy, I. P., & Ruckman, K. (2017). Licensing speed: Its determinants and payoffs. *Journal of Engineering and Technology Management, 46,* 52–66.

Munyoki, J., Kibera, F., & Ogutu, M. (2011). Extent to which university-industry linkage exists in Kenya: A study of medium and large manufacturing firms in selected industries in Kenya. *Business Administration and Management (BAM), 1*(4), 163–169.

Mwiria, K. (1995). *Enhancing linkages between African universities, the wider society, the business community and governments. 'The University in Africa in the 1990s and beyond'.* AAU Colloquium.

Neary, J., & Osborne, M. (2018). University engagement in achieving sustainable development goals: A synthesis of case studies from the SUEUAA study. *Australian Journal of Adult Learning, 58*(3), 336–364.

Ogunwale, S. (2021). *Industry-academia collaboration tested during COVID-19.* Accessed May 01, 2022, from https://www.universityworldnews.com/post.php?story=20210201045446705

Perkmann, M., & Walsh, K. (2007). University–industry relationships and open innovation: Towards a research agenda. *International Journal of Management Reviews, 9*(4), 1006–1021.

Prigge, G. W. (2005). University-industry partnerships: What do they mean to universities? A review of the literature. *Industry and Higher Education, 19*(3), 221–229.

Prigge, G. W., & Torraco, R. J. (2006). University-industry partnerships: A study of how top American research universities establish and maintain successful partnerships. *Journal of Higher Education Outreach and Engagement, 11*(2), 89–100.

RAND Corporation. (2001). *Science and technology collaboration: Building capacity in developing countries?* Accessed May 01, 2022, from http://www.rand.org/content/dam/rand/pubs/monograph_reports/2005/mr1357.0.pdf

Robertson, J., McCarthy, I. P., & Pitt, L. (2019). Leveraging social capital in university-industry knowledge transfer strategies: A comparative positioning framework. *Knowledge Management Research and Practice, 17*(4), 461–472.

Ruwoko, E. (2021). *University-industry collaboration cannot be delayed.* Accessed May 01, 2022, from https://www.universityworldnews.com/post.php?story=20210616220921125

Saad, M. & Girma, Z. (2011). *Theory and practice of the triple helix system in developing countries: Issues and challenges.* Accessed April 20, 2022, from https://www.routledge.com/Theory-and-Practice-of-the-Triple-Helix-Model-in-Developing-Countries-Issues/SaadZawdie/p/book/9780415475167

Salleha, M. S., & Omar, M. Z. (2013). University-industry collaboration models in Malaysia. *Procedia - Social and Behavioral Sciences, 102*, 654–664.

Sparks, D. L., & Barnett, S. T. (2010). The informal sector in sub-Saharan Africa: Out of the shadows to foster sustainable employment and equity? *International Business and Economics Research Journal, 9*(5), 1–11.

Team, T., Perrault, M. F., Kilo, O. M. M., & Hettinger, L. M. P. (2013). *African development bank group.* Kenya Bank Group.

The World Bank. (2009). *University-industry partnership.* Accessed May 01, 2022, from https://www.worldbank.org/en/events/2019/09/03/university-industry-partnership

The World Bank. (2010). Annual World Bank conference on development economics. *The World Bank.* Accessed May 10, 2022, from https://elibrary.worldbank.org/doi/abs/10.1596/978-0-8213-8060-4

The World Bank. (2012). Africa's pulse. *The World Bank.* Accessed May 10, 2022, from http://siteresources.worldbank.org/INTAFRICA/Resources/Africas-Pulsebrochure_Vol6.pdf

The World Bank. (2013). Promoting university-industry collaboration in developing countries. *Policy Brief.* Accessed April 20, 2022, from https://www.researchgate.net/publication/278961909PromotinguniversityindustrycollaboratinindevelopingcountriesInnovationPolicyPlatformOECDandWorldBank

Tumuti, D. W., Wanderi, P. M., & Lang'at -Thoruwa, C. (2013). Benefits of university-industry partnerships: The case of Kenyatta university and equity Bank. *International Journal of Business and Social Science, 4*(7), 26–33.

Vedovello, C. (1997). Science parks and university-industry interaction: Geographical proximity between agents as a driving force. *Technovation, 17*(9), 491–502.

Volmink, J. (2005). Addressing inequality in research capacity in Africa. *British Medical Journal, 331*(7519), 705–706.

Welch, A. (2005). Challenge and change: The academic profession in uncertain times. *Dynamics (Pembroke, Ont.), 7*, 1–19.

Yusuf, S., Saint, W., & Nabeshima, K. (2009). *Accelerating catch-up: Tertiary education for growth in sub-Saharan Africa*. The World Bank.

Higher Education as an Engine of Development: *Sites of Domination, Contestation, and Struggle in Africa*

John Kamwi Nyambe and Ngepathimo Kadhila

1   INTRODUCTION

The argument over the purpose of higher education (HE) is set in the context of recent developments in addressing socioeconomic inequities in the world including Africa, as well as relationship to the mass model of HE and associated policy considerations for continued participatory growth (Kromydas, 2017). Consequently, HE has become the subject of growing interest on the African continent, and internationally. Among others, the growing interest has been fuelled by optimistic beliefs about HE as the engine power that will not only propel economic and social development

---

J. K. Nyambe
School of Education, University of Namibia, Windhoek, Namibia
e-mail: jnyambe@unam.na

N. Kadhila (✉)
University of Namibia, Windhoek, Namibia
e-mail: nkadhila@unam.na

© The Author(s), under exclusive license to Springer Nature
Switzerland AG 2024
P. Neema-Abooki (ed.), *The Sustainability of Higher Education in Sub-Saharan Africa*, Sustainable Development Goals Series,
https://doi.org/10.1007/978-3-031-46242-9_12

in Africa but also serve as a social equaliser. Yet, despite the growing interest, HE in Africa has remained a highly contested space: a site for domination, contestation and HE, as competing forces engage in an almost trench warfare over the roles and purposes of HE. At the core of the tensions and struggles are two major competing forces, directly opposed to each other, namely: the African development agenda and the Western neoliberal hegemonic interests.

The tensions and contestations notwithstanding, HE has been the most highly sought-after antidote for solving the challenges related to societal development in Africa, and lately, the attainment of knowledge-based economies, and the Sustainable Development Goals (SDGs) (Lawrence, 2015; Woldegiorgis & Doevenspeck, 2013). Framed within the neoliberal Western ideology, development has been conceptualised to mean the attainment of the capitalist state, which is regarded as the natural, inevitable, and essentially unchangeable best kind of society that could ever be (Martinez-Vargas, 2020). Thus, on the social front, vision statements of many countries in Africa view societal development as upliftment of lifestyles of their citizens to levels similar to their western counterparts. Advancing a hegemonic European perception of development, the neoliberal agenda has constructed and positioned the Western world as the ultimate yardstick for measuring attainment of development in Africa. Yet, several lessons from the booming Asian economies dating back to the 1990s (Woldegiorgis & Doevenspeck, 2013) present a diversity of alternatives for Africa to learn from. The hegemonic Eurocentric view of the Western world as the ultimate state of development has been reinforced by, among others, the intensified shrinking of the world in terms of time and space as recently experienced through advancements in information and communication technologies (ICTs), and globalisation.

This chapter contributes to the global debate on the role of HE as an engine of societal development in Africa. The chapter problematizes the role of HE by illuminating the underlying tensions, conflicts, and contestations as competing agendas seek to advance their own interests through HE in Africa. In particular, the chapter seeks to:

- Analyse the dominant neoliberal theory of development, and its attendant human capital theory, that shapes and informs mainstream conceptualisation of development and the role of HE in Africa
- Analyses the extent to which HE in Africa serves a transformative role that promotes societal development

- Examine the underlying tensions, conflicts, and struggles on the African HE landscape in executing the role of societal development
- Propose an alternative paradigm of development that positions HE as a catalyser of development in the African context

The chapter interrogates the dominant conceptualisation of development on the African continent, its underlying assumptions, and how this conceptualisation has informed and shaped HE in Africa, and the African university in particular. It then examines the different roles and the tensions, contestations, and struggles; and concludes by proposing an alternative paradigm of development that is likely to support genuine societal development in Africa.

2 Theoretical Framework

This chapter is informed by the human capital theory supported by the neoliberal ideology as a lens to conceptualise the present-day view of the purpose for higher education. The human capital theory informs the mainstream view in the world including Africa that education in general, and HE in particular, is a high yielding investment, and that the main reason why someone spends time and money to pursue higher levels of education is the high returns expected from the corresponding wage premium when they enter the labour market (Kromydas, 2017). In today's fast-paced world, the primary argument centres on the many kinds of institutional reforms in HE. Kromydas (2017) divides HE into three categories. The first is the elite form, whose primary goal is to develop and mould the minds of pupils from the most powerful social class. The second type is mass HE, which transfers information and skills gained in HE to the technical and economic tasks that students later fill in the labour market. Finally, there is the universal form, which has as its primary goal the adaptation of students and the wider public to rapid social and technological changes. This chapter locates the human capital theory within the dominant neoliberal development model outlined below.

3 The Conceptualisation of Development Under Neoliberal Ideology

Several (e.g., Martinez-Vargas, 2020; Lawrence, 2015; Heaney, 2015) have pointed out the extent to which neoliberalism is at all levels of social life and informing what many now accept as everyday thinking, including in HE. Davis and Bansel (2007, p. 247) observe that:

> It is sometimes assumed neoliberalism is peculiar to the northern hemisphere but ... the South has been far from immune. The advent of neoliberalism extends to those capitalist countries participating in the global economy, and its impacts are more widely geographically dispersed through the activities of such groups as the World Bank and the IMF.

For the purposes of this chapter, neoliberalism is interpreted to mean the transformation of a state that was previously responsible for human well-being, as well as for the economy, into one that gives power to global corporations. Underpinned by the market ideology, neoliberalism advances "the belief that the market should direct the fate of human beings rather than that human beings should direct the economy" (Davies & Bansel, 2007, p. 253). Profit maximisation, instead of human wellbeing, is therefore the driving motive under neoliberalism:

> For neo-liberals, 'profit is God', not the public good. Capitalism is not essentially, kind. In Capitalism it is the insatiable demand for profit that is the motor for policy, not public or social or common weal, or good. Thus, privatised utilities approximating as railway system, health and education services, free and clean water supply are run to maximise the shareholders' profits, rather than to provide a public service (Hill, 2003, p. 6).

Following on these scholars, we offer that neoliberalism has been the dominant ideology that has shaped and informed not only mainstream conceptualisation of the development process over the past decades but also the role HE plays in development. In particular, the neoliberal modernisation theory of development (Allahar, 1995; Sklair, 1991) has over the decades dominated mainstream development thinking and the role of HE. Yet, far from advancing African development interests as purported by proponents, namely, the World Bank, IMF, and several development agencies, neoliberal development theory has served to pave the way for the unrestrained global expansion of private capital. Modern day development thinking has been at the service of global capital, ensuring that it flourishes in all corners of the globe.

Three key assumptions underpin neoliberal development thinking. First, is the Eurocentric assumption that South nations, Africa included, are underdeveloped because of their lack of, or their insufficient development of certain internal values and characteristics corresponding to capital, technology, cultural attitudes, institutional arrangements or social organisations and entrepreneurial spirit (Allahar, 1995; Sklair, 1991; Janos, 1986). For instance, one of the dominant arguments that have been advanced over the years concerns the purported values African societies have about families, and the population growth, which has apparently led to hunger and starvation across the continent. Thus, to create a global hegemony and legitimise the unequal capitalist hoarding of global resources, neoliberal theories of development have advanced bizarre arguments about population growth in Africa, for instance, blaming it on the lack of appropriate attitudes and values:

> In the West, the usual attitude toward the population explosion is that the poor multiply because they are ignorant of biology, illiterate, and don't know any better. So, they have more children than they actually want ... when the sun goes down ... the people are left in the darkness. They have no books, no movies, no television. There is only one thing to do—go to bed. There they find their sole source of recreation and amusement ... at the root of [the population explosion] is copulation (George, 1977; p. 38).

Dictated by the apparent deficit of cultural values, Africa has over the years witnessed massive importation of European and North American cultures, educational models, technology, capital, and advice. As elaborated elsewhere in this chapter, massive importation of foreign culture and foreign advice into Africa has benefited the exporting Western countries more than it has done to Africa.

Neoliberal assumption is based on the unilinear and Eurocentric belief that underdeveloped South nations *ought to* and *must* follow the similar path taken by, and blazed for them by the now developed societies, with capitalism as the highest stage of development. A developed country, it is tussled, only shows to the underdeveloped one, the image of its own future. Martinez-Vargas (2020, p. 116) refers to this assumption as:

a Eurocentric interpretation of the world that represents a linear evolution of societies, determining the Western block as the most advanced position of all, and for us in the South to follow.

Yet, as alluded to earlier, several best practices now exist for Africa to learn from, especially in China and other Asian countries. But the West delegitimises these achievements and pushes the agenda for the Western capitalist state. While presenting the capitalist state as the final, inescapable stage of development, several strategies have been adopted through neoliberal development theories to ensure similar conditions exist in the South to ease expansion and operation of global capital. These strategies have included, for instance, the 'development partners', study programmes abroad, staff and student exchange programmes, foreign aid, and several donations, all geared towards transforming the underdeveloped South nations along the path blazed for them by the Western world. Soft strategies, foreign aid, for instance, have been used to entice South nations to emulate their 'Western mentors'. However, where South nations have chosen to become stubborn and play mischievous games, Western and North American nations have not been short of military invasions, raiding of nations, and supporting the overthrow of democratically elected governments (Chomsky, 1992).

The third neoliberal assumption is that, because they are way far ahead on the universal road to development, Western nations have *a positive role* to play in fostering development in the underdeveloped South nations through trade, investment, and aid. Highly celebrated are the global transnational corporations, originating from the West, which are believed to be the engines of development. African leaders, for instance, have gone an extra mile to entice these corporations as a way of promoting foreign investment in Africa and thus purportedly boosting local economy and development. Yet, Barnet and Muller's seminal observation is very instructive in this regard:

> Most poor countries appear to be so eager to entice global corporations to their territory, so eager in fact to create a good investment climate for them, that they are generous with tax concessions and other advantages. *The unfortunate role of the global corporation in maintaining and increasing poverty around the world is due primarily to the dismal reality that global corporations and poor countries have different, indeed conflicting interests, priorities, and needs. The primary interest of the global corporation is worldwide profit*

maximization ... it is often advantageous for the global balance sheet to divert income from poor countries [Emphasis, added] (Woldegiorgis & Doevenspeck 2013, p. 151).

Thus, far from the belief that global corporate giants are engines of development that spread goods, capital, and technology around the world, the key objective of such corporations is worldwide profit maximisation. When these global corporate giants are enticed by African leaders, the argument is that they will create millions of jobs and employ locals. However, one reason, among others, why these global corporations choose to come to Africa is mainly due to cheap labour that they can exploit under government protection.

As will be elaborated in the next sections, HE occupies a central position at the core of the mainstream development thinking. Through human capital development, research, and knowledge production, among others, HE, it is believed, will serve as the engine power that will ignite and propel the development process in Africa, and elsewhere. Due to this perceived role, together with the human rights drive, HE on the African continent has witnessed unprecedented massification. Suffice to reiterate, our argument is that despite its perceived role in promoting national development, HE has remained a site for domination, contestation, and struggle, with competing interests almost drawing each other into trench-warfare. In the next section, we examine the role of HE as an engine of societal development, together with the underlying tensions, and conflicts that surround higher education institutions (HEIs).

4   THE PURPOSE OF HIGHER EDUCATION IN AFRICAN SOCIETIES

HE's purpose and role in modern society is still a contentious philosophical issue with significant practical and policy repercussions. This chapter contributes to the discourse by giving a synthesis of the relevant literature, which may be used as a foundation for future theoretical and empirical study in understanding contemporary developments in HE as they relate to larger socioeconomic changes. According to Martinez-Vargas (2020), education should include both inculcation and emancipation in order to support both individual intellectual growth and societal advancement. Ntshoe (2020) highlights the importance of HEIs providing a public purpose beyond simple self-serving objectives, as well as enforcing societal change in order to reflect the desired nature of a society. Ntshoe (2020) challenged the purpose of education and its significance in current western societies, claiming that education has become a paradoxical concept that allows no room for liberation since it does not allow students to improvise. As a result, rather than feeling emancipated, the students feel trapped within the system. Bo agrees with Mokyr, who emphasised the importance of revisiting ancient philosophers' essential conceptions of education, namely, that education should be combined by both inculcation and liberation in order to serve both individual intellectual growth and societal advancement. Two theoretical perspectives, in particular, have been considered. The mainstream viewpoint is primarily concerned with assisting individuals in increasing their income and improving their relative position in the labour market. The intrinsic idea, on the other hand, focuses on understanding its aim in terms of ontological and epistemological issues. Individual creativity and liberation are in conflict with current institutional contexts associated to power, domination, and economic logic within this philosophical framework. This conflict has the potential to influence people's conceptions of HE's purpose, which can either maintain or revolutionise societal connections and roles.

Kromydas (2017) refutes that in a modern world where monetary costs and benefits are the foundation of policy arguments, the massification and broad diffusion of HE to a much larger population implies marketization and commercialisation of its purpose, and thus its inclusion on an economy-oriented model where knowledge, skills, curriculum, and academic credentials inevitably presuppose a monetary value and have a financial purpose to fulfil. Despite its dismal success in relieving wealth and social disparities, policy tendencies towards an economy-based-knowledge, through a rigid instrumental reasoning, rather than the supposed knowledge-based-economy, appear to remain and dominate. However, in a global context of persistent economic stagnation and deterioration of society's democratic reflexes, a shift towards a model in which knowledge is not subjugated to economic reasoning can inform a new societal paradigm of a genuine knowledge-based economy, in which the economy becomes a means rather than an end goal for human development and social progress.

Based on the arguments offered in the literature about HE, this chapter argues that the current policy focus on labour market-driven policies in HE has resulted in an ever-increasing competition, transforming this social institution into a regular market, where attainment and degrees are viewed as a currency that can be converted to a labour market value. Education has shifted from its initial mission of providing context for human growth to being a tool for economic advancement. As a result, HE has become prohibitively costly, and even if regulations promote openness, only a select few can afford it. A transition towards a hybrid paradigm, in which HE's intrinsic as well as instrumental purposes are equally recognised, should be considered by policymakers as the way ahead in creating more inclusive educational institutions and more knowledgeable and fair societies.

5   SITES OF DOMINATION
AND COUNTERHEGEMONIC DISCOURSES

A glaring observation in the literature on HE in Africa (Martinez-Vargas, 2020; Ntshoe, 2020; Padayachee et al., 2018; Altbach, 1978; Mazrui, 1978) pertains to African HE as a site for domination, and at the same time, a site for counterhegemonic discourses. Thus, in the process of carrying out the role of an engine of societal development, African HE has been drawn into a struggle for domination and counter-hegemony. We support that the domination aspect of HE is largely attributed to the role apportioned to HE through the mainstream neoliberal development ideology, which is linked to the colonial agenda. The aspect of HE as a site for domination is rooted in the genesis of HEIs on the African continent and the purposes for which HE was initially created. This is elaborated as follows:

> European HEIs in colonial Africa were not created to meet the needs of African societies of the time. Instead, they were instruments of executing colonial agendas in Africa. Thus, as of independence, the new African governments inherited institutions which did not have local autonomies and did not reflect African interests (Woldegiorgis & Doevenspeck, 2013, p.43)

Although the existence of HEIs on the African continent predated the colonial period, albeit the numbers were less, the growth and development of such institutions was stalled by the colonial system. The European model of HE was thus established on the African continent as an instrument for propagating Western cultural artefacts and values, attuned to the neoliberal development ideology outlined above. To date, HE on the

African continent has remained a key instrument for facilitating the neoliberal model of development, despite that the neoliberal driven cuts on public spending have also affected HEIs. Despite undergoing transformation at independence for the purposes of serving as engines of societal development, Africanisation, and nation building, African HE, has, by and large, remained an instrument for advancing the Western epistemic system. Way back in the 1970s, the problem of HEIs serving as instruments of domination was already observed thus:

> The African university became the clearest manifestation of cultural domination. By the 1950s it had replaced primary schools and churches as the prime symbol of cultural penetration. The university is a cultural corporation with political and economic consequences (Mazrui, 1978).

While one would have expected that the situation has improved since Mazrui's observations in the 1970s, the instrumentalisation of HE as a tool for Western hegemonic cultural domination has continued. Similar observations as those made by Mazrui were made by Martinez-Vargas (2020, p. 113) thus:

> These institutions are in contradiction with the cultural capital that students bring with them to their institutions, therefore, generating many challenges as educational failure, emotional distress, or identity problems. They do alienate its university student body from their cultural background.

As catapulted in the literature, HEIs have remained colonial sites in terms of teaching a Western oriented curriculum, pedagogical strategies, textbooks, and the entire epistemic system, which has largely remained Western biased. In terms of serving as an engine of societal development in Africa, it would have been expected that while learning what is appropriate from the West, African HE students would have been exposed to curricular that recognise the role played by student cultures, languages, values, and worldviews in learning. Instead, these appear to have been subordinated and dominated by the Western epistemic system.

Apart from the Western epistemic system resulting into what Padayachee et al. (2018, p. 291) term "epistemic violence", African HEIs are to a larger extent replication of the Western institutions in terms of organisation structure, operation, and climate or milieu. Thus, with their predominantly colonial architecture and the Eurocentric academic model, HEIs have alienated their students rather than serving as engines of societal development. Therefore, as a space for foreign domination, HEIs in Africa have continued to impose a Eurocentric worldview and have negated the importance not only of African epistemologies but also of African languages, which are important aspects of learning (Ngugi Wa Thiongo, 1986). Hence, while they may have facilitated development, HEIs have played a role in creating and perpetuating an inferiority of local cultures and a superiority of Western culture.

While serving as sites of domination, HEIs also served as sites of counterhegemonic resistance. The role played by HEIs in the liberation of African colonies can never be disputed. In the same vein, the most recent events in South Africa, namely, the student protests of 2015 and 2016 highlighted the competing agendas on the African HE landscape. The *#FeesMustFall* and the *#RhodesMustFall* student movement underscored the contestations, struggles and conflicts on the HEIs landscape as opposing discourses faced each other. The counterhegemonic discourse advanced by the study body battled not only the neoliberal commodification and marketization of HE into a commodity that is sold and bought on the HE market, but also the colonial heritage marked by domineering symbols and statutes of colonisers that still formed part of the campus architecture. The decolonial discourse (Nyamnjoh, 2020; Mbembe, 2016), has recently become an important feature of the HE landscape, centring not only on the epistemic aspects of HE, but also on issues ranging from pedagogy to leadership and management of these institutions, which still largely resemble the Western model. We illustrate that the decolonial discourse is one steps towards turning African HE around to serve its purpose as an engine of societal development. In the next section, we examine the tensions and conflicts surround the role played by HE in producing human capital that is needed to drive the development process in the African context.

6   Investing in Human Beings: Developing a Nation?

As elaborated in the theoretical framework for the Chapter, one of the key arguments central to the role HE holds in the societal development is that of human capital development. As an investment in human beings, HE leads to higher returns in terms of national development (Davies & Bansel, 2007). Liable to the drive for national development, and global competitiveness, education in general, and HE plays in particular, is expected to produce citizens with the appropriate knowledge, skills, and attitudes. Tabulawa (2009, p. 87) observes that HE is expected to produce the new kind of graduate: one with human capital attributes: "creativity, versatility, innovativeness, critical thinking, problem-solving skills, and positive dispositions towards team-work". Lately, HEIs have had to reform their curricular in pursuit of human capital attributes, namely, the twenty-first century skills, or skills oriented towards the fourth and the fifth Industrial revolution. Heaney (2015, p. 294) cites Theodore Schultz's elaboration of the concept of human capital as originator of the concept:

> I propose to treat education as an investment in man and to treat its consequences as a form of capital. Since education becomes part of the person receiving it, we shall refer to it as human capital.

Therefore, the knowledge, skills, and attitudes that students acquire from their HE Studies are a form of capital, and since these skills become part of the person receiving it, Theodore's concept of human capital applies. By and large, from the African development perspective, these human capital attributes are viewed as public good, intended to serve the nation, and to contribute to the public interest of national development. For Africa, attainment of national development is perceived as hinging upon investment in human capital. It is the human capital as a public good that is perceived to drive the development process.

Yet, far from being interested in advancing public good per se, neoliberal ideology poses yet another source of tension, conflict, and struggle on the HE landscape, namely: shifting the emphasis from human capital as a public good to human capital as a private good. Neoliberal ideology positions an individual as an entrepreneur of oneself, an economic competitive individual who develops and sells one's own capital. Human capital is positioned more as an individual private good, a view which creates tension with the public good that is needed to facilitate national development. The neoliberal reconstruction of human capital is elaborated as follows:

> The individual then is an enterprise-unit. *Homo oeconomicus*, economic man, is an entrepreneur of *himself*: being for himself his own capital, being for himself his own producer, being for himself the source of his earnings …

education investment is of course an important aspect of this investment in the self (Heaney, 2015, p. 294)

Therefore, from a neoliberal development perspective, human capital is viewed as solely a private good as opposed to also serving as a public good. Dwelling on the interest in human capital, neoliberals are therefore likely to exert some form of control in terms of the nature of graduates that are produced by HEIs. Following Hill (2003)'s observations, it can be concluded that HEIs:

> are dangerous because they are intimately connected with the social production of labour power, equipping students with skills, competences, abilities, knowledge, and attitudes, and the attributes and personal qualities that can be expressed and expended in the capitalist labour process ... [higher education institutions] are guardians of the quality of labour power (Hill, 2003, p. 9).

Therefore, apart from privatisation of human capital, neoliberal development perspectives are likely to exercise interest and control in terms of the quality and personal attributes of the HE graduates. As will be elaborated in the subsequent sections, forms of pedagogy that are likely to produce graduates that are perceived to be militant are not encouraged by the neoliberal ideology.

The view of human capital as a private good whose production should be closely controlled is also tied to the neoliberal invasion of HE space as a corporate space which should be privately owned, with students responsible for their own studies as private individuals. It is related to the commodification and marketization of HE, which was referred to in the previous sections. The provision of HE is increasingly seen as a market opportunity (Hill, 2003). It is observed, for instance, that:

> The corporate presence in HE cannot be denied as bespoke by the prolific rise in enrolment rates in nations, Brazil s n example, where between 1995 and 2008, the growth rate of public sector enrolment was 259.3%, while public sector growth was around 81% (Lawrence, 2015, p. 262)

As human capital is a private good and not a public good, governments shy away from their moral responsibility for providing HE and abandon students to carry the tuition fees on their own. This neoliberal human capital view contributed to the *#FeesMustFall* student movement of 2015 and 2016 in South Africa. In the next section, another area of contestation is discussed, namely: the drive towards knowledge-based economies.

7   HIGHER EDUCATION AS THE ENGINE POWER FOR THE DRIVE TOWARDS A KNOWLEDGE-BASED ECONOMY

In the drive for attaining national development, there is increasingly a shift of emphasis in many African nations towards knowledge-based economies. For instance, in Botswana, the country's Vision 2036, and its National Development Plan 11, outline national aspirations to transform to a knowledge-based economy and knowledge-based society by 2036 (Government Republic of Botswana [GRB], 2016). In the same vein, Namibia's Vision 2030 aspires to transform the country into a knowledge-based economy by 2030, "joining the ranks of high-income countries and affording all its citizens a quality of life that is comparable to that of the developed world" (Government of the Republic of Namibia [GRN], 2004, p. 5). Elsewhere on the African continent, the aspiration for many nations is to transform to knowledge-based economies.

One of its key functions is knowledge production and knowledge dissemination. HE is placed high on the pedestal of the knowledge-based economy. Through research, HE will produce valuable knowledge that will be used to drive knowledge-based economies. Thus, as knowledge becomes central to the economy, so is HE (Lawrence, 2015; Woldegiorgis & Doevenspeck, 2013). While indeed, the contribution that HE can make in terms of knowledge creation for a knowledge-based economy cannot be disputed, there is a need to take heed of what many scholars have termed the geopolitical imbalances in knowledge production and knowledge dissemination, as well as the interests leading to these geopolitical imbalances. At the same time, it is important to also take heed of Apple (1990, p. 20)'s observations about knowledge as a form of ideology that can create "false consciousness which distorts one's picture of social reality and serves the interests of the dominant classes in society".

We submit that the knowledge production function of HE within the drive towards knowledge-based economies in the African setting is potentially rife with conflicts, tensions, and contestations. Fundamental questions about knowledge and knowledge production need to be raised. These include, for instance: who produces knowledge in the global epistemic space? Whose knowledge is regarded as legitimate? Whose and which knowledge is de-legitimised? How relevant is the knowledge to the African development agenda? Whose interests does the knowledge serve and to what extent does it serve as a form of epistemic oppression or alienation for the African context?

Martinez-Vargas (2020, p. 116), for instance, refers to "the geopolitics of knowledge … the knowledge power structures—whose knowledge matters and why" as central to understanding global knowledge inequalities. The geopolitical inequalities in knowledge production render HE a site for domination, conflict, and struggle. Already in the 1970s, Altbach (1978, p. 302) observed:

> The Third World suffers from an unfavourable balance of intellectual payments. It imports more knowledge products than it exports. The Third World is beholden to the industrialised nations for books and journals and also for knowledge in most scientific and technical fields, for applied research findings, and often for the results of research about the Third World itself.

He further elaborates:

> The world's leading universities, research institutions, publishing houses, journals, and all the elements that constitute a modern technological society are concentrated in the industrialised nations of Europe and North America. As a result, the industrialised nations dominate the world's research production, mass media, information systems, and advanced training facilities. In fact, the Third World dependency is perhaps increasing because of the impressive research and development activities in the West in recent years. This dependency means that the Third World relies on the old colonial centres of power for expatiates, licenses for technological innovation, books, and many other artefacts of model culture.

Therefore, as Africa shift towards knowledge-based economies, it is important to interrogate the issue of knowledge itself. The question is whether or not African HE will serve its role as generator of knowledge needed for the knowledge-based economy, give the current geopolitical epistemic imbalances and the domination by the Western world and North America. Under the current geopolitical inequalities, many African scholars struggle to contribute to knowledge production and dissemination in the European-based platforms like academic journals that have gained hegemonic superiority as impact journals. The peer review process, whether legitimate or not, tends to throw many African scholars out of publishing in these journals. Publication in the local African journals carries less currency and legitimacy on the global scale. Research in Africa that is funded by neoliberal institutions as the World Bank and the International Monetary Fund (IMF) is likely to import researchers from Western universities. Where Africans are involved in the research exercise, their roles are limited to rudimentary tasks while major research decisions, and indeed acknowledgement for having conducted the research, are apportioned to the researcher from Europe or North America.

The role of HE as producer of knowledge for the knowledge-based economy is therefore fraught with tensions and conflicts. It is a space for oppositional forces, with different forces pursuing conflicting agendas. Although dominating the knowledge production process, the Western World may not necessarily have research interests related to African development. Much of the knowledge produced may be of lesser use to the African context, and much more of use to the neoliberal global corporations that fund most of the research. In the next section, the role of HE in shaping and creating citizens that can drive the development process is examined.

8   Higher Education as a Producer of Subjects

As an engine for societal development, HE in the African context, and elsewhere, is viewed as a site for the production of subjects or citizens, individuals with the appropriate attributes to drive national development agenda. It is a space for creativity, insight, and the expanding of intellectual and social horizons (Heaney, 2015). It is a site for producing citizens who are public intellectuals (Kovacs, 2008) or transformative intellectuals (Freire, 1973) who do not only question received assumptions and institutions but are also capable of engaging not only a language of critique but also taking actions to realise possibilities. Such citizens would have achieved deepened and awareness of the socio-political and economic structures which shape their lives and those of their fellow citizens, particularly the poor and marginalised. Hence, citizens who are capable of engaging in ethical and ideological questioning and debate in the public space, and are able to take appropriate transformative actions, are likely to drive the national development agenda. Such citizens are individuals who use intellect to problematize power, oppression, deceit, dominance, and injustice.

Ntshoe (2020), Lawrence (2015), Kovacs (2008) and Freire (1973) maintain that various pedagogical approaches ranging from pedagogy to social justice pedagogy and critical conscientisation pedagogy are available in the HE space to produce active citizens who will drive the national development agenda. However, like other roles HE occupies, the role of producing active citizens is never free from the tensions, conflicts and struggles in an HE system operating under neoliberal ideology. For the neoliberal ideology, the interest is to ensure that pedagogies that work against labour-power production do not exist. It is, for instance, observed:

> the capitalist state will seek to destroy any forms of pedagogy that attempt to educate students regarding their real predicament—to create an awareness of themselves as future labour-powers and to underpin this awareness with insight that seeks to undermine the smooth running of the social production of labour power (Hill, 2003, p. 9).

Therefore, there is a tendency to supress space and critique which is perceived as being hostile to neoliberal ideology. However, the fact that this suppression may face resistance should be acknowledged. In the drive to supress and banish space, the neoliberal ideology seeks to produce what Kovacs (2008) has termed neo intellectuals who are willing tools of neoliberal interests. It thus observed:

> Neo-intellectuals serve power in the face of people, using pseudoscience and fear to forward corporate orthodoxy and dogma, sweeping people and issues under the rug as they work to maintain the present social order: high capitalism ... hundreds of intellectuals have indeed become the willing tools of big economic interests. Housed in neoconservative/neoliberal think tanks, institutes, and foundations and in universities nationwide, individuals who claim to be researchers, scientists, and scholars generating working papers, policy briefs, and journal articles in order to build public and political support for neoconservative and/or neoliberal legislation (Kovacs, 2008, p. 2).

The foregoing discussion presupposes that the role of HE in producing citizens that can drive the national development agenda is a highly contested site. It is a space for struggle and tension between divergent and conflicting interests. In the next section, lessons for Africa are explored.

9 HIGHER EDUCATION AND THE SUSTAINABLE DEVELOPMENT GOALS

In addition to national development agenda, HE, particularly university education, is accorded a central role as a key driver of the global development agendas. The most recent global development agenda, Agenda 2030, launched by the United Nations (UN) in 2015, views HE as a key lynchpin to the attainment of its 17 Sustainable Development Goals (SDGs). In particular, Sustainable Development Goal number 4 (SDG4), nested within the Education 2030 global agenda, encapsulates the vision for HE: "By 2030, ensure equal access for all women and men to affordable and quality technical, vocational and tertiary education, including university" (UNESCO, 2015: 6).

HE has an irreplaceable role in research, innovation and development, and human capital development. HE therefore serves as an enabler across all the 17 SDGs.

We surmise that the SDGs present yet another avenue where the role of HE as an engine of development is a highly contested space: a site of domination, contestation, and struggle. Thus, while on the one hand HE is expected to play a decisive role in enabling the attainment of the SDGS, on the other hand, it serves as a space for advancing the linear and Eurocentric neoliberal view of development where the Western world, inclusive of its epistemic and cultural aspects, is seen as the inevitable and ultimate state of development for the underdeveloped South nations. This is evident in the Education 2030 agenda where a clarion call is made for "developed countries, traditional and emerging donors, middle income countries and international financing mechanisms" to support the implementation of Agenda 2030 in the underdeveloped world. The call was further extended on developed countries to "provide technical advice, national capacity development and financial support" (UNESCO, 2015: 9).

While this is not to gainsay the support of developed nations towards the underdeveloped ones, what is not evident in these calls for support is how to guard against the neoliberal tendencies where the development interventions have tended to benefit global corporations and the Western world. In the previous sections, we highlighted how massive import of foreign technical advice and financial support has tended to benefit the exporting world more than it has benefitted Africa. Yet, the clarion call at the launching of Agenda 2030 and the SDGs did not put in place mechanisms to avoid a repeat of previous experiences where the exporting countries and their global corporations bent on worldwide profit maximisation were the beneficiaries.

The chapter insists that while HE holds a central role in the attainment of the SDGs, there is a need to put mechanisms in place to guard against practices which have in the past tended to benefit global corporations and the Western world to the disadvantage of the underdeveloped world.

10 Way Forward and Lessons for Africa

In terms of moving forward, we propose adoption of an alternative model of development, which reconceptualises development in the South context and the role of HE. The alternative model radically departs from the dominant neoliberal view that Africa is underdeveloped due to its own internal inefficiencies and lack of certain internal characteristics, values, capital, technology, and cultural aspects. Instead, the alternative model endorses the observation by scholars from the South, who submit that the very same process that led to development in the Western and north American worlds is responsible for the underdevelopment currently experienced in the South (Allahar, 1995). This is explained in terms of the colonial and neocolonial economic relationship between the Western world and Africa over the past decades. While the merits and de-merits of the dependency theory have largely been debated, in this chapter we focus on the most recent discourse in the alternative paradigm of development, namely, the decolonisation discourse. We defend that decolonisation is one of the key processes to make HE in Africa more relevant to serve as an engine of societal development. Decolonisation as a counter-hegemonic discourse that will help Africa to recognise, understand, and challenge the ways in which the African consciousness is shaped by colonial and Eurocentric epistemic systems. Decolonisation provides a vantage point for redefining the HE space, culture, and institutional decision-making structures in a manner that addresses the competing interests that seek to sway African HE away from its intended purpose of societal development.

While decolonisation of HE is broader, four key focus areas are discernible, namely: knowledge or epistemic system, pedagogy, assessment, and the broader university itself. In terms of epistemic decolonisation, fundamental questions should be raised pertaining to what epistemic changes should be effected in the present curriculum and what needs to remain, as well as how we can embed African and other epistemologies into programmes and curricular. There is a need to examine the knowledge in terms of its relevance to the diverse student body, gender, race, and nationality. Regarding pedagogy, a decolonisation model should raise questions pertaining to the extent to which pedagogy is being emancipatory of students, as well as how it addresses issues of power and control. Examples of transformative pedagogy or social justice pedagogy abound in the literature (Ntshoe, 2020). In the same vein, we need to interrogate the extent to which we use alternative forms of assessment that are empowering of students as well as examining the values and beliefs underpinning our assessment practices. Lastly, there is a need to forensically examine the extent to which African HEIs are transnational institutions modelled on institutions in the North, by and large, perpetuating a foreign culture. There is a need to re-centre the African university and curriculum (Mbembe, 2016) in such a manner that Africa becomes part of the history, and HE ceases to alienate the African student.

While we do not advocate for decolonisation as a process of substituting one hegemonic order with another, we propose for Africa to draw lessons from many scholars who have advanced a pluralistic approach to decolonisation that accommodates different perspectives. Martinez-Vargas (2020, p. 120), in particular, offers insightful advice thus:

> The point as mentioned in previous sections is to not radicalise our position and think that everything produced by the Western system is terrible, thus a colonial imposition, or that every knowledge excluded by the Western system is good, and thus glorified. Both the Western epistemic system with its wrongs and goods as well as the excluded epistemic systems need to be analysed. Not all practices and thinking rooted in the North are automatically an attempt to colonise us, despite that might be in many cases.

Therefore, in the decolonisation drive, Africa should be cautious not to put in place new hegemonic systems that will repeat the very same mistakes committed by the Western world that are supposed to be decolonised. Decolonisation entails moving away from what Martinez-Vargas (2020) terms the "uni-verse" to a "pluri-verse" system. The "uni-verse", as contained in the nomenclature of the "university", communicates the idea of a single, universal truth or a single universal epistemic system, which as shown throughout the chapter is the Western epistemic and cultural system. "Pluri-verse", on the other hand tips an epistemic system that is pluralistic and is inclusive of diverse worldviews. Decolonisation of HE should therefore entail:

HIGHER EDUCATION AS AN ENGINE OF DEVELOPMENT: *SITES...* 301

A transformation from a uni-verse which is totalised and homogeneous—the Western project—to a pluri-verse that is able to preserve the heterogeneity of the world, its worldviews, and epistemic systems (Martinez-Vargas, 2020, p. 120).

In decolonising from a "uni-verse" to a "pluri-verse", considerations can be adduced about curricular in HE, pedagogy, and several other constituent parts of the learning and teaching environment which have so far been found to be imposing and dominating.

11 Conclusion

While the positive role played by HE in facilitating societal development in Africa cannot be disputed, we have denoted throughout this chapter that the role HE plays in societal development has been executed in a space characterised by tensions, conflicts, domination, and struggles. To a larger extent, Eurocentric values, cultures, and epistemic systems have tended to dominate and colonise the HE pace. The university, in particular, has been the most sophisticated instrument of domination and violence meted out on Africa in terms of language, knowledge and many other aspects. While on the one hand this hegemonic Eurocentric interest constituted a daring force, fighting to use HE to its own advantage, on the other hand, the African development interest has been another force, also fighting to use HE to achieve its interests. The current policy focus on labour market-driven policies in HE, according to this chapter, has resulted in an ever-increasing competition, transforming this social institution into a regular m HE market-place, where attainment and degrees are seen as a currency that can be converted to a labour market value. Education has shifted from its initial mission of providing context for human growth to being a tool for economic advancement. As a result, HE has become prohibitively costly, and even if regulations promote openness, only a select few can afford it. With the two forces pulling in two different directions over the same subject, contestation, conflict, and struggles have been the result. The chapter thus proposes an alternative development model that hinges upon decolonisation from a "uni-verse" that advances a universal European worldview to a "pluri-verse" that is pluralistic and inclusive of diverse worldviews. A shift towards a hybrid model, in which HE's intrinsic as well as instrumental purposes are equally recognised, should be seen

302   J. K. NYAMBE AND N. KADHILA

by policymakers as the way forward in creating more inclusive educational systems and more knowledgeable and just societies.

REFERENCES

Allahar, A. L. (1995). *Sociology and the periphery: Theories and issues.* Garamond Press.

Altbach, P. G. (1978). The distribution of knowledge in the Third World: A case study in neo-colonialism. In P. G. Altbach & G. P. Kelly (Eds.), *Education and colonialism.* Longman.

Apple, M. (1990). *Ideology and curriculum.* Routledge.

Chomsky, N. (1992). *Deterring democracy.* Hill and Wang.

Davies, B., & Bansel, P. (2007). Neoliberalism and education. *International Journal of Qualitative Studies in Education, 20*(3), 247–259.

Freire, P. (1973). *Education for critical consciousness.* Seabury.

George, S. (1977). *How the other half dies: The real reasons for world hunger.* ALLANHELD.

Government Republic of Botswana. (2016). *National Development Plan II, April 2017–March 2023.* Gaborone.

Government Republic of Namibia. (2004). *Namibia Vision 2030: Policy framework for long term development.* Office of the President.

Heaney, C. (2015). What is the university today? *Journal for Critical Education Policy Studies, 13*(2), 287–314.

Hill, D. (2003). Global neo-liberalism, the deformation of education and resistance. *Journal for Critical Education Policy Studies, 1*(1), 1–93.

Janos, A. C. (1986). *Politics and paradigms: Changing theories of change in social science.* Stanford University Press.

Kovacs, P. (2008). Neointellectuals: Willing tools on a veritable crusade. *Journal for Critical Education Policy Studies, 69*(1), 2–28.

Kromydas, T. (2017). Rethinking higher education and its relationship with social inequalities: Past knowledge, present state and future potential. *Palgrave Communications, 3*(1), 33–49. https://doi.org/10.1057/s41599-017-0001-8

Lawrence, M. (2015). Beyond neoliberal imaginary: Investigating the role of critical pedagogy in higher education. *Journal for Critical Education Policy Studies, 13*(2) Accessed April 24, 2022, from http://www.jceps.com/archives/2648

Martinez-Vargas, C. (2020). Decolonising higher education research: From a university to a pluri-versity of approaches. *South African Journal of Higher Education, 34*(2), 112–128.

Mazrui, A. A. (1978). The African university as a multinational corporation: Problems of penetration and dependency. In P. G. Altbach & G. P. Kelly (Eds.), *Education and colonialism.* Longman.

Mbembe, A. J. (2016). Decolonising the university: New directions. *Arts and Humanities in Higher Education, 15*(1), 29–45.

Ngugi Wa Thiongo. (1986). *Decolonising the mind: The politics of language in African literature.* Heinemann.

Ntshoe, I. M. (2020). Ontological social justice in decolonised and post-apartheid settings. *South African Journal of Higher Education, 34*(3), 263–280.

Nyamnjoh, F. B. (2020). *Decolonising the academy: A case for convivial scholarship.* Basler Afrika Bibliographien.

Padayachee, K., Matimolane, M., & Gans, R. (2018). Addressing curriculum decolonisation and education for sustainable development through epistemically diverse curricula. *South African Journal of Higher Education, 32*(6), 288–304.

Sklair, L. (1991). *Sociology of the global system.* The John Hopkins University Press.

Tabulawa, R. T. (2009). Education reform in Botswana: Reflections on policy contradictions and paradoxes. *Comparative Education Review, 45*(1), 87–107.

UNESCO. (2015). *Incheon declaration: Education 2030.* Republic of Korea.

Woldegiorgis, E. T., & Doevenspeck, M. (2013). The changing role of higher education in Africa: A historical reflection. *Higher Education Studies, 3*(6), 35–45. https://doi.org/10.5539/hes.v3n6p35

Energy Sustainability in African Higher Education: *Current Situation and Prospects*

Alfred Kirigha Kitawi ⓘ *and*
Ignatius Waikwa Maranga ⓘ

1   INTRODUCTION

Sustainability refers to how our present actions, in relation to meeting current needs, affect future generations. In recent years, higher education institutions (HEIs) have been forced to be explicit on sustainability issues. This entails not only their processes, which relate to teaching, research, and community service, but also the structures and infrastructures that support HEIs mission. HEIs spur efforts towards a circular economy (CE) through reusing, recycling, and transformation of waste into wealth by narrowing flows (using less), slowing flows (use longer), regenerating

---

A. K. Kitawi (✉)
School of Humanities and Social Sciences, Centre for Research in Education, Strathmore University, Nairobi, Kenya
e-mail: akitawi@strathmore.edu

I. W. Maranga
Energy Research Centre (SERC), Strathmore University, Nairobi, Kenya
e-mail: imaranga@strathmore.edu

© The Author(s), under exclusive license to Springer Nature Switzerland AG 2024
P. Neema-Abooki (ed.), *The Sustainability of Higher Education in Sub-Saharan Africa*, Sustainable Development Goals Series,
https://doi.org/10.1007/978-3-031-46242-9_13

flows (make cleaner) and cycling flows (use again) (Walzberg, 2021). HEIs are agents in promoting practices that decouple economic activity from the consumption of finite resources and designing waste out of the system (Yu et al., 2022). The recent Coronavirus pandemic has brought to the fore a need to build sustainable practices that can enable HEIs to overcome wicked problems. In addressing sustainability issues, HEIs must place the human being at the centre of sustainability concerns (Rio Declaration 1992, Principle 1). They, together with other stakeholders, need to ensure that their sustainability approach is anchored on economic, social, and environmental concerns, which some other entities refer to the 3Ps (Profit, People, Planet). This implies that the economic models and processes which they choose to navigate and ensure their own existence or profitability, need to be socially sustainable within the contexts they operate in and include environmental concerns. Since HEIs educate the growing and diverse youth population, they are at the centre of assisting to achieve Agenda 21 and principle 21 of the Rio Declaration which focusses on mobilization of the youth of the world to forge a global partnership to achieve sustainable development.

2 Sustainability in Higher Education

The agenda on sustainability received more prominence in September 2015 when 193 Member States of the United Nations General Assembly outlined a vision for economic, social, and environmental development for the next 15 years. The 17 Sustainable Development Goals (SDGs) and 169 targets were framed to set a collective vision and act as a standard for the implementation framework. The SDGs were set at a macro level and institutions at the meso level and micro level need to work together jointly to meet these goals and targets.

Kitawi and Mumbi (2023) provide an indication on how the discourse on sustainability has evolved from the Magna Charta of European Universities, the Taillores Declaration of University Presidents for a Sustainable Future, the Luneberg Declaration on for Sustainable Development (2001), the Ubuntu Declaration on Education and Science and Technology for Sustainable Development (2002) to the recent Rio+20 Treaty on Higher Education which over 100 institutions signed. The authors further highlight the tools for assessing sustainability in HEIs. These included environmental management systems; assessments of teaching and learning processes that embed Education for Sustainable

Development; and Social innovations that are not only economically viable but respect the diversity of values and norms of individuals in societies with an emphasis on social enterprises.

Sustainability should be the lifeblood of HEIs. These institutions should see themselves as institutions that actively shape and transform society towards a sustainable future. The actions which HEIs can take are putting in place ecological policies that support waste management, energy efficiency, and use of materials parallel to paper. It can also set up specialized courses, extension programmes, research, and community driven initiatives to influence society to harness the skills and knowledge to act sustainably and to create solutions that address the deeply embedded sustainability challenges (Kitawi & Mumbi, 2023). Some HEIs have chosen to incorporate sustainable reporting thus being able to show positive or negative contribution towards the goals of sustainable development. This form of reporting was developed as a standard of reference from the Global Reporting Initiative (GRI) and it allows for measurement, accountability, and disclosure of an organization's performance. Research to sustainability performance which is mainly focused on corporate entities focusses on size, profitability, financial leverage, corporate governance structure, ownership structure, firm's age, sector, firm posture and board qualifications and experience (Farisyi et al., 2022).

There have been several initiatives that have mapped out how the topic of sustainability has been embedded into teaching, research, and social innovation. An example is the Times Higher Education University rankings which measured the contribution of universities to SDGs. The universities that ranked top in the ranking embedded and built sustainability into their processes and programmes. The University' Sustainable Development Report 2022, provided a range of initiatives of the University of Auckland has engaged in to exhibit how the 17 SDG goals were being achieved in both teaching and research. Other universities like Macquarie University and Monash University have also mapped out the topic of sustainability in relation to the third mission of universities. In relation to innovation, it is important to embed sustainability into design thinking, including the ideation processes, whether it follows a lean methodology or agile. In relation to lean thinking and Lean Six Sigma, it may focus sustainability on four aspects quality, productivity, cost, and profitability within the define, measure, analyse, improve and control (DMAIC) processes (Erdil et al., 2018). Lean in this case refers to a balanced use of materials, resources, and personnel (Marhani et al., 2013). Agile processes and sustainability imply embracing sustainability both at the individual level and working in teams during product development and its iterations. At individual level, they embrace the aspects of environment, social and economic sustainability in their individual projects and bring this on board to team discussions when improving products. This implies that they promote not only individual capacities and capabilities, and thereby expanding individual's freedoms (Sen, 1999), but also in maximizing their efforts in commonality when working on an innovation while at the same time taking into consideration future needs. Embracing sustainability in the teaching process of a university requires an interaction of students (being at the centre of teaching and learning with considerations of inclusivity and social justice), students' competencies (integration of sustainability knowledge, skills and attitudes through communication, collaboration, reflective thinking and creativity), alliances (between different stakeholders), lecturers and teaching methodologies (mainly constructive in nature) (Zamora-Polo & Sanchez-Martin, 2019).

HEIs can further involve themselves in ensuring sustainable policies by being resolute agents in agenda setting; this implies that they insist on the implementation of CE methods, participating as stakeholders in policy formulation, providing evidence basis that informs decision making, following up policy implementation and providing feedback on the positive and negative externalities created and policy evaluation. Since HEIs are involved in research, i.e., knowledge generation and knowledge sharing, it can provide predictive indicators for other institutions to incorporate within their practices. These predictive indicators can be applied to civic organizations (society), private institutions, governments and academia (Quadruple Helix Model) (Carayannis & Campbell, 2009).

3   Ways Higher Education Institutions Can Employ Sustainability through Energy Self-sufficiency

Research into alternate sources of energy dates back to the late 1990s when many economies were affected by oil price hikes. There was thus greater attention into renewable resources to achieve sustainability (Owusu & Asumadu-Sarkodie, 2016). The developments in the energy sector have been linked to the different industrial revolution phases. In the first industrial revolution (B.C-eighteenth century), the main source of energy was natural energy. During the second industrial revolution (late nineteenth century-twentieth century), the main source was industrial energy. In the third industrial revolution (1970s–1990s), there was a new emphasis, as mentioned earlier, on renewable energy. This set the stage for the sustainable development phase (late 1990s–2010s). This is also the time earmarked as the third industrial revolution, with the emergence of sustainable energy sources. We are currently in the fourth industrial revolution, marked by digitalization (Strielkowski et al., 2021), together with an insistence on sustainable and abundant energy sources and more powerful computing power that can allow artificial intelligence. Sustainable energy is becoming an increasingly important issue in HEIs, which have a responsibility to reduce their environmental impact and prepare the next generation of leaders to address sustainability challenges. There are myriad ways in which HEIs can promote sustainable energy. We shall focus on: (1) sustainability plans; (2) energy sources which are underpinned in such a plan; (3) how these energy sources can be a source of revenue and reduction of institutional expenditures.

3.1   Development of a Sustainability Plan

A plan is as good as its realization. HEIs can set targets for reducing carbon footprint and develop a sustainability plan that includes sustainable energy initiatives: the use of renewable energy sources, energy-efficient buildings, and sustainable transportation. Setting targets and developing a sustainability plan is an important first step for HEIs to reduce their environmental impact and promote sustainable energy. The development of sustainability plans involves a number of steps. Calhoun (2014) explicate a three-step process: use of the sustainability assessment tool to measure the programmes sustainability, use of results from sustainability tool to inform and prioritize the planning process, and implementation of plan and keeping track of the progress.

The first step is to conduct a sustainability assessment of the institution's operations, including energy consumption, water usage, waste management, transportation, and procurement practices. This will help identify areas where the institution can reduce its environmental impact and develop specific sustainability targets. Based on the sustainability assessment, the institution can establish specific sustainability goals, plus reducing energy consumption by conducting energy audits, increasing the use of renewable energy, reducing water usage, reducing waste, and promoting sustainable transportation. Such sustainability goals will require some capital investment; hence the important to rationalize the expenditure vis-a-vis savings that will accrue.

Once sustainability goals have been established, the institution can develop an action plan outlining specific steps to achieve the goals. This could include initiatives similar to installing solar panels, implementing energy-efficient lighting and Heating, Ventilation, and Air Conditioning (HVAC) systems, promoting sustainable transportation options, and reducing waste through recycling and composting. The sustainability plan should identify who is responsible for implementing each initiative and establish timelines for completion. This will help ensure that the plan has been implemented effectively and efficiently. Regular monitoring and evaluation of the sustainability plan's progress will help the institution to identify areas where it needs to improve and make adjustments to the plan accordingly. This should take cognisance of the changing micro and macro institutional dynamics. Engaging students, faculty, staff, and the local community, is essential for the success of the sustainability plan. The institution can involve stakeholders in the planning process and seek their feedback and support for sustainability initiatives.

The implementation of plans will focus on specific renewable energy sources. We shall focus these renewable energy sources to sources available within the African continent. These include solar, wind, geothermal, hydroelectric and biomass sources

3.2 Renewable Energy Sources as a Source of Sustainability

The use of renewable energy sources in higher learning institutions is a great way to reduce their carbon footprint and promote sustainable energy. Renewable energy systems are technologies that harness and convert naturally replenishing resources: sunlight, wind, water, geothermal heat, and biomass into usable forms of energy: electricity, heat, and fuels (National Renewable Energy Laboratory, n.d.). Unlike non-renewable energy sources as coal and oil, renewable energy systems do not deplete finite resources and do not produce greenhouse gases and other harmful emissions that contribute to climate change and air pollution (Lund, 2014). Examples of renewable energy systems include solar photovoltaic (PV) systems, wind energy systems, hydroelectric energy systems, geothermal energy systems, and biomass energy systems.

3.2.1 Solar Photovoltaic (PV) Systems as a Source of Sustainability

Solar PV systems, also known as solar panel systems, are a type of renewable energy technology that converts sunlight into electricity. The system consists of one or more solar panels, which are made up of photovoltaic cells that capture the energy from the sun and convert it into direct current (DC) electricity (NREL, n.d.). The DC electricity produced by the solar panels is then converted into alternating current (AC) electricity through an inverter, making it compatible with the electrical grid or suitable for use in buildings. Solar PV systems are commonly used in residential, commercial, and industrial settings, as well as in large-scale solar power plants (NREL, n.d.). They are an environmentally friendly +alternative to traditional fossil fuels, as they do not emit greenhouse gases or other pollutants during operation. Solar PV systems can be designed to be connected to the electrical grid or operate independently, providing electricity in remote areas without access to electricity grids. Solar PV systems are often cost-effective in the long term, as they can provide electricity at a lower cost than traditional sources over the lifetime of the system. Sustainability of Solar PV systems is tied to its configuration as grid-tied solar PV systems, off-grid solar PV systems, hybrid solar PV systems, or building integrated solar PV systems (International Energy Agency, 2020).

Grid-tied Solar PV systems are connected to the electrical grid and do not have battery storage. The solar panels generate electricity during the day, which is consumed in the building or sent back to the grid (NREL, n.d.).

Universities in Africa are exposed to long hours of sunshine. Solar panels can be placed on buildings and land and therefore generate electricity during the day that can limit consumption of electricity from the paid grid. A cost-benefit analysis needs to be conducted to evaluate the payback period from the initial investment and to make a solid case for it.

Off-grid solar PV systems operate independently of the electrical grid and typically include battery storage to provide electricity at night or during periods of low solar generation (NREL, n.d.) Off-grid systems are commonly used in remote areas without access to the grid or in situations where grid electricity is unreliable. This implies that during the evening and night hours, specific operations, depending on the amount of energy available, can continue. Computers and other low energy consumption devices like inkjet dot printers, energy saving overhead lighting and fax machines can be operated during the night quite efficiently (Kawamoto et al., 2001).

Hybrid solar PV systems combine solar PV with another source of energy as wind or diesel generators, to provide reliable power in locations with variable renewable energy resources or unstable grid connections (NREL, n.d.). These systems can ensure consistency in power supply in institutions where stable power supply is an important issue for continued operations, in terms of teaching, research, or consultancy.

The last type of solar PV system is the building-integrated solar PV system which is integrated into the building structure in the form of solar shingles or panels incorporated into the building's facade or roof (International Energy Agency, 2020). This implies other than the materials serving aesthetic purposes, it can also be used to generate energy to meet some institutional needs.

HEIs can install solar panels on rooftops, parking lots, or other open areas on campus to generate electricity from the sun. This can significantly reduce the institution's reliance on non-renewable energy sources and provide a clean source of energy. Solar-powered outdoor lighting on campus can be used to provide lighting for pathways, parking lots, and other areas. Solar-powered outdoor lighting is an energy-efficient and sustainable alternative to traditional lighting systems that rely on non-renewable energy sources. An example of HEIs that have used this form of third-stream revenue generation activity is Strathmore University, Kenya (Strathmore University, 2014). The grid tie solar photovoltaic system meets 65–75% of the electricity needs in the university and exports 25–35% of the electricity generated to the electricity distribution utility company, Kenya Power. This 600 kilo-watts (kW) system at Strathmore University has led to electricity bills savings of up to 30–40%. In February 2023, the University of Mauritius installed a 14.5 kW solar PV system on the roof of one of its buildings to reduce its dependence on fossil fuels and lower its energy costs as part of a solar laboratory in the university funded by Huawei. The innovative solar laboratory aims to facilitate the government to meet its objectives to generate 60% of Mauritius' energy from renewable energy sources (Government of Mauritius, 2023).

With the onset of electric vehicles, solar-powered charging stations can provide a sustainable source of energy for electric vehicles while also reducing the institution's carbon footprint. These charging stations can be installed in universities and power vehicles belonging to the campus or communities around the campus, and therefore earn extra revenue. The University of Lagos in Nigeria implemented a solar powered electric vehicle charging system in 2019 (University of Lagos, 2019). The charging station served Nigeria's first 100% electric vehicle and encouraged the use of electric vehicles in the university. HEIs can also use solar energy for classroom and laboratory equipment. This way, the HEI can reduce the institution's energy consumption and promote sustainable energy practices.

3.2.2 Wind Energy Systems as a Source of Sustainability

Wind energy systems convert the kinetic energy of wind into usable electricity. Wind energy systems typically consist of wind turbines, which are tall, rotating structures that are designed to capture the energy from the wind and convert it into electricity (Manwell et al., 2010). The blades of the wind turbine are connected to a generator, which converts the rotational energy of the blades into electrical energy. Wind energy systems can be either small wind energy systems (capacity of 100KW) or large wind energy systems (several Megawatts). They are in areas that experience constant and strong wind resources (Burton et al., 2011). This implies if a HEIs is considering this source, it is important to measure the strength of wind currents. Wind energy systems are considered a sustainable and environmentally friendly source of electricity because they do not produce greenhouse gas emissions or other pollutants. They also have the potential to reduce our dependence on fossil fuels and help to mitigate the effects of climate change.

Small wind energy systems can supplement electricity supplied to the grid and can generate enough power for small business use and to power base transceiver stations located within institutions (Musgrove, 2011). Generally, such systems are less expensive to install and maintain, and are more accessible to individual homeowners and small businesses. Once a sustainability assessment has been done within HEIs, it can assess whether this type of investment makes sense and can generate additional third-stream revenue.

Large wind energy systems are typically found in large commercial wind farms and provide electricity to the grid (Musgrove, 2011). I HEIs is located in a very windy area and it has large tracts of land, it can consider have a public-private partnership agreement with a wind energy company to host such a facility since it requires more investment and expertise to install and operate. Purchase of wind energy from off-campus sources to generate electricity provides a clean and renewable source of energy while also supporting the development of wind energy projects. South Africa has been in the forefront of wind energy deployment due to its robust policy and structural environment. It has produced a number of white papers on renewable energy which includes wind, an independent power producer's procurement programme and integrated energy plans.

Stellenbosch University in South Africa has installed a three megawatt (MW) wind turbine on its campus as part of its renewable energy initiatives. The wind energy system provides about 30% of the university's electricity needs and has contributed to the reduction of its electricity bills.

3.2.3 Hydroelectric Energy Systems as a Source of Sustainability

Hydroelectric power systems are systems that generate electricity using the power of water. They use the energy generated by falling or flowing water to turn turbines, which in turn drive generators that produce electricity (Pandey & Kark, 2010). Hydroelectric power systems are a type of renewable energy source, as the water that drives the turbines is typically replenished by rainfall or melting snow. Hydroelectric power systems typically consist of a dam or other water control structure that creates a reservoir, a powerhouse where the turbines and generators are located, and a transmission system that carries the electricity generated to users. The size and capacity of hydroelectric power systems can vary widely, from small systems that generate power for a single home or small community to large-scale systems that generate power for entire cities or regions (Pandey & Kark, 2010). One of the advantages of hydroelectric power systems is that they are clean and produce no greenhouse gas emissions or air pollution. They are also reliable, as long as there is a steady supply of water. However, the construction of dams and other water control structures can have significant environmental impacts, including changes to river ecosystems, displacement of people and wildlife, and alterations to water flow and quality (Gulliver & Arndt, 1991). A number of governments, including Kenya, have recently expressed the wish to create at least 100 dams which can have the potential of also generating hydroelectric energy. Nevertheless, the potential of hydroelectric energy systems is not as powerful as solar PV systems.

Installation of hydroelectric power plants on campus to generate electricity from flowing water can provide a clean and sustainable source of energy. There are several different classes of hydroelectric power systems, which can be classified based on their size, location, and design (Gulliver & Arndt, 1991). Large-scale hydroelectric power systems (more than 30 Megawatts) may be part of a national or regional power grid and can provide power to thousands or millions of people. Small-scale hydroelectric may be used to provide power to individual homes or small communities, or to supplement other sources of electricity in remote areas (Sharma, 2003). Small-scale hydroelectric power systems may be run-of-river systems (which use the natural flow of the river) or storage systems (which use a reservoir to store water). Micro hydropower systems on campus are a sustainable way to generate electricity from small-scale water sources compared to streams or rivers. Irrigation systems for campus gardens or green spaces are a possible load that can be powered by hydroelectric systems (Sharma, 2003). This can provide a sustainable energy source for watering plants. A five KW hydroelectric plant in the University of Ilorin, Nigeria was built in the 1980s and was upgraded in 2012 to increase its efficiency. The small hydroelectric power plant generates electricity to power the campus (Henry et al., 2010).

3.2.4 Geothermal Energy Systems as a Source of Sustainability

Geothermal energy is a form of renewable energy that comes from the Earth's natural heat. It involves extracting heat from the Earth's crust to generate electricity or provide heat for buildings and industrial processes. The heat is obtained from hot water and steam that exist in rock formations beneath the Earth's surface. The heat can be harnessed using geothermal power plants that use heat exchangers and turbines to convert the thermal energy into electricity. The steam is used to power a turbine which generates electricity, and the water is re-injected back into the reservoir to be reheated. Geothermal energy is a clean and sustainable source of energy that can help reduce greenhouse gas emissions and provide a reliable source of electricity and heating. This implies that this type of energy source can only be derived if an HEIs is in areas with registered volcanic and seismic activities, i.e., after Geophysical data, Geological data, and surface manifestation analyses have been done. Geothermal systems are highly efficient and can provide a reliable source of heating, cooling, and electricity with minimal environmental impact. However, the installation and maintenance of geothermal systems can be expensive, and they require specific geological conditions to be effective. The "United Nations Environment Program and the Infrastructure Consortium for Africa estimate a geothermal potential capacity of more than 20 GW of geothermal energy across Eastern Africa, which has encouraged countries, including Comoros, Eritrea, Djibouti, Rwanda, Uganda, and Tanzania, to conduct preliminary investigations for geothermal resources. Ethiopia has a future plan to reach 1 GW production from geothermal energy by 2021.

Additionally, Uganda, Burundi, and Zambia are working to establish new small-scale geothermal power plants" (Elbarbary et al., 2022). The use of geothermal energy to provide hot water for buildings, swimming pools, or other facilities is a way to provide a sustainable source of energy for heating water in HEI. Geothermal energy can also be used to provide heat for greenhouse facilities on campuses that have agricultural activities.

3.2.5 Biomass Energy Systems as a Source of Sustainability

Biomass energy systems are renewable energy systems that utilize organic materials in the form of wood, crops, and waste materials, to generate heat or electricity. These systems convert the chemical energy stored in biomass into thermal or electrical energy, through a range of processes including combustion, gasification, and anaerobic digestion. They can be classified into two main categories: direct combustion systems and advanced conversion systems. Direct combustion systems are the simplest and most common biomass energy systems. They burn biomass directly to produce heat, which is then used to generate steam or hot air to drive a turbine and generate electricity. Advanced conversion systems use a combination of heat, pressure, and chemical reactions to transform biomass into energy-rich gases or liquids (De Jong & Van Ommen, 2014).

Biogas can be used to generate electricity or heat, or can be processed further to produce transportation fuels or chemicals. Biomass is abundant and can be sourced locally, making it a cost-effective option for HEIs. Additionally, biomass energy systems can help reduce greenhouse gas emissions by displacing fossil fuels, and can help divert waste materials from landfills. The University of Nigeria, Nsukka has implemented a 10 MW biomass gasification power plant that utilizes wood chips from sustainable sources to generate electricity for the university and its surrounding communities (Samson et al., 2011). However, these systems also have some drawbacks, including potential environmental impacts associated with land use and transportation, as well as air pollution from combustion processes. HEIs can use biomass to support sustainable agriculture practices by using organic waste, example being food scraps, as a source of biomass.

An example of where a university setup a bio-digester system is the UW-Oshkosh. It is a dry fermentation bio-digester system. It generates energy from plant and food waste. Accordingly, "it enhances the university's revenue stream in three ways: producing energy that can be sold back to the grid, creating marketable carbon credits, and leaving waste products for which there is also a market (Pelletier, 2012)."

3.3   Energy Sources as a Source of Revenue and Reduction of Institutional Expenditures

Institutions should be more strategic in expanding revenue streams in addition to cost-cutting measures. This section will elaborate how HEIs can reduce institutional expenditures through energy efficiency practices, adoption of sustainable transportation and education/outreach activities as well as third stream funding initiatives available to HEIs as a source of additional revenue.

3.3.1   Reduction of Institutional Expenditures through Energy Efficiency Practices

Central in the use of energy sources which can lead to reduction of institutional expenditures is energy efficiency. Energy efficiency refers to the practice of using less energy to perform the same task or achieve the same outcome (Goswami & Kreith, 2015). This involves using technologies, products, and practices that consume less energy while still providing the same level of service. Energy efficiency can be applied to various sectors, including buildings, transportation, industry, and agriculture in HEIs. In buildings, for example, energy-efficient technologies such as insulation, efficient lighting, and HVAC systems can help reduce energy consumption and costs. The benefits of energy efficiency are reduction in energy costs, lowering greenhouse gas emissions, improvement of air quality, and increase in energy security. It can also create jobs and promote economic growth by stimulating innovation and investment in new technologies and practices.

Another aspect of energy efficiency which can and should be employed in HEI is energy-efficient lighting; efficient HVAC systems; use of energy-efficient appliances (Goswami & Kreith, 2015).

Energy-efficient lighting systems are lighting technologies and products that consume less energy while still providing the same or better lighting performance as traditional lighting systems. They are designed to use less electricity, produce less heat, and have longer lifetimes compared to traditional lighting technologies. There are three main technologies used for energy-efficient lighting applications. These are light emitting diodes (LEDs), compact fluorescent lamps (CFLs) and halogen incandescent lamps. LEDs are highly energy efficient and have a longer lifespan than traditional incandescent bulbs. They consume up to 75% less energy and last up to 25 times longer than incandescent bulbs (Watson & Labs, 1983). It is common for HEIs in the United States of America (USA) to retrofit their lighting sources to LEDs in a bid to achieve energy sustainability. There are at least five HEIs in the USA that have implemented LED lighting technology. One of them is the University of California, Berkeley. The university has implemented a lighting retrofit programme that replaced over 55,000 light fixtures with LED lighting, resulting in an estimated annual energy savings of 10 million kilowatt-hours (kWh) (Laboratory, 2014). Arizona State University has implemented an LED lighting retrofit in its parking garages, resulting in an estimated annual energy savings of 3.3 million kWh (News, 2016). CFLs are an efficient alternative to traditional incandescent bulbs, using up to 75% less energy and lasting up to 10 times longer. Halogen Incandescent bulbs are more efficient than traditional incandescent bulbs, using about 25% less energy and lasting up to three times longer (Goswami & Kreith, 2015).

HVAC systems are systems designed to regulate the indoor environmental conditions of a building or space, including temperature, humidity, and air quality (Goswami & Kreith, 2015). These systems typically consist of a network of ducts, filters, thermostats, and other components that work together to control the heating, cooling, and ventilation of a building. Heating components in an HVAC system may include furnaces, boilers, and heat pumps, while cooling components may include air conditioners or refrigeration units. Ventilation components may include exhaust fans and air ducts that provide fresh air from outside the building (Goswami & Kreith, 2015). These systems play a weighty role in maintaining comfortable and healthy indoor environments, promoting energy efficiency, and ensuring proper airflow and ventilation. HEIs can implement this technology in a number of ways to achieve sustainability. Energy-efficient HVAC systems that consume less energy can be implemented in HEIs, while maintaining optimal indoor environmental conditions. The Technical University of Denmark University has implemented a low-energy ventilation system that uses natural ventilation and heat recovery to reduce energy consumption and greenhouse gas emissions. The system has reduced energy consumption by 80% compared to traditional HVAC systems (Architects, 2012). Such systems could include geothermal heat pumps, variable refrigerant flow systems, and energy recovery ventilation systems. The University of British Columbia has implemented a geothermal system for heating and cooling several buildings on campus. The system uses renewable energy and has reduced greenhouse gas emissions by 1800 tonnes per year (Handschuh, 2011). Regular maintenance and upgrades of HVAC systems can help ensure they are running at peak efficiency. Upgrades could include the installation of more efficient equipment, like high-efficiency air filters, programmable thermostats, and smart controls that allow for more precise temperature and humidity control. The University of California Irvine has implemented a district energy system that includes a central plant with high-efficiency boilers and chillers, energy recovery systems, gas and steam turbines and a building automation system that optimizes HVAC performance. The gas and steam turbine system meets 85% of the university's electricity needs, and the entire system has led to significant energy savings (Ivrine, n.d.). The New York University has reduced energy consumption by 30% in some buildings by implementing a building automation system that optimizes HVAC performance based on occupancy, temperature, and weather conditions. In Africa, the American University in Cairo, Egypt has implemented a building automation system that optimizes HVAC performance based on occupancy and weather conditions (AUC, n.d.). Monitoring and optimizing HVAC systems can help institutions identify and address inefficiencies and malfunctions. This could involve the use of building automation systems (BAS) to monitor HVAC equipment, control indoor environmental conditions, and optimize energy consumption (Watson & Labs, 1983). HEIs can integrate their HVAC systems with renewable energy sources: solar panels, wind turbines, and geothermal systems. Finally, HEIs can educate students, faculty, and staff about the importance of sustainable HVAC practices: using energy-efficient equipment, reducing energy waste, and maintaining HVAC systems regularly. This can help create a culture of sustainability and encourage individuals to make sustainable choices in their everyday lives.

Energy-efficient appliances in HEIs are an effective way to promote sustainability and reduce energy consumption. Some of the initiatives that can be implemented by HEIs to achieve sustainability with regard to the use of energy-efficient appliances include using energy-efficient light sources like LEDs, and using premium efficient pumps, laptops, computers, and kitchen appliances. The University of Massachusetts, Amherst, has implemented LED lighting, efficient HVAC systems, and smart building controls (The Sustainability Tracking, 2019). Reduction of institutional expenditures through sustainable transportation initiatives is another method.

Sustainable transport is an approach that aims to reduce the carbon footprint and pollution caused by transportation and to promote social and economic development. Examples of sustainable transportation include walking, cycling, public transport (buses, trains, and trams, electric vehicles, carpooling, and shared mobility services like bike-sharing and car-sharing). These modes of transport are typically designed to be more energy efficient, emit fewer pollutants, and reduce greenhouse gas emissions, compared to traditional transport methods. They are also designed to prioritize safety, accessibility, and affordability for all users. Sustainable transport promotes a healthy and equitable society by improving accessibility, reducing traffic congestion, and creating more liveable cities. Additionally, it supports economic growth by reducing transportation costs and improving the efficiency of the transportation system (Thapa & Vispute, 2017).

HEIs can foster the reduction of institutional expenditures through enhancing and encouraging sustainable transport initiatives. Bike-sharing or bike rental programmes can make cycling a more accessible and convenient transport option. Providing secure bike parking spaces and repair stations on campus can also serve to encourage cycling on and to campus and by implication a reduction of the budget allocated to transport (Thapa & Vispute, 2017). The University of Warwick, UK has implemented a car-sharing programme for staff and students (Liftango) (Estates, 2022a, 2022b), as well as a bike-sharing programme under the West Midlands Cycle Hire Programme. The campus has over 3500 bike parking spaces, specified cycling routes, and shower facilities (Estates, 2022a, 2022b). HEIs can offer discounted or free public transit passes to students and staff to encourage public transit as a transport option. Though this may sound more remote than possible, within Africa the adoption of electric vehicles can be one way to reduce institutional expenditures. This implies that HEI will need mechanisms and infrastructure to support (both human and technical) not only the acquisition but also maintenance of such vehicles. Electric vehicle (EV) charging stations on campus can encourage staff and students to adopt the use of EVs. The University of California, Davis has been a leader in promoting sustainable transport for decades. It has implemented electric vehicle charging stations on campus among other incentives to promote sustainability (University of California, 2018).

3.3.2 Reduction of Institutional Expenditures through Education and Outreach Activities

Education and outreach activities play a significant role in promoting energy sustainability in HEIs. Raising awareness, encouraging behaviour change, creating a culture of sustainability, and providing resources that support sustainability initiatives are initiatives that HEIs can implement to achieve sustainability. In America, the Arizona State University has offered classes in the design of photovoltaic systems with different configurations including stand-alone and integrated (grid-connected) solar systems and hybrid systems while in the same context the University of Arizona offer classes in computer energy analysis that includes modelling, simulation and design of green buildings and passive solar systems (Mahmoud, 2008). Strathmore University in Kenya offers a renewable energy training programme that provides participants with the skills and knowledge needed to design, install, and maintain renewable energy systems. These programmes have had a significant impact on promoting energy sustainability by equipping individuals with the knowledge and skills needed to design and implement sustainable energy solutions. This includes adoption of a solar PV system that operates in hybrid form together with the energy from the main grid. There are over 2000 licensed solar photovoltaics technicians by the Energy Petroleum and Regulatory Authority (EPRA) in Kenya (Energy Petroleum and Regulatory Authority, 2019). More than half of the licensed solar photovoltaic technicians in Kenya have been trained at Strathmore University, at the Energy Research Center (SERC) (Maranga, 2022). Awareness campaigns aimed at promoting energy sustainability among students, staff, and the wider communities around HEIs have had a significant impact on promoting energy sustainability by raising awareness about the importance of energy conservation and sustainability.

3.3.3 Third Stream Funding Initiatives Available to HEIs as a Source of Additional Revenue

Third stream funding refers to a form of revenue for HEIs that comes from sources other than government funding or tuition fees. Some potential sources for third-stream funding in HEIs are through energy savings, capacity building in sustainable energy, offering consultancy services in sustainability, and carrying out research in sustainable energy systems (Hatakenaka, 2014).

Energy savings can be achieved through various means, related to reducing energy consumption, optimizing building systems, and implementing renewable energy sources. HEIs can leverage these savings to generate revenue through third-party financing arrangements same as energy performance contracts (EPCs) or power purchase agreements (PPAs). PPAs allow institutions to purchase or sell renewable energy at a fixed price over a long-term contract; a good example is Strathmore University in Nairobi that has a PPA with Kenya Power, the electricity distribution utility in Kenya—to sell excess energy generated from a grid-tied solar PV system in the university (Strathmore University, 2014). Under an EPC, an energy services company (ESCO) provides the upfront capital to implement energy efficiency measures and renewable energy systems, and the institution repays the investment from the savings achieved over a set period of time (Taylor et al., 2008).

Sustainable energy capacity building refers to activities that help institutions develop the knowledge, skills, and infrastructure needed to support sustainable energy systems. This may include training programmes for staff and students, research activities to advance sustainable energy technologies, and infrastructure investments in renewable energy systems and energy efficiency measures (Blewitt & Cullingford, 2004). These training and consultancy programmes offer additional revenue to HEIs. In addition, knowledge in sustainable infrastructure investments does not only support revenue generation but has a ripple effect in the infrastructure and buildings for other industries and companies, and therefore overall in minimizing the carbon footprint.

HEIs can leverage third stream funding through various means including grants, partnerships with industry and government, and revenue from energy sales or carbon credits (Blewitt & Cullingford, 2004). Country-country partnerships with the HEI being at the focal point is a way to transmit knowledge from the global north to the global south and can form other avenues including better energy-efficient practices from common research done in these energy efficiency aspects.

A university could develop a training programme for staff and students on sustainable energy technologies and sell this programme to other institutions or organizations. The university could also partner with industry or government to undertake research projects on sustainable energy technologies and receive funding for this work. Additionally, the university could invest in renewable energy systems and sell excess energy generated back to the grid or earn revenue from carbon credits.

However, it is important to note that sustainable energy capacity building activities can be costly and require significant investment upfront. Therefore, institutions must carefully assess the potential benefits and risks of these activities and develop a comprehensive strategy to ensure they are financially sustainable in the long term (Blewitt & Cullingford, 2004). HEIs should adopt an intentional commitment to imbue sustainability into its agenda if it is to adopt energy-efficient and effective practices that can reduce expenditure and generate revenue. It should be intentional in creating policies that speak to the sustainability agenda from infrastructure, buildings, its processes (financial, technological, and human), and the interaction of its internal and external constituents (Taylor et al., 2008). HEIs which have adopted suitable practices for third stream funding through sustainable measures should participate in a broader national agenda of policy formulation. Continuous evidence bases should be provided by such institutions on how the sustainability agenda has not only generated economic savings, but benefited the environment, the society and improved institutional governance plus the positive externalities created. In terms of a broader national, regional and transnational agenda, such HEIs should be at the vanguard of preferring broader indicators of quality mechanisms in HEI that include the environmental, social and governance aspects. These indicators can also be useful in the quintuple helix agenda of the interaction of academia, civic organizations, private institutions, government, and environment.

4 Conclusion

The adoption of sustainable energy practices in HEIs is key to reduce greenhouse gas emissions and to mitigate the impacts of climate change. The use of renewable energy sources analogous with solar, wind, geothermal, hydropower, and biomass can help HEIs achieve their sustainability goals while also providing educational opportunities for students and staff. Moreover, implementing sustainable energy practices can lead to cost savings, improved energy efficiency, and enhanced institutional reputation. It is therefore important for HEIs to set targets and develop sustainability plans that prioritize the use of renewable energy sources, energy-efficient appliances and implement energy efficiency practises. By doing so, HEIs can exemplify leadership in sustainability, support the transition to a low-carbon economy, and contribute to a more sustainable future.

REFERENCES

Architects, C. (2012). *Revolutionary hybrid ventilation system, for new research building. Technical University of Denmark* (Online). Accessed February 7, 2023, from https://archello.com/project/revolutionary-hybrid-ventilation-system-for-new-building-technical-university-of-denmark

AUC. (n.d.). *AUC campus planning and design standards*. AUC.

Blewitt, J., & Cullingford, C. (2004). *The sustainability curriculum: The challenge for higher education*. Routledge.

Burton, T., Sharpe, D., Jenkins, N., & Bossanyi, E. (2011). *Wind energy handbook*. John Wiley & Sons.

Calhoun, A. (2014). Using the program sustainability assessment tool to assess and plan for sustainability. *Preventing Chronic Disease, 11*.

Carayannis, E., & Campbell, D. F. (2009). Mode 3' and 'Quadruple Helix': Toward a 21st century fractal innovation ecosystem. *International Journal of Technology Management, 46*(3–4), 201–234.

De Jong, W., & Van Ommen, R. J. (2014). *Biomass as a sustainable energy source for the future*. Wiley.

Elbarbary, S., Saibi, H., Fowler, A.-R., & Saibi, K. (2022). Geothermal renewable energy prospects of the African continent using GIS. *Geothermal Energy, 10*(8). https://doi.org/10.1186/s40517-022-00219-1

Energy Petroleum and Regulatory Authority. (2019). *Solar photovoltaic regulations*. Energy Regulation and Regulatory Authority.

Erdil, N., Aktas, C., & Arani, O. (2018). Embedding sustainability in lean six sigma efforts. *Journal of Cleaner Production, 198*(10), 520–529.

Estates, U. O. W. (2022a). *Carpooling* (Online). Accessed March 3, 2023, from https://warwick.ac.uk/services/estates/transport/carpooling/

Estates, U. O. W. (2022b). *Cycling* (Online). Accessed March 7, 2023, from https://warwick.ac.uk/services/estates/transport/cycling/

Farisyi, S., Al Musadieq, M., Utami, N. H., & Damayanti, R. C. (2022). A systematic literature review: Determinants of sustainability reporting in developing countries. *Sustainability, 14*, 1–18.

Goswami, Y., & Kreith, F. (2015). *Energy efficiency and renewable energy handbook (mechanical and aerospace engineering series)* (2nd ed.). CRC Press.

Government of Mauritius. (2023). *All Africa* (Online). Accessed February 28, 2023, from https://allafrica.com/stories/202302230338.html

Gulliver, J. S., & Arndt, R. E. A. (1991). *Hydropower engineering handbook*. McGraw-Hill, Inc..

Handschuh, D. (2011). *Harnessing nature's energy to heat an entire campus* (Online). Accessed February 9, 2023, from https://news.ubc.ca/2011/11/03/harnessing-natures-energy-to-heat-an-entire-campus/

Hatakenaka, S. (2014). *Development of third stream funding.* Higher Education Policy Institute.

Henry, D., Adekunle, A., & Idehai, O. (2010). Retrofitting a hydropower turbine for the generation of clean electrical power. *Journal of Research Information in Civil Engineering, 7*(2).

International Energy Agency. (2020). *IEA PVPS annual report 2020.* International Energy Agency.

Ivrine, U. (n.d.) *UCI microgrid* (Online). Accessed March 1, 2023, from http://www.apep.uci.edu/UCI_Micro_Grid.html

Kawamoto, K., Koomey, J., Nordman, B., & Brown, R. (2001). *Electricity used by office equipment and network equipment in the U.S.: Detailed report and appendices.* Energy Analysis Department, Environmental Energy Technologies Division, University of California.

Kitawi, A., & Mumbi, M. (2023). Re-imagining the sustainable African university post-Corona virus pandemic. In *Quality in higher education in Africa: Development perspectives.* s.l.: s.n.

Laboratory, L. B. N. (2014). *Berkeley lab completes UC Berkeley lighting retrofit program.* University of California, Berkeley.

Lund, H. (2014). Introduction. In *Renewable energy systems: A smart energy systems approach to the choice and modeling of 100%* (pp. 1–14). Academic Press.

Mahmoud, W. H. (2008, January). Higher education for sustainability. *Innovative Higher Education.*

Manwell, J. F., McGowan, J. G., & Rogers, A. L. (2010). *Wind energy explained: Theory, design and application.* John Wiley & Sons.

Maranga, I. (2022). *Status of electric vehicle charging and battery swapping infrastructure in Kenya.* UNEP.

Marhani, M. A., Jaapar, A., Bari, A. N. A., & Zawawi, M. (2013). Sustainability through lean construction approach: A literature review. *Procedia- Social and Behavioural Sciences, 101,* 90–99.

Musgrove, P. (2011). Offgrid wind energy systems. In *Renewable energy origins and flows.* Routledge.

National Renewable Energy Laboratory. (n.d.) *Energy basics* (Online). Accessed February 6, 2023, from https://www.nrel.gov/research/learning.html

News, A. (2016). *ASU parking and transit services recognized for sustainability efforts.* Arizona State University.

NREL. (n.d.) *Solar energy basics* (Online). Accessed February 6, 2023, from https://www.nrel.gov/research/re-solar.html

Owusu, A. P., & Asumadu-Sarkodie, S. (2016). A review of renewable energy sources, sustainability issues and climate change mitigation. *Cogent Engineering, 3,* 1–14.

Pandey, B., & Kark, A. (2010). *Hydroelectric energy: Renewable energy and the environment.* CRC Press.

Pelletier, S. G. (2012). Rethinking revenue. *Public Purpose*, 2–5.

Samson, E. I., Ofoefule, A., Uzodinma, E. & Okoroigwe, C. (2011). *Characterization and performance evaluation of 11M3 biogas plant constructed at National Center for energy Research and Development, University of Nigeria, Nsukka.*

Sen, A. (1999). *Development as freedom.* Oxford University Press.

Sharma, S. (2003). *Water power engineering* (2nd ed.). Vikas Publishing House.

Strathmore University. (2014). *Strathmore University News* (Online). Accessed February 3, 2023, from https://strathmore.edu/news/strathmore-university-and-kenya-power-sign-solar-energy-purchase-deal/

Strielkowski, W., Civín, L., Tarkhanova, E., & Tvaronaviciene, M. (2021). Renewable energy in the sustainable development of electrical power sector: A review. *Energies, 14,* 8240.

Taylor, R., Govindarajalu, C., Levin, J., Anke, M., & William, W. (2008). *Financing energy efficiency: Lessons from Brazil, China, India, and beyond.* World Bank.

Thapa, B., & Vispute, S. (2017). Green energy practices in higher education institutions: A review of the literature. *Renewable and Sustainable Energy Reviews, 1*(68), 1–14.

The Sustainability Tracking, A. & R. S. (2019). *University of Massachusetts Amherst OP-5: Building energy efficiency* (Online). Accessed March 11, 2023, from https://reports.aashe.org/institutions/university-of-massachusetts-amherst-ma/report/2020-03-06/OP/energy/OP-5/

University of California, D. (2018). *UC Davis transportation services - Electric vehicles* (Online). Accessed March 9, 2023, from https://taps.ucdavis.edu/parking/ev

University of Lagos. (2019). *Unilag general news* (Online). Accessed February 7, 2023, from https://unilag.edu.ng/?p=9135

Walzberg, J. (2021). *Solar PV reuse & recycling: How human behaviour affects the fate of aging solar panels.*

Watson, D., & Labs, K. (1983). *Climatic building design: Energy-efficient building principles and practices.* McGraw-Hill.

Yu, Y., Junjan, V., Yazan, M. D., & Iacob, E. (2022). A systematic literature review on circular economy implementation in the construction industry: A policy making perspective. *Resources, Conserving and Recycling, 183,* 1–17.

Zamora-Polo, F., & Sanchez-Martin, J. (2019). Teaching for a better world. Sustainability and sustainable development goals in the construction of a change-maker university. *Sustainability, 11,* 1–15.

Quality Assurance of Higher Education in a Neo-liberal Context: *Towards Transformative Practices in Africa*

Joel Jonathan Kayombo **ⓘ**, *Mjege Kinyota* **ⓘ**, *and Patrick Severine Kavenuke* **ⓘ**

1 INTRODUCTION

Quality Assurance (QA) in modern higher education (HE) in Africa is predominantly influenced by neo-liberal ideologies. QA philosophies and practices have as well evolved into an increasingly professionalised power mechanism. The QA philosophies and practices have advocated compliance and an imbalance between power and responsibility. This chapter analyses and critiques the dominance of neo-liberal-driven QA regimes in HE. The chapter first analyses the intersection between QA regimes in Africa and the five defining features of global neo-liberalism as identified by Martinez and Garcia (1997). Then, the chapter proceeds by arguing for change from concern-neo-liberal-driven QA regimes to quality culture

---

J. J. Kayombo (✉) • M. Kinyota • P. S. Kavenuke
University of Dar es Salaam, Dar es Salaam, Tanzania
e-mail: joel.kayombo@udsm.ac.tz; kmjege@yahoo.com;
patrickkavenuke@gmail.com

© The Author(s), under exclusive license to Springer Nature
Switzerland AG 2024
P. Neema-Abooki (ed.), *The Sustainability of Higher Education in Sub-Saharan Africa*, Sustainable Development Goals Series,
https://doi.org/10.1007/978-3-031-46242-9_14

327

to sustainably improve HE in Africa. Neo-liberalism is a highly contested term, at least in part because the term is used by different people to mean different things. For instance, Gertz and Kharas (2019) identify three distinct but related conceptualisations of neo-liberalism namely; neo-liberalism as thought collection; neo-liberalism as academic theory, and neo-liberalism as a policy practice.

As a thought collection, neo-liberalism refers to an organised intellectual and political movement, propagated by a specific group of people (Gertz & Kharas, 2019). This group of thinkers emerged in the 1930s, as market economies were challenged by the emergence of Nazism and communism. This group of liberal thinkers mostly in Europe and America felt the need to promote an alternative discourse championing the priority of the price mechanism, private free enterprise, competition, and a strong and impartial state (Gertz & Kharas, 2019). From this perspective, neo-liberalism constitutes a debate over what ideas these self-proclaimed neo-liberal individuals believed in and supported (Gertz & Kharas, 2019).

As a thought collection, neo-liberalism represents an ideology that promotes markets over the state, and regulation and individual advancement/self-interest over the collective good and common well-being (Ball, 2012). Indeed, the collapse of the Soviet Union and the Berlin wall is considered a marking point of the changes that ended communism. As a political movement, ending communism (and thus replacing it with neo-liberalism) dominated many policies in Europe and America. For instance, President Harry S. Truman of the United States of America (USA) in 1947 promised support to any nation that would engage in preventing the spread of communism.

Neo-liberalism as academic theory refers to the academic study of economics using neoclassical models. These models are neo-liberal in the sense that they are grounded in individual choices on what to consume and produce (Harvey, 2007). Besides, neo-liberalism maintains that countries remove obstacles to free market capitalism and allow capitalism to spawn development. The argument is that, if allowed to work freely, capitalism will generate wealth which will trickle down to everyone. In other words, neo-liberalists believe that private enterprises or companies should take the lead in development.

When individual and firm preferences are combined, they lead to supply and demand curves that constitute markets. It is assumed that when individuals and firms optimize their decisions, economists can identify stable, Pareto-optimal equilibria. Thus, the theory stresses opening up the market and capital and limiting protectionism (Mintz, 2021). It is further insisted that the governments should also embrace a role through taxes and spending that are modelled to maximize some social welfare functions. Generally, in this perspective, neo-liberalism is bounded by the discourse of neoclassical economics and the proper interpretation of neoclassical microeconomic and macroeconomic models (Gertz & Kharas, 2019).

Lastly, neo-liberalism as a policy practice constitutes a set of economic policies that have been executed by governments adopting the same principles of individualism and markets. From the late twentieth century onwards, neo-liberalism as a policy practice stressed the hymn of "stabilization, privatization, and liberalization"(Gertz & Kharas, 2019). It demanded governments adopt a friendly approach toward regulation, avoid the industrial policy, and use the logic of market competition to allocate resources wherever possible, including the area of education (Harvey, 2007).

That is to say, neo-liberalism can be considered as a complex, often incoherent, unstable, and even contradictory set of practices that are organised around a certain imagination of the market (Ball, 2012). The practice and its related discourses have triggered the universalisation of market-based social relations that has percolated in almost every single aspect of our lives which emphasizes commodification, capital accumulation, and profit-making (Ball, 2012).

The conceptualisation of neo-liberalism is constant contestation not only because it is commonly applied to these three overlapping yet not identical conceptualizations, but also because in each of these cases, there is no one single, narrowly-defined idea, but rather considerable opacity and flexibility (Gertz & Kharas, 2019). Despite the complexities in defining the term, there have been several attempts to compile the defining features of neo-liberalism. For instance, Martinez and Garcia (1997) identified five defining features of global neo-liberalism that the current HE system in Africa is not immune to, namely (i) the rule of the market, (ii) cutting public expenditure for social services like education and health care, (iii) deregulation, (iv) privatization, and (v) eliminating the concept of "the public good" or "community" and replacing it with "individual responsibility".

Similarly, Wrenn (2015) advanced very similar categories of neo-liberal narrative consisting of three well-defined tropes: (i) privatisation of state-provided goods and services, (ii) deregulation of industry and (iii) retrenchment of the welfare state. She confirms that all three tropes reinforce a central premise: the locus of control is the individual exercising agency through a (free) market operation.

Generally, neo-liberalism embodies the ideological shift in the purpose of the state from one that has a responsibility to ensure full employment and protect its citizens against the exigencies of the market to one that has a responsibility to ensure the protection of the market itself (Harvey, 2005). We acknowledge that as policy practice, neo-liberalism has been adapted to local contexts and conditions as it was implemented by local elites, leading to a considerable disparity across place and time (Gertz & Kharas, 2019). Nevertheless, there are core defining features of neo-liberalism, which can be used to determine both, what it includes and what it excludes, and it remains a useful analytical concept for understanding the contemporary world. For this chapter, we focus primarily on the last meanings of neo-liberalism-neo-liberalism as a policy practice.

2   MULTIPLE LEVELS ANALYSIS: A CONCEPTUAL FRAMEWORK

To understand the intersection between neo-liberalism and QA in HE, we use three levels of analysis; macro level, meso-level, and micro levels of analysis (Kentikelenis & Rochford, 2019). The conceptual framework presented here (see Fig. 1) seeks to clarify how neo-liberalism manifests at different levels of analysis, thereby structuring the complexity of the QA practices and processes.

Firstly, at the macro institutional level is where complexes of routines, rules, roles, and meanings reside, simultaneously shaping and being shaped by social action. Simplistically, institutions have regulative, normative, and cognitive dimensions that guide the behaviour of organisations and individuals (Kentikelenis & Rochford, 2019). This level of analysis focuses on elaborating on how institutions came to be in the first place, how they change, and how they structure social and policy environments. Neo-liberalism runs through macro–micro–macro processes.

Secondly, QA is substantially shaped by meso-level processes that carry imprints of global neo-liberalism. This meso-level includes various types of actors who employ distinct forms of decision-making to develop an array of QA policies. As to the terms of types of actors involved, the neo-liberal era has promoted the increased prominence of the private sector in HE,

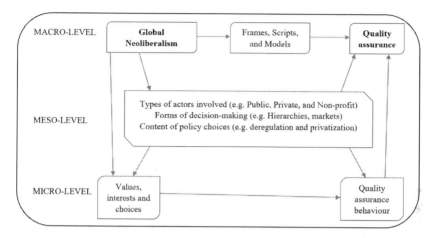

Fig. 1 Multiple levels analysis of neo-liberalism and quality assurance. Adapted from Kentikelenis and Rochford (2019)

with a shift away from purely public solutions. Also, the forms of organizational decision-making reflect neo-liberal approaches. This can be amplified by the introduction of market-based approaches within public h HE. Besides, the emergency of the logic of efficiency, profit-maximisation, and managerialism in HE have impacted QA systems.

Third, neo-liberalism also manifests at the level of individuals, or the "micro" level (Kentikelenis & Rochford, 2019). Some individuals have more powerful resources, case in point, money, networks, or positions of authority than others to advance their values and interests. At this level, individuals must not be viewed as all-powerful free-floating actors, but rather, as representatives of the dominant institutional orders. At this level, neo-liberal frames become internalized by individuals around the world and are in turn reflected in their ideas, values, and choices. Ideas and values are important, as they are tools that people employ to make sense of the world. For example, neo-liberalism's focus on the individual can turn collective policy problems into questions of individuals. More generally, neo-liberalism creeps into individual behaviours towards QA.

To comprehend how neo-liberalism intersects with QA practices, analyses must pay attention to processes operating at various levels of analysis: from individual-level behaviours to organizational decision-making, and right up to the institutional structures that frame policy thinking and individual actions. At the macro level, global neo-liberalism produces frames, scripts, and models that shape our understanding of the world including QA issues. At the meso level, neo-liberal doctrines determine the type of actors, forms of decision-making, and the content of policy choices in HE. And finally, at the micro level, neo-liberalism shapes individual values, interests, and choices. For instance, policy elites may wish to advance neo-liberal values when designing QA policies or engaging in the internationalisation of HE.

We rule that whoever is practising QA in a neo-liberal context at any level among the three levels, s/he has to understand that QA needs to be treated as an engaging and transformative endeavour. In that regard, practitioners of QA need to freely reflect and take action by removing all the social arrangements that are usually created by the existing unfair social hierarchies in HEIs. Eventually, the practice may lead to transformative QA systems.

3   Understanding Quality Assurance Using the Lens of Educational Policy Transfer

QA in HE has been considered important attesting to its contribution towards increasing organisational excellence and productivity. However, there have been some key questions that people have kept asking when they discuss QA. Such questions include: what is quality? What is QA? How do institutions conduct QA? What are the quality indicators in any HEI? Who are the customers of HEIs? (Jingura & Kamusoko, 2018; Bwalya, 2023). The other key question that has been asked is how to build a quality culture in HEIs. (Legemaate et al., 2021).

Any product in an organisation needs to meet customer expectations, and set standards. In that regard, a product is thought to be of good quality if it complies with the QA requirements specified in the industry and meets the needs of the customers (Bwalya, 2023). To have quality outputs, there is a need to monitor and evaluate the process —that is, conduct as a form of formative evaluation, other than waiting for the outputs (Jingura & Kamusoko, 2018; Bwalya, 2023). In the process of conducting QA, it is advised that the task is made collegial and the purpose should be to identify the gaps and fix them before things go astray. For the QA to be conducted smoothly there is a need to build a quality culture in the institutions. The details on how to build a quality culture are provided in the next sections.

QA systems have become inevitable in the context of neo-liberalism which focuses on free market capitalism, competition, and university ranking (Jarvis, 2014a, 2014b). In that regard, in this neo-liberal era, there is no way one can talk of free market capitalism, competitions, and university rankings without talking about QA systems. Therefore, as we reiterate throughout the chapter, neo-liberalism ideals have dominated QA in HE across the globe. About Africa, the framings of QA as we know it today are very much connected and thus can be understood using the lens of educational policy transfer, or simply the transfer of "best practices" in education.

In the case of QA, "best practices" can be reflected in almost similar regulatory regimes that seek to manage, and direct, the HE sector across many countries. According to Steiner-Khamsi (2006) educational policy transfer or the diffusion of "best practices" takes place in different forms which represent a continuum of autonomy. First, neo-liberal informed "best practices" in QA might have diffused in Africa through deliberate or autonomous efforts to learn from other countries. In this case, strategies—let's say. Sending a team of experts to go and learn from other countries could have taken place. Secondly, the diffusion of "best practices" in QA might be an automatic spread of the discourse through globalisation, publications, attendance at international conferences, and notably through Information and Communication Technologies (ICT) (Arnove, 2007). Finally, the transfer of "best practices" in QA might have been imposed through conditionality by donors or being forced to conform to international standards to compete in global university rankings and other league tables.

No matter which form of policy transfer was involved, there is a need to understand how "best practices" in QA were received and internalised in Africa. Without this kind of analysis, we will be denying the distinct and unique cultures of Africa, within which such best practices are implemented. While many African academics have been globalised and thus conform to international standards, it is an undeniable fact that they are a product of a unique socialization process that may have a tremendous influence on how they perceive, receive, and implement policies in QA.

In the same sense, even a shift towards promoting quality culture as we propose in the next sections should be analysed in terms of how it intersects with unique cultures in Africa. In doing so, we recommend a framework of policy borrowing in education by Phillips (2004) who proposed four stages-cross-national attraction, decision-making, implementation, and domestication/indigenisation through which policies or "best practices" transfer across localities.

The first stage of cross-national attraction is concerned with how countries are attracted to borrow (sometimes lend in the case of imposition) educational practices from elsewhere. Policy attraction, in this case, reforms regarding best practices in QA, may be fuelled by regime change, negative external evaluation, suppose that, not conforming to international standards and in which the existing practices are scandalised, systemic collapse, and internal dissatisfaction by stakeholders within Africa. Such factors may spark the transfer of policy aspects, to illustrate, QA guiding philosophy, strategies, goals and processes as well as the "enabling structures" (Phillips, 2004, p. 58) which are necessary for QA practices to function in a new context.

Decision-making is the second stage in which countries or institutions decide to borrow an educational practice or policy. According to Phillips (2004), the decision may be "theoretical" (expectations that certain viewpoints/practices will work in the home contexts) or "practical" (in which careful analysis of the home context has been made and implementation proved to be promising) or decision can be a "quick fix" where the practice is not compatible with the new context but it must be somehow implemented to combat internal criticism or satisfy influential actors in the name of donors. (p. 58).

The third stage of implementation involves efforts to put into practice the borrowed practices in which case the speed of implementation will depend on support/resistance from implementers (e.g. do academics view QA as a good thing that will enhance the quality or as policing and thus a nuisance?) as well as the extent of compatibility in terms of context (are unique cultures and institutional framework in Africa compatible with neo-liberal ideals?). Overall, successful implementation will depend on how the new practices are adapted to a new context. The last stage involves "internalization" (Phillips, 2004, p.59) and "indigenisation" (Steiner-Khamsi, 2006, p. 162) under which evaluation is made to gauge the effect of a borrowed practice on existing systems and the general performance in relation existing structures. Findings of evaluation may yet lead to another cross-national attraction, repeating the cycle once again.

3.1 Global and African Perspectives on Quality Assurance as Policy Transfer

Historical legacies of colonialism, aid, and the effects of top-down programmes, to illustrate structural adjustment plans (SAPs), have all influenced African countries' perceptions of development relations with the "global North", and developed countries' perceptions of the "global South" (Constantine & Shankland, 2017). While reforms in many countries have moved on from the days of the SAPs and another coercive neoliberal "policy solutions" imposed on (and accepted by the elites in) developing countries, their echoes are present in the politics of North-South development "knowledge transfer" and internationalisation of HE.

The emphasis on knowledge transfer as a central element of development cooperation had its origins in the late 1990s, a period when criticism of the social impacts of SAPs and increasing political resistance to explicit conditionality were beginning to force Northern-dominated development reforms to consider a change in approach.

There is no contradicting argument that the global education reforms have impacted education policies in Africa in a significant manner. The mode of transfer, the motives and the resources available for the implementation of those reforms have influenced and will certainly continue to influence the form and content in varying degrees in different universities, countries and regions of Africa claiming to harmonise and elevate HE to the global standards. Indeed, some governments in Africa and international organisations are proactive in promoting harmonisation and convergence or exporting/importing policy lessons.

The African perspective on QA in HE is a unique one. Nonetheless, this is not to assume that Africa is an entity that can be described with a precise level of homogeneity. We acknowledge that Africa is very big in geographical size as well as in terms of the number of countries which have diverse cultures, traditions, and philosophies (including those which guide HE and QA practices). That being said, some common features unite Africa. Notably, the African philosophy of Ubuntu can be said to unite African social, cultural, and economic lives. As a philosophy, Ubuntu entails an aspiration to build and maintain a just society that is based on the pillars of dignity, reciprocity, harmony, compassion, and humanity (Assié-Lumumba, 2017). These pillars, one can say, are not very much aligned with pillars of, for example, individualism and competitiveness which dominate much of HE and its QA practices when conceived in their neo-liberal leanings.

Secondly, Africa has a unique history which may distinguish it from the Global North. In particular, HE reflects on the struggles for independence from the 1950s to the mid-1970s. This implies that many HEIs in Africa are just celebrating 60 years since their establishment, making them comparatively very young. There is also the issue of why these institutions were HEIs in the first place. In this case, the majority of HEIs were established as part of larger efforts to rebuild new post-colonial countries notably in the area of human capital development. It is also important to note that, some of these institutions hoped to build socialist countries in a way that contradicts the aims of the present–day neo-liberal HEIs. We use Tanzania as an example to illustrate this assertion. Tanzania embarked on transforming her education system to prepare a learned community that would lead post-colonial Tanzania (Nyerere, 1967). The University of Dar es Salaam (formally University College, Dar es Salaam) was established in 1961 to serve the aforementioned purpose. Accordingly, Education for Self-Reliance (ESR) (Nyerere, 1967) was launched to prepare a socialist society which would be characterised by social and economic equality (Nyerere, 1968). However, later in the 1980s, following the introduction of SAPs as a result of intensified globalisation, there was a major shift from an education policy based on socialism to an education system based on a market-driven economy. This time, the country's focus had changed to that of preparing citizens who could compete in the science-and-technology-driven world. These shifts in philosophy not only imply the immaturity of many African countries in terms of coping with the neo-liberal principles of competition and standardisation but also the fact that these countries are still struggling to manage these transitions.

Finally, many HEIs in Africa are state-owned when compared to private institutions. The institutions tend to be mainly financed by the government to build a workforce that will save national (often not global) interests. As a consequence, many stakeholders in HE, except the management cadre, are not under much pressure to compete in ways congruent with many neo-liberal institutions.

4 Neo-liberalism, Quality Assurance, and the Sustainable Development Goals

The Sustainable Development Goals (SDGs) were adopted by more than 150 nations at a special UN General Summit in September 2015 as an ambitious programme to promote shared prosperity and well-being for all over the next 15 years. Goal 4 of the programme aims to ensure inclusive and equitable quality education and promote lifelong learning opportunities for all. The emphasis here is placed on the "quality" of education, and not simply on access. Whether such education leads to greater employment in "decent jobs", will depend on how the global over-supply of workers is managed through radically different labour market policies than those erected by the last forty years of neo-liberal economics (Belda-Miquel et al., 2019; McCloskey, 2019; Weber, 2017).

The debate around the goals goes to the heart of contemporary HE policies. Does the attainment of the goals remain a largely technical, managerial, and depoliticised discourse? Or, does it widen its ambit for debate into the need for systemic change and political influence addressing the quality crisis in education? It is difficult to anticipate a scenario where the managerial path alone, which the SDGs appear to represent, will provide the transformative change needed to achieve inclusive and equitable quality education.

Although the 17 SDGs and their 169 targets aim at improving the living environment of populations for a more sustainable future, their universal and inclusive claim has been challenged, as well as the methods and processes used to achieve them (Begashaw, 2019; Belda-Miquel et al., 2019; Shettima, 2016). For instance, the fact that the Agenda 2030 is formulated as "goals to be achieved" and not as "the rights of persons to be guaranteed" is a limitation. Similarly, SDGs are blamed for not questioning the unequal power relations at the root of the issues that the SDGs claim to tackle (Belda-Miquel et al., 2019). Those voices advocate for the need to value local knowledge and schemes of thought, and to decolonise the practices of international development to establish a real dialogue for bottom-up development and the respect of people (Meusburger et al., 2018). Others doubt that the stated goals could be achieved without a challenge of norms, regulations, and logic at play. The degradation of contexts in a variety of fields (environmental, economic, social, and political) also questions the reality of progress and throws some doubt on its sustainability (Belda-Miquel et al., 2019).

The SDGs are framed as a universal project, with quite substantial institutional monitoring mechanisms aimed at ensuring the successful implementation of aligned policies. Indeed, the implementation of highly contested neo-liberal policies are themselves explicit goals of the SDG agenda (Weber, 2017). In that case, as universities institute QA mechanisms, they are impliedly responding to neo-liberal demands.

5   Neo-liberalism, Higher Education, and Quality Assurance

In this chapter, we recognise that neo-liberalism in all its forms has percolated in HEIs as it is in all other spheres of life as in political and economic spheres. As Harvey (2007) advises, neo-liberalism has become a hegemonic discourse with pervasive effects on ways of thought and political-economic practices to the point where it is now part of the common sense way we interpret, live in, and understand the world. The argument is also supported by Torres (2011) who postulated that although neo-liberalism has utterly failed as a viable model of economic development, the politics of culture associated with neo-liberalism are still in force, becoming the new common sense shaping the role of government and education.

With the influence of neo-liberal ideologies, education policies are being reformed and reworked in different parts of the world including Africa. Policies are flowing and converging to produce a singular vision of 'best practice' based on the methods and tenets of the 'neo-liberal imaginary' (Ball, 2012). Indeed, the networks through which neo-liberal ideals diffuse across national education policies are so complex that one cannot easily trace where they originate (Ball, 2012). Organized around the imagination of the rule of the market, cutting government expenditure for social services, deregulation, and privatization, neo-liberalism has profound social effects including the functioning of HEIs (Ball, 2012; Harvey, 2007; Kayombo, 2019; Mintz, 2021). As stakeholders in HE, it is high time that we accept the bitter truth that neo-liberalism is not out there, it is with us. Different stakeholders in HE are faced with deciding in 'conditions of undecidability' as well as being positioned in all of this, being complicit, imbricated and compromised (Ball, 2012).

Neo-liberalism has a considerable impact on creating a market-oriented HE system (Toyibah, 2020). Consequently, HE is greatly subjected to the realm of accountability, QA, and efficiency. As noted earlier, neo-liberalism in HE has been implemented in principles of commodification, privatisation, internationalisation, and QA to respond to market demands. As they pursue these principles, the universities shift into providing services, research, and labour to the industrial sector of the economy. This is highly emphasised in global, regional, national and local (university) policies. With the consolidation of neo-liberalism in HE, universities are seen as a key driver in the knowledge economy and as a consequence, HEIs have been encouraged to develop links with industry and business in a series of new venture partnerships (Olssen & Peters, 2005; Mgaiwa, 2021).

Although universities in Africa have changed since the dawn of their existence, the speed of change started to accelerate remarkably in the 1960s after independence. Exponential growth in the number of students and staff was immediately followed by administrative reforms aimed at managing this growth and managing the demands of students for democratic reform and societal relevance (Lorenz, 2012). Since the 1980s, however, an entirely different wind has been blowing along the academic corridors. The fiscal crisis of the welfare states and the neo-liberal course of the governments made the battle against budget deficits and government spending a political priority.

Towards the end of the 1980s and the beginning of the 1990s, the neo-liberal agenda became more pervasive and the governments in many African countries were forced to systematically reduce public expenditure by privatising public services and introducing market frameworks. Such reforms led to a market-driven HE system which employs modes of governance based on a corporate model (Enright et al., 2017). Along with the strained business models, universities in the neo-liberal context need to manage diverse competing needs within and outside academia (Kayombo & Misiaszek, 2022). While on one hand, university activities are now gauged and audited against value for money criteria, on the other hand, activities are as well ought to respond to the market needs and ensure customer satisfaction (Kayombo & Misiaszek, 2022).

In an attempt to problematize QA in the context of neo liberalisation in Africa, this chapter was stirred by Jarvis's (2014a, 2014b) theorisation of QA regimes in the governance approaches to HE. Jarvis elucidates that the emergence and spread of QA regimes impose quasi-market, competitive-based rationalities premised on neo-liberal managerialism using a policy discourse that is often informed by conviction rather than evidence. Besides, maintains that QA regimes must not only be understood as managerial instruments but must also be understood as part of a broader series of agendas associated with neo-liberal policy prescriptions that valorise market rationality.

Although the purpose of QA in HE has been a topic of inconclusive discussion among stakeholders, its practices have become a basic principle in HE. Under neo-liberal university management, the universities are subjected to mechanisms under performance-based evaluation. Evaluation and assessment relating to the performance of a university and its academic staff are considered strategies to ensure the quality of HE (Toyibah, 2020). Academic life is shaped and framed by performance evaluation and assessment of research and the outputs of academic activities as awards, grants, consultation and the impact of research, and the quality of teaching. These aspects of academic performance are fundamental to determine the recruitment process, academic career advancement, and promotion.

Historically, universities have been regarded as institutions that provide academic freedom, freedom of thought and expression, heterodoxy and exploration to create new knowledge. Initially, universities were self-regulating and the quality they offered was never assessed as it was taken for granted to be excellent. Nevertheless, with the spread of neo-liberalism, universities are mandated to adapt to a series of regulatory regimes that seek to manage and direct the sector in ways that serve the interests of the neo-liberalist economy, and the denationalised state by applying principles of efficiency, value, and performance.

6 Towards a Quality Culture in a Neo-liberal University Context

The analysis of the neo-liberal narratives as they intersect with QA in HE is not meant to put in two cents that there is a comprehensive and complete 'neo-liberal agenda' that is actively enforced by maniacal powers. Rather, we are advancing an argument that the neo-liberal narratives consist of a central ideological construct of hyper individualism. As a result, market activities are legitimised and prioritised above socially integrative activities. Through the socialisation processes, neo-liberal ideologies advocate that each individual should be accountable to oneself and in so doing, each individual's responsibility to others and the collective society is eroded. As a result, society is made up of self-interested and atomistic individuals carrying their agendas. Besides, the emphasis on individual accountability and responsibility within neo-liberalism naturally shifts powers to the individual acting alone rather than collective responsibility.

On the other hand, quality culture required shared values, beliefs, expectations, and commitment QA. Within this context, we categorise that if universities are to institute sustainable QA practices, they have to begin by instituting a quality culture. It is however very important to note that there is no full and common understanding about what constitutes quality culture as well as how it can be built in an institution. In this case, there is a need to create a shared understanding among the stakeholders, so that the promotion of quality culture is undertaken by a group of people with a shared understanding (Legemaate et al., 2021).

Indeed, Legemaate et al. (2021) have castigated the lack of ownership of the QA process by members of the organisation which per se creates a great challenge to promoting quality culture. In that regard, they hold the view that a quality culture should be built by ensuring ownership of QA processes by members of the organisation, setting rules and regulations, encouraging teamwork, developing proper QA policies and implementing them, conducting regular staff training, providing effective mentorship and support, as well as implementing total quality management.

The implementation of neo-liberal-driven quality assurance regimes represents one of the major challenges facing today's globally operating universities. HEIs face an increasingly competitive environment, both locally and globally, leading to elevated demands for quality assurance in teaching, research as well as in service and administration (Sattler & Sonntag, 2018). Quality assurance has therefore been a central policy issue in HE for the past three decades. Extensive debates on quality assurance have served as a starting point for introducing the concept of quality culture.

In the context of quality culture, it is no longer only a question of assessing quality using hard facts, for instance, the number and quality of publications or the amount of funding attracted, but also the question of assessing the extent to which quality is subscribed to and lived by members of a HEIs (Sattler & Sonntag, 2018). All along, academics have been in disagreement with the commonly used quality criteria, for example, bibliometric indicators. With the institutionalisation of quality culture, quality could well become a concept with which they can all identify, regardless of their discipline.

We have to come to terms that culture by design is a transformative process that never ends and this is evidenced by an institution's commitment to continuous delivery and integration of quality systems that begin with people, values, and mission alignment. As a set of shared attitudes, values, goals, and practices, culture characterizes an institution or organization. In HEIs, culture signalises that administration, faculty, students, community, and partners must desire a quality culture and make a commitment to organizational excellence. In this regard, we understand institutions and academic programmes that signify quality first, commit to an excellence journey, and secondly, integrate quality as a core value.

The quality culture approach requires an understanding that certain things in groups are shared or held in common. The major categories of observables that are associated with culture in this sense were identified by (Schein & Schein, 2016) as observed behavioural regularities when people interact; group norms; espoused values; formal philosophy; rules of the game; climate; embedded skills; habits of thinking, mental models, and linguistic paradigms; shared meanings; "root metaphors" or integrating symbols; and formal rituals and celebrations. Schein and Schein (2016) further classified these observables into three distinct levels namely: Artefacts (Visible organizational structures and processes); Espoused beliefs and values (Strategies, goals, philosophies); and Underlying assumptions (Unconscious, taken-for-granted beliefs, perceptions, thoughts, and feelings).

Schein and Schein's categorisation provides a valuable framework for different levels that need consideration when developing a quality culture in an organisation. In this case, to assess quality culture, it is essential not only to take into account the visible quality artefacts within an organization (e.g. quality assessment tools) but also to analyse its quality values and shared basic assumptions (e.g. commitment) about quality. The quality culture approach thereby goes far beyond classic ranking procedures, which are limited primarily to the assessment of artefacts that distinguish quality (Sattler & Sonntag, 2018).

Scholars (Loukkola & Zhang, 2010; Adina-Petruţa, 2014; Vilcea, 2014; Sattler & Sonntag, 2018) have certified that the purpose of incorporating quality culture in an organizational culture is to sustain and enhance quality permanently. To achieve that, two distinct elements have to be taken on board. On the one hand, a cultural/psychological element of shared values, beliefs, expectations, and commitment towards quality and, on the other hand, a structural/managerial element with defined processes that enhance quality and aim at coordinating individual efforts need to be considered.

In other words, QA should not only objectively focus on tangible aspects: tools and procedures (artefacts) of quality management but also the quality culture which encompasses organisational-psychological aspects: espoused values, expectations, and commitment to quality which are rather difficult to capture in the traditional QA practices.

Experience has it that QA in HE has mainly focused on academic issues leaving the administrative part ignored (Boateng, 2014). In the quality culture, quality assurance needs to focus on both academics and administration. It is quality in the administrative part that necessitates quality in the academic part. Despite its pivotal role, the administrative part has tended to be ignored by QA in HE mainly because it is perceived as somehow disconnected from the core functions of teaching, research, and consultancy. Nonetheless, it is important to note that the administrative part of HE performs duties that have a direct impact on the quality of academics. This is executed from recruitment and management of human resources (including academic staff), management of funds through proper budgeting, and planning to policy development, implementation, and evaluation. Thus, the role of the administrative part cannot be overlooked in promoting quality in organisations.

Since neo-liberal-driven QA are inherently a technical process, we advocate that QA in HE should change from being a technical process to a political process. We draw reasons for this argument from Skolnik (2010) who favours three reasons why QA should, in many cases, be viewed as a largely political process:

(i) The considerable differences of opinion among different stakeholders about the definition of quality.
(ii) The likelihood that QA serves as a vehicle for the pressures towards conformity within the academe. The tendency to exclude some stakeholders, particularly faculty and administrative staff, from any significant role as in the design and implementation of quality assurance processes.

Viewing QA as a political process reminds practitioners in quality assurance to consider it as a democratic undertaking where people are engaged in developing and implementing quality assurance policies. In this way, the employment of a participatory approach as opposed to an audit approach to quality assurance for ensuring the sustainability of the quality of HE in Africa is indispensable. The participatory approach starts with the claims, concerns, and issues put forth by all stakeholders. Also, the use of a participatory approach reminds practitioners of QA of the need to do total QA by considering the participation of all members of the organisation (for instance, lecturers, students, administrators, drivers, games and sports tutors) to ensure quality.

7   Conclusion

From the discussion, we conclude that QA should be an integral part of any institution including HE, particularly in the era of neo-liberalism where the systems need to be monitored, evaluated, and regulated regularly to bring about relevant outputs and overall organizational excellence. Since, QA is a socially constructed domain of power and its philosophies, design, and practices are political, we propagate that QA practices and philosophies must transform oppressive practices and social arrangements that are created by unfair social hierarchies, white supremacy, and unregulated capitalism. Therefore, QA philosophies and practices have to be emancipatory where the main focus is praxis— reflection and action- as put forward by Paulo Freire (Freire, 2010). As QA does reflection and takes action to the problems identified, in the process the institutions are transformed and through that transformation, the institutions respond to the neo-liberal forces.

References

Adina-Petruţa, P. (2014). Quality culture - a key issue for Romanian higher education. *Procedia - Social and Behavioral Sciences, 116*, 3805–3810.

Arnove, R. F. (2007). In R. F. Arnove & C. A. Torres (Eds.), *Comparative education the dialectic of the global and the local* (3rd ed.). Rowman & Littlefield Publishers, Inc.

Assié-Lumumba, N. T. (2017). The Ubuntu paradigm and comparative and international education: Epistemological challenges and opportunities in our field. *Comparative Education Review, 61*(1), 1–21. https://doi.org/10.1086/689922

Ball, S. J. (2012). *Global education inc.: New policy networks and the neo-liberal imaginary*. Routledge.

Begashaw, B. (2019). Africa and the sustainable development goals: A long way to go. *Brookings.* https://www.brookings.edu/blog/africa-in-focus/2019/07/29/africa-and-the-sustainable-development-goals-a-long-way-to-go/

Belda-miquel, S., Boni, A., & Calabuig, C. (2019). SDG localisation and decentralised development aid : Exploring opposing discourses and practices in Valencia' s aid sector. *Journal of Human Development and Capabilities, 20*(4), 1–17. https://doi.org/10.1080/19452829.2019.1624512

Boateng, J. K. (2014). Barriers to internal quality assurance in Ghanaian private tertiary institutions. *Research on Humanities and Social Sciences, 4*(2), 1–9.

Bwalya, T. (2023). Quality assurance in higher education and its implications on higher education institutions and challenges in Zambia. https://doi.org/10.20944/preprints202301.0049.v1.

Constantine, J., & Shankland, A. (2017). From policy transfer to mutual learning?: Political recognition, power and process in the emerging landscape of international development cooperation. *Novos Estudos CEBRAP, 36*(1), 99–122. https://doi.org/10.25091/s0101-3300201700010005

Enright, E., Alfrey, L., & Rynne, S. B. (2017). Being and becoming an academic in the neo-liberal university: A necessary conversation. *Sport, Education and Society, 22*(1), 1–4.

Freire, P. (2010). *Pedagogy of the oppressed (30th Anniv)*. Continuum.

Gertz, G., & Kharas, H. (2019). *Beyond neo-liberalism: Insights from emerging markets* (pp. 1–107). Global Economy and Development Program.

Harvey, D. (2005). *A brief history of neo-liberalism*. Oxford University Press.

Harvey, D. (2007). Neo-liberalism as creative destruction. *Annals of the American Academy of Political and Social Science, 610*(1), 22–44. https://doi.org/10.1177/0002716206296780

Jarvis, D. S. L. (2014a). Policy transfer, neo-liberalism or coercive institutional isomorphism? Explaining the emergence of a regulatory regime for quality assurance in the Hong Kong higher education sector. *Policy and Society, 33*(3), 237–252. https://doi.org/10.1016/j.polsoc.2014.09.003

Jarvis, D. S. L. (2014b). Regulating higher education: Quality assurance and neo-liberal managerialism in higher education-a critical introduction. *Policy and Society, 33*(3), 155–166. https://doi.org/10.1016/j.polsoc.2014.09.005

Jingura, R., & Kamusoko, R. (2018). A framework for enhancing regulatory cooperation in external quality assurance in southern Africa. *Quality in Higher Education, 24*(2), 154–167. https://doi.org/10.1080/1353832 2.2018.1480343

Kayombo, J. J. (2019). Becoming, doing, being and belonging into academics: Career trajectories of early-career academics at the University of Dar es Salaam. *Paper in Education and Development, 2*(37), 180–200.

Kayombo, J. J., & Misiaszek, L. I. (2022). *The Palgrave handbook of imposter syndrome in higher education*. Springer International Publishing. https://doi.org/10.1007/978-3-030-86570-2

Kentikelenis, A., & Rochford, C. (2019). Power asymmetries in global governance for health: A conceptual framework for analyzing the political-economic determinants of health inequities. *Globalization and Health, 15*(Suppl 1), 1–10. https://doi.org/10.1186/s12992-019-0516-4

Legemaate, M., Grol, R., Huisman, J., Oolbekkink-Marchand, H., & Nieuwenhuis, L. (2021). Enhancing a quality culture in higher education from a socio-technical systems design perspective. *Quality in Higher Education, 28*(3), 345–359. https://doi.org/10.1080/13538322.2021.1945524

Lorenz, C. (2012). If you're so smart, why are you under surveillance? Universities, neo-liberalism, and new public management. *Critical Inquiry, 38*(3), 599–629. https://doi.org/10.1086/664553

Loukkola, T., & Zhang, T. (2010). *Examining quality culture: Part 1 – Quality assurance processes in higher education institutions.* European University Association.

Martinez, E., & Garcia, A. (1997). *What is neo-liberalism? A brief definition for activists.*

McCloskey, S. (2019). The sustainable development goals, neo-liberalism and NGOs: Its time to pursue a transformative path to social justice. *Policy and Practice: A Development Education Review, Autumn, 29,* 152–159.

Meusburger, P., Heffernan, M., & Suarsana, L. (2018). Geographies of the university. In P. Meusburger, M. Heffernan, & L. Suarsana (Eds.), *Raumforschung und Raumordnung* (Knowledge) (Vol. 77, 3). Springer.

Mgaiwa, S. J. (2021). Fostering graduate employability: Rethinking Tanzania's university practices. *SAGE Open, 11*(2), 215824402110067. https://doi.org/10.1177/21582440211006709

Mintz, B. (2021). Neo-liberalism and the crisis in higher education: The cost of ideology. *American Journal of Economics and Sociology, 80*(1), 79–112. https://doi.org/10.1111/ajes.12370

Nyerere, J. K. (1967). *Education for self-reliance* (Vol. 19, p. 382). Oxford University Press.

Nyerere, J. K. (1968). *Ujamaa - Essays on Socialism.* Oxford University Press.

Olssen, M., & Peters, M. A. (2005). Neo-liberalism, higher education and the knowledge economy: From the free market to knowledge capitalism. *Journal of Education Policy, 20*(3), 313–345. https://doi.org/10.1080/02680930500108718

Phillips, D. (2004). Toward a theory of policy attraction in education. In G. Steiner-Khamsi (Ed.), *Lessons from elsewhere: The politics of educational borrowing and lending* (pp. 54–67). Teachers College Press.

Sattler, C., & Sonntag, K. (2018). Quality cultures in higher education institutions—Development of the quality culture inventory. In P. Meusburger, M. Heffernan, & L. Suarsana (Eds.), *Geographies of the university* (Knowledge) (pp. 313–327). Springer.

Schein, E. H., & Schein, P. A. (2016). *Organizational culture and leadership* (5th ed.). Jossey-Bass. https://doi.org/10.12968/indn.2006.1.4.73618

Shettima, K. (2016). Achieving the sustainable development goals in Africa: Call for a paradigm shift. *African Journal of Reproductive Health, 20*(3), 19–21. https://doi.org/10.29063/ajrh2016/v20i3.2

Skolnik, M. L. (2010). Quality assurance in higher education as a political process. *Higher Education Management and Policy, 22*(1), 1–20. https://doi.org/10.1787/hemp-22-5kmlh5gs3zr0

Steiner-Khamsi, G. (2006). The economics of policy borrowing and lending: A study of late adopters. *Oxford Review of Education, 32*(5), 665–678. https://doi.org/10.1080/03054980600976353

Torres, C. A. (2011). Public universities and the neo-liberal common sense: Seven iconoclastic theses. *International Studies in Sociology of Education, 21*(3), 177–197. https://doi.org/10.1080/09620214.2011.616340

Toyibah, D. (2020). Neo-liberalism and inequality in higher education. *Icri, 2018*, 1590–1597. https://doi.org/10.5220/0009932415901597

Vilcea, M. A. (2014). Quality culture in universities and influences on formal and non-formal education. *Procedia - Social and Behavioral Sciences, 163*, 148–152. https://doi.org/10.1016/j.sbspro.2014.12.300

Weber, H. (2017). Politics of ' leaving no one behind ': Contesting the 2030 sustainable development goals agenda. *Globalizations, 14*(3), 399–414. https://doi.org/10.1080/14747731.2016.1275404

Wrenn, M. V. (2015). Agency and neo-liberalism. *Cambridge Journal of Economics, 39*(5), 1231–1243. https://doi.org/10.1093/cje/beu047

The Role of Higher Education and the Future of Work in Africa's Fourth Industrial Revolution

Kenneth Kamwi Matengu ⓘ, *Ngepathimo Kadhila* ⓘ, *and Gilbert Likando* ⓘ

1 INTRODUCTION

The world has not seen the amount of upheaval in the workforce since the industrial revolution in the seventeenth and eighteenth centuries, and the information age that followed in the previous century (Ramos, 2019). The fourth industrial revolution or popularly known as "Industry 4.0" or "4IR" marks a significant shift in how we live, work, and interact with one another. It is the start of a new era in human progress made possible by incredible technological advancements unparalleled to those of the first, second, and third industrial revolutions.

Avis, cited by Fataar (2020), conceptualised the 4IR as an ideological construct with specific material interests and that it has implications for

---

K. K. Matengu • N. Kadhila (✉) • G. Likando
University of Namibia, Windhoek, Namibia
e-mail: kmatengu@unam.na; nkadhila@unam.na; glikando@unam.na

© The Author(s), under exclusive license to Springer Nature
Switzerland AG 2024
P. Neema-Abooki (ed.), *The Sustainability of Higher Education in Sub-Saharan Africa*, Sustainable Development Goals Series,
https://doi.org/10.1007/978-3-031-46242-9_15

education and training. The "4IR socio-technical imaginary is based on recurrent themes as artificial intelligence (AI), robotisation, digitisation, and smart machines" (p. 7). As a result, 4IR is acclaimed as holding the promise of addressing and resolving long-standing developmental challenges associated with health, welfare, and climate change to which higher education (HE) should begin to a play a pivotal role, especially in Africa (Fataar, 2020).

Recent developments show that economic sectors and vocations are booming and contracting at an alarming rate, and the skills required to stay current in practically any employment are churning at an even quicker rate. In the global economy, knowledge is regarded as a primary engine of economic growth. Today, nations' competitive advantages are generated from technological innovation and the application of knowledge, rather than natural resources or cheap labour, as they were in the past (Jung, 2020). The use of the robot, for example, lodged the "idea that education and skills acquisition would have to respond decisively to the impact of technology on economic change and a rapidly changing labour market" (Fataar, 2020). Consequently, the world is undergoing a major social and economic change, aided by astonishing breakthroughs in automation and artificial intelligence, as well as unparalleled access to data and computing (Kayembe & Nel, 2019). These technologies have permeated almost every facet of our economy, impacting a diverse variety of professions in healthcare, banking, transportation, energy, and manufacturing, among others. As was the case with past industrial revolutions, these advancements have the potential to provide great opportunities for humanities, ultimately resulting in enormous wealth (Kayembe & Nel, 2019). Innovations will reshape the future of employment, aggravate the skill bias that has existed for decades, and lead to a widening divide between the best and least educated members of society (Morsy, 2020).

Amidst all these socioeconomic and technological transformations, education in general, and HE in particular, is imperative for navigating a future marked by enormous social changes. However, recent discourse about the fourth industrial revolution has exposed some concerns in Africa, including the need for significant reforms in the industrial structure and higher education institutions (HEIs) to prepare the continent for possible times of technological unemployment. According to the World Economic Forum (WEF), with more mechanised, digitalised, and fluid labour markets, today's HEIs are rapidly becoming incompatible with the future we want (WEF, 2019). Accordingly, the WEF (2019) attests that the world is two decades into the twenty-first century, yet HE is still mostly oriented towards success in the twentieth century. Institutions of higher learning themselves voice reservations about their capacity to adapt to future changes. Globally, while the majority of discussions about the future of HE centre on the skills required for the future and the necessity of reskilling, it is as important to address the inevitable structural changes to HE (WEF, 2020a).

In Africa, despite some accomplishments, there are rising concerns about the continent's knowledge production system due to the economic downturn, a lack of new industry and job creation, and a gap between HE and industry. There are questions, in particular, regarding whether the continent is ready for the rapid challenges that the fourth industrial revolution brings. The purpose of this chapter is to explore the role that Africa's HE plays to prepare the continent for the fourth industrial revolution and the future of work.

2   THEORETICAL FRAMEWORK

The proper conceptualisation of how the school system should be structured in response to the new models of education in the 4IR requires that countries understand the schools of the future (WEF, 2019). This chapter used the 4IR ecosystem for HEIs of the future (Kayembe & Nel, 2019) as a conceptual framework to examine HE aspects of Africa's readiness for the 4IR by analysing programme offerings and labour demands for the future of work. Figure 1 provides a framework for how the HE in the African education system was analysed to determine whether it is ready to provide skills for the future.

There are key elements presented in the 4IR ecosystem for schools of the future that should be considered when ascertaining whether HE in Africa is ready for the 4IR adaptation. These tenets referred to as 4IR dynamism are as follows: (1) labour force, education, and skills; (2) enabling infrastructure; (3) innovation systems; and (4) regulatory and investment climates. Although the ecosystem is premised on schools of the future, the 4IR tenets within the ecosystem have been used in this chapter as a model to examine the readiness of HE in Africa in providing skills for the future.

Using these tenets, we maintain that although African countries have been presented with wide opportunities to reform and transform their education systems because of the 4IR boom, there are challenges that still

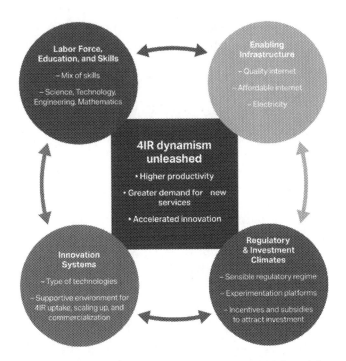

Fig. 1 4IR ecosystem for schools of the future. Source: Adapted from ACET (2018, p. 4)

require attention. These challenges and the mitigation strategies are examined in the subsequent sections of this chapter.

3   Conceptualising the Fourth Industrial Revolution

The term "fourth industrial revolution" refers to recent rapid advances in technology, industry, and society. It was inspired by Germany's "Industry 4.0," which referred to the strategic direction of industrial change required to digitise manufacturing processes and develop purchasing networks (Wilkesmann & Wilkesmann, 2018). After the World Economic Forum used the term in 2016, it gained traction and became part of the national agenda of several advanced countries. The fourth industrial revolution assumes the atomisation and connection of industry and society, and it

THE ROLE OF HIGHER EDUCATION AND THE FUTURE OF WORK... 353

alludes to a future world based on technological advancements commensurate with robotisation, artificial intelligence, and smart factories (Jung, 2020).

According to Kayembe and Nel (2019), technology powered the world's four industrial revolutions, and the use of various technologies aids the government and the private sector in experiencing rapid growth. Using technology, different new concepts and ideas are brought to life in today's world. Virtual worlds, smart cities, big data, the Internet of Things (IoT), and artificial intelligence (AI) have all taken centre stage in propelling development in the new era. Another commonality between these revolutions is the improvement of people's lives as well as the simplicity with which they may conduct business and provide services. Unlike past industrial revolutions, the pace of development has been substantially faster, resulting in abrupt rather than gradual changes in entire societal systems (Schwab, 2017). The consequences of a fourth industrial revolution are likely to be immeasurable, but much will depend on the countries' economic, social, and cultural contexts, as well as their willingness to alter and adapt. According to the Africa Centre for Economic Transformation (ACET), countries with flexible economic frameworks, for example, will gain from the reforms, while those without would suffer far greater difficulties (ACET, 2018).

4   THE GENESIS OF THE FOURTH INDUSTRIAL REVOLUTION

Industrial revolutions, according to Ramos (2019), are periods of technological upheaval accompanied by larger societal changes. These phases are about more than the invention and deployment of technology; they are about the transformation of whole power systems. Industrial revolutions are new ways of looking at the world that causes economic and social institutions to transform (Jung, 2020). The emergence of the industrial revolution is summarised in Table 1.

As shown in Table 1, the first industrial revolution (1IR) covered the period between the eighteenth and nineteenth centuries. During this period, human communities developed from agricultural activities to the use of mechanisation. The steam engine was invented in the 1IR, which changed the means available for production (Ramos, 2019). The second industrial revolution (2IR) began in the early nineteenth century as a continuation of the preceding period. Steel, chemicals, electricity, and a

354   K. K. MATENGU ET AL.

Table 1 Emergence of industrial revolutions

Industrial revolutions	Timelines	Main characteristics
First industrial revolution (1IR)	1760–1840: late eighteenth–early nineteenth century	• Mechanisation, water and steam engine/power *(machines replace animal and manual labour)* • Improvement in standard of living (including transportation, communication, and banking); increase in manufactured goods; the growth of industries in coal, iron, and textile
Second industrial revolution (2IR)	1870–1914: late nineteenth–mid-twentieth century	• Mass production, assembly line, and electricity *(mass manufacturing, machines, and processes)* • Electric railroad and electric cars; radio communications; radio wave transmission; inventions of the elevator, the telephone, refrigerator, typewriter, phonograph, washing machine, and diesel engine
Third industrial revolution (3IR)	1950s–1970s: second half of twentieth–early twenty-first century	• Computer, automation, and information technology *(digital revolution and globalisation; analogy to digital technology; worldwide web)* • Computer, digital mobile phones, and the Internet; digital communication (cellular phones), digital camera, CD-ROM, including automated teller machines, industrial robots, electronic bulletin boards, and video games
Fourth industrial revolution (4IR)	Early twenty-first century–present	• Cyber-physical systems *(automation/robotics, artificial intelligence, analytics, Internet of Things)* • Self-driving cars; drones; virtual realities; software that translates, invests, analyses, and identifies; social media; nanotechnology

Source: Adapted from Ramos (2019: 3)

variety of other sectors witnessed significant scientific advancements during the 2IR (Jung, 2020). The introduction of electricity was a pivotal milestone since it allowed numerous industries to operate and grow their operations. Mineral prospecting became feasible as a result of this technical breakthrough. The employment of machines, which were largely powered by electricity, was a defining feature of the 2IR (Berchin et al., 2021). In the mid-nineteenth century, the third industrial revolution (3IR) began. The 3IR was pushed by technology developments in manufacturing, distribution, and energy considerations, according to Ramos (2019). The development of nuclear power and the widespread usage of electronics were two of the most significant achievements of the 3IR period.

Kayembe and Nel (2019) avow that technological advancements characterised the preceding three industrial revolutions, although not at the same rate as they are now. For numerous reasons, technology has witnessed fast development, implementation, and application in recent years. Technology has become an integral element of people's lives. Technology is transforming lives and offering civilisations new talents and capacities. As a result, society is presently heading towards the 4IR. The 4IR is about the digital revolution that is now taking place. For society, the 4IR has opened up new possibilities and opportunities. It is based on the foundations of numerous prior revolutions' accomplishments. The twenty-first century presents a variety of issues that will necessitate novel solutions. The 4IR is a collection of developing technologies. New ideas, new possibilities, new creations, and innovations are all part of the new revolution. It is the goal of this new revolution to tear down barriers. The next revolution is "marked by a lot more pervasive and mobile internet, smaller and more powerful sensors that have grown cheaper, and artificial intelligence (AI) and machine learning," according to Kayembe and Nel (2019). The popularity of mobile devices skyrocketed at the turn of the century.

5   Characteristics of the Fourth Industrial Revolution

The 4IR is the most recent industrial revolution, with a greater emphasis on information and communication technology, technological advancement, innovation, and creativity. The characteristics of the 4IR, including big data, AI, robotics, ICT, 3D printing, and quantum computing, were identified in this chapter. According to Kayembe and Nel (2019), the IoT, Cyber-Physical Systems, Internet of Services, and Smart Factory are the components of Industry 4.0. AI, three-dimensional (3D) printing, robotics, blockchain technology, cryptocurrency, quantum computing, nanotechnology, and biotechnology are among the many additional features and elements of the 4IR. These technologies are transforming how resources, products, and services are created and consumed. Technologies that drive the 4IR are summarised in Table 2.

Table 2 Technologies that drive the 4IR

Technology	Concept/definition
Internet of Things (IoT)	It refers to the integration of Internet connectivity and the incorporation of electronics, software, sensors, actuators, and communication capabilities into electronic devices, vehicles, structures, buildings, and other devices, allowing them to send, transmit, and process data with fewer human interventions. It also allows things to be remotely monitored and controlled, improving efficiency, accuracy, and productivity. Smart power networks, virtual power plants, intelligent transportation systems, and automated houses are just a few examples (a component of smart cities).
Big data	It refers to the creation and processing of extremely big datasets. Volume (size), velocity (streams over time), and variety (complexity) are three typical definitions, which include variability and complexity. Electronic gadgets, social media, search engines, sensors, and tracking devices (GPS) all contribute to this data. Descriptive analytics (e.g., obtaining customer profiles and behaviour from social media and customer transaction databases), predictive analytics (forecasting future events), and even prescriptive analytics (using simulation and optimisation methods) are all methods for extracting value from digital data.
Robotics and automation	Mechanical, electrical, computer science, and other engineering-based subjects are all included in this multidisciplinary engineering and science field. It encompasses the science of robot design, construction, operation, and implementation, as well as computer systems for control, feedback, and data processing. Robotics attempts to replace humans in laborious and dangerous activities that are more acceptable in a human setting, as well as completing repetitive human duties. Humanoid robots emulate human walking patterns, object lifting, communication, and cognitive skills, as well as behaviour and motions found in nature, particularly animal movements.
Artificial intelligence (AI)/machine intelligence	These are computer-based programs that perform things that humans do, such as visual perception, decision-making, and speech recognition. It is the application of machine reasoning and thinking skills to replicate human or animal intelligence. The design and execution of intelligent agents (units that perceive the environment and make decisions) are also classified as AI in the computer science discipline. Cognitive (mental) skills research is concerned with the study of people's ability to learn and solve issues.

(*continued*)

THE ROLE OF HIGHER EDUCATION AND THE FUTURE OF WORK... 357

Table 2 (continued)

Technology	Concept/definition
Three-dimensional (3D) printing	The historical patterns of 3D printing growth are similar to those seen in the late 1970s with the rise of home computing. 3D printing, or additive manufacturing, is the method of creating solid 3D objects from a digital file. This innovative technology has found widespread application in a variety of industries, including medical, automobile production, and many others, for varied objectives. Aeroplane parts and artificial organs made from human cells are two examples of 3D printing applications. This technology has been in use since the 1980s, but it has gone through several changes. 3D printing presents a much faster and cheaper way to create objects.
Quantum computing	Quantum computing can solve problems that would take billions of years on today's systems in days or hours. It also allows for discoveries in healthcare, energy, environmental systems, smart materials, and other fields. Quantum computers could aid in the development of new scientific breakthroughs, life-saving medicine, machine learning methods to diagnose illnesses faster, materials to make more efficient devices and structures, retirement financial strategies, and algorithms to quickly direct resources like ambulances.

Source: Adapted from Ramos (2019)

Modern societies are defined in part by the priority placed on education. For countries to benefit from the 4IR, a fundamental transformation in HE is needed to enable digital social engagement. Education, on the other hand, is prone to "wicked difficulties." Wicked problems have various interpretations for essential parts and frequently comprise mutually dependent and interconnected problems. The next sections discuss the implications of the 4IR for HE.

6 IMPLICATIONS OF THE 4IR FOR HIGHER EDUCATION IN THE GLOBAL CONTEXT

According to Ramos (2019), any 4IR education strategy must build on the 3IR's development of in-person instruction and asynchronous educational resources. In the fourth industrial revolution, education is a sophisticated, dialectical, and thrilling opportunity that can change society for the better. The 4IR has ramifications in a variety of other areas of life. As a result, it presents educational opportunities as well as obstacles. The education sector could be significantly altered by incorporating several components of the 4IR—IoT, 3D printing, quantum computing, and AI—to provide solutions to new challenges. Unlike past industrial revolutions, the pace of development for the fourth industrial revolution has been substantially faster, resulting in abrupt rather than gradual changes in whole societal systems (WEF, 2020a). Jung (2020) warns that the consequences of the fourth industrial revolution are likely to be immeasurable, but much will rely on nations' economic, social, and cultural contexts, as well as their willingness to alter and adapt. As such, countries with flexible economic systems, for example, will gain from the changes, while those without would suffer far greater difficulties (WEF, 2020b). In general, considerable changes in the knowledge production system will occur, including changes in industry and labour market structures, as well as changes in HE (Jung, 2020). To begin with, the industrial structure, which includes technology, production, and trade procedures, may be significantly impacted. Traditional businesses will be challenged by technology breakthroughs like information and communications technology (ICT), AI, robotics, the Internet of Things (IoT), big data, and digital printing, and there will be no distinction between online and offline trading.

Changes in industrial structure have a direct impact on the labour market. The majority of job creation estimates for the fourth industrial revolution are gloomy, implying that robots and artificial intelligence would replace many employments. Although new employment will be generated in new industries, the number of jobs that will be lost will significantly outnumber new occupations. For example, labour-intensive businesses will either go out of business or be overtaken by businesses that use cheaper robots or clever software. Artificial intelligence and social media will become increasingly vital knowledge service businesses (Berchin et al., 2021).

According to Africa Centre for Economic Transformation (ACET) the fourth industrial revolution will result in the polarisation of the labour market (ACET, 2018). Machines will replace low-skilled professions that do not need complex knowledge, resulting in technological unemployment (Schwab, 2017), whereas highly trained and experts will be in greater demand. Middle-skill occupations, regular work requiring certain skills, will be reduced. Job instability will rise since employees' labour levels can be readily managed and monitored after automation, and their performance can be measured. Figure 2 paints a picture of the jobs of the future.

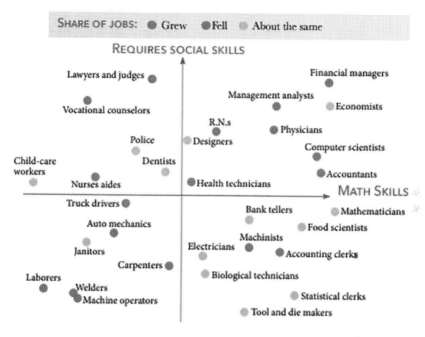

Fig. 2 Jobs of the future. Source: http://quantitative.emory.edu/news/articles/Future-of-Work-Part-I.pdf

In comparison to the ecosystem for the schools of the future, the figure provides a second model for assessing the readiness of HE to prepare graduates for future jobs.

The data in Fig. 2 show the typical characteristics of HE, with a high concentration on hard skills that will be easily replaceable by machines in the future. While we note that the rate at which the fourth industrial revolution is taking place is determined by the state of development of a particular country or region, the pace of technological advancement is alarming such that skills currently in use may become redundant in the future. The implication is that the fourth industrial revolution has had a significant impact on HE, both in terms of what knowledge is created and in terms of how highly skilled people, for instance, postgraduate students, are trained, as well as what curricula are provided for undergraduate students to prepare them for their labour market transition. As a result, HE's purpose and structure will need to alter to keep up with changes in the workforce. This requires a paradigm shift in the education system rather than following an industrial model of education, in which students follow a prescribed curriculum delivered primarily in formal classroom settings. HE in the future will need to equip students with collaborative, problem-solving skills to self-direct their lifelong learning in a way that complements rather than competes with technology.

Masindei and Roux (2020) assert that it is undeniable that the 4IR will 'revolutionise' HE, potentially leading to Higher Education Revolution 4.0 (HE 4.0). Some of the implications of the 4IR in the education sector, according to Ramos (2019), are curricula, teaching, learning teaching, research, and innovation. In other words, cross-sector teaching and learning are required. The role of HE in any society cannot be overemphasised; the sector's role in emerging technologies is even more substantiated.

Figure 3 is a summary of some of the ways that 4IR will influence the three main functions of HEIs, particularly universities.

As regards teaching, every university has a primary responsibility to educate the youth. As a result, proper teaching strategies must be implemented, and work must be organised in a way that promotes learning. This has implications for flexible learning programs, improved learning

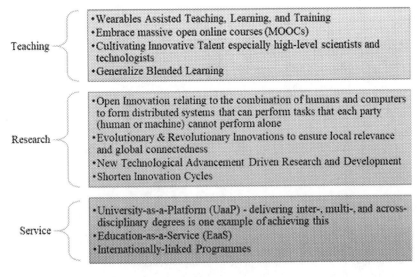

Fig. 3 Ways that 4IR will influence the three main functions of higher education institutions. Source: Masindei and Roux (2020: 36)

experiences, and a commitment to lifelong learning. Students and educators from numerous disciplines must understand the various variables that go into implementing the 4IR successfully. Students studying the basic and applied sciences, as Berchin et al. (2021) expound, must understand the political and social nature of the world in which they live, and students studying the humanities and social sciences must understand at least the foundations of which AI is based and how it operates.

The 4IR promotes a multidisciplinary field in which humanities and social sciences collaborate with technology to solve problems. The 4IR, as well as advances in biotechnology and artificial intelligence, fundamentally questions beliefs about people and their relationship with nature. To adjust for the social dislocations caused by the 4IR, liberal arts programmes should be constructed. In general, the 4IR curriculum should adapt to political and social difficulties caused by rapid technological innovation (Fataar, 2020). Online instruction and the growing usage of AI in teaching and learning demand new principles to give a theoretical foundation for digital pedagogy (Berchin et al., 2021). Digital literacy is a necessary skill for students to obtain to engage in the global digital society, benefit from the digital economy, and gain new chances for employment, innovation, creative expression, and social inclusion (Morsy, 2020).

Any digital education approach should consider the educational system's impact. This is a difficult situation. If students are not effectively prepared and resources are not adequately invested, changes may have an impact on the quality of graduates (Morsy, 2020). Education is especially vulnerable to wicked challenges, particularly when it comes to quality measures (Fataar, 2020). In a disputed area of educational transformation and strategy, conceptualising and operationalising quality measurements, performance indicators, and educational outcomes become increasingly difficult (Berchin et al., 2021).

As for research, the most essential driving force for research and development (R&D) is generally new technology developments (Berchin et al., 2021). Technology-assisted research has many favourable results. Technology-driven research and development, according to Berchin et al. (2021), can take many forms, including using mobile capabilities to improve data acquisition accuracy, advanced big data analytics to spot hidden statistical patterns, and AI techniques to retool information search, collection, organisation, and knowledge discovery, to mention a few. To compete globally in HE, universities must devote significant resources to research and development (R&D). These influences, according to experts, include anything from new technology deployment to global cooperation and collaboration.

Xing and Marwala (2017) withstand that as far as service is concerned HEIs must drastically strengthen educational services to maintain their competitive HE position in the global HE system. HEIs need to encourage much more innovation and competition in education, in particular. Because of the speed and scope of the aforementioned megatrends and the consequent fourth industrial revolution for HE, it is a duty for nations to comprehend the influence of these developments on all aspects of human life, including HE.

7   Implications of the 4IR for Higher Education in the African Context

HEIs command a role in society's transformation for a more sustainable future. HEIs provide the way for sustainable development through knowledge generation and dissemination, research, education, and outreach. This highlights the need for a stronger role of universities to promote sustainable development paths, expand and disseminate knowledge, build capacity through training, and work with local communities to increase their resilience (Ramos, 2019). HE is one of the sectors on which the 4IR has a significant impact. In African HE, the fourth industrial revolution promises the emergence of artificial intelligence and big data to bring together cyber-physical systems (CPS), also known as the Internet of Things (IoT), to produce mass automation (4.0), building on the previous ages of steam (1.0), electricity (2.0), and information (3.0) (Matthews et al., 2021). The fast growth of technology is creating challenges regarding the purpose of HE. The HEIs of the future will have to adapt to these developments, which have the potential to have far-reaching implications not only for HE but also for society at large. Therefore, the biggest challenge for HE in the 4IR is to prepare students for an uncertain and unpredictable future.

The 4IR will have a significant impact on the skills that are required in the work market in Africa. For example, the demand is shifting away from regular activities and restricted abilities connected to specific occupations and towards adaptive social, behavioural, and non-repetitive cognitive talents all around the world. Software engineers, marketing professionals, authors, and financial counsellors are in high demand, whereas mechanical technicians, administrative assistants, and accountants are in short supply (Morsy, 2020).

Kayembe and Nel (2019, p. 79) postulate that

> [t]he 4IR is the current and developing environment in which changing technologies and trends unvarying as the Internet of Things (IoT) and artificial intelligence (AI) are changing the way we live and work. The 4IR presents several implications for skills development and education. Some of these implications include reinventing education systems and strategic approaches to increase creativity and innovation.

The sentiments of the immediate quoted duo imply that the 4IR requires drastic changes in African HE systems due to the eminence of smart technologies including big data, artificial intelligence, the Internet of Things, and automation. For the inadequate IT infrastructure in most countries on the continent, the disruption caused by emerging technologies is happening at an unprecedented speed. As a result, Africa needs to put systems in place to be able to leverage opportunities and address challenges brought by the 4IR. In other words, the education system in Africa should adopt the 4IR ecosystem as a model for transforming the HE system. Using the four tenets of the 4IR ecosystem (labour force; education and skills; enabling infrastructure innovation systems; and regulatory and investment climates) as a readiness model, the preparedness of African HE system that many countries in Africa lag is assessed.

8   African Higher Education and Future Skills

World Economic Forum (2020) declares that Sub-Saharan Africa is only capturing 55% of its human capital potential, compared to a worldwide average of 65%. Sub-Saharan Africa is the world's youngest area, with more than 60% of its inhabitants under the age of 25. The continent's working-age population is expected to double by 2030, from 370 million people in 2010 to over 600 million by 2030. The largest long-term advantages of ICT-heavy occupations in the region are anticipated to be in digital design, innovation, and engineering, rather than the lower-skilled delivery of digital products or services. To create a pipeline of future skills, African educators should create future-ready curricula that promote critical thinking, creativity, and emotional intelligence, as well as accelerate the acquisition of digital and science, technology, engineering, and mathematics (STEM) skills to match the way people will work and collaborate in the 4IR. Of recent it has been observed that arts have become attuned to the 4IR, hence the emphasis on science, technology, engineering, arts, and mathematics (STEAM) skills.

While significant progress has been achieved in HE in Africa there are still numerous obstacles to overcome. The UN Conference on Trade and Development's "Technology and Innovation Report 2021," published earlier in 2022, reveals that Africa as a whole was the world's least equipped area for 4IR technologies. Education is also widely cited as a major roadblock, both in terms of the region's often inadequate educational systems, which result in a small number of people enrolling in HE, and in terms of a skills mismatch, with many people lacking the necessary training to take advantage of the 4IR. South Africa is the continent's best-prepared country, but it is behind the BRICS countries (Brazil, Russia, India, and China) in terms of preparedness. The Democratic Republic of Congo, Gambia, and Sudan are among the continent's most vulnerable countries (Kayembe & Nel, 2019).

According to Fataar (2020), "[U]nderstanding the nature of labour demand is essential for an accurate depiction of the capitalising logics of 4IR, its productive dynamics and the questions that this raises for education" (p. 15). In other words, African countries should understand the demand for future skills to inform curricula to respond to job losses and the labour market in the digital economy. The 4IR (IR 4.0), as Eleyyan (2021) has observed,

> will accelerate the rate of disruption in jobs which we are already experiencing, and it is necessary to empower individuals to take charge of their education and career strategies. (p. 24)

In this understanding, there should be a strong synergy between HE and training and the job market.

HE 4.0 combines industry and education by providing authentic real-world and virtual-world learning experiences, and preparing students for a mixed-reality future. Numeracy, literacy, digital, social, critical thinking, moral, and creative problem-solving capacities are all goals of education 4.0, as are emotional intelligence, flexibility, and adaptability, as well as a lifelong learning attitude. The curriculum is non-linear, interdisciplinary, decolonised, learner-centred, personalised, and of international quality. As such, education 4.0 reflects a paradigm shift from the previous industrial revolution's educational paradigm, employing emerging pedagogies close to heutagogy, paragogy, and cybergogy (Ally, 2019) within new, increasingly relevant, practical, and immersive learning environments.

9   Challenges for Higher Education in the Fourth Industrial Revolution in Africa

The 4IR requires significantly more investment in HE, especially within developing countries. Technology is emerging so rapidly that educational systems, institutions, educators, and other resources cannot keep pace. Curricula and pedagogical approaches that focus on specific technical skills are inadequate. Kayembe and Nel (2019) reason that the 4IR presents some implications for skills development and education. Some of these implications include reinventing education systems and strategic approaches, and increasing creativity and innovation. However, the current HE system in Africa seems not to be ready to create jobs for the future when assessed using the 4IR ecosystem, model propounded by the traditional approach to curriculum development. With the development and use of new technology, the 4IR has its own set of risks. To reduce these risks, careful planning is required. New risk management systems and processes will be required. However, the risks and negative consequences of these new technical developments should not be overlooked. Inequality is the first risk of the 4IR for education. In Africa, societal problems of inequality and wealth distribution are prevalent, like the social challenges of high crime rate, gender violence, and unemployment.

An assumption exists that an effective education system, with proper curricula, qualified teachers, and good infrastructure, will provide the required vocational and digital skills. However, the traditional nature of curricula and the "socio-economic conditions that structure learners" are often ignored. In other words, the 4IR discourse should take into consideration educational inequalities as well as the deeper implications because of socioeconomic conditions. "Educational outcomes can be improved only if widespread poverty is addressed, and the social conditions of the majority improved" (Fataar, 2020, p. 8). This is true in the African context where a strong relationship exists between educational inequality and socioeconomic conditions.

Inequality in Africa's HE system is a contentious issue, and new technological breakthroughs have the potential to bolster this view. There is a possibility that only the wealthiest members of society will be able to acquire new technology for educational reasons, leaving the poor behind. The execution of the preceding three industrial revolutions exemplifies this. Today, many people in African communities lack access to safe drinking water, transportation, power, and the Internet. As a result, a wider gap between the "haves" and "have nots" would increase alienation, distrust, and social instability. The human situation as well as social fairness should be considered. At different socioeconomic levels, the effects of technological advancement and shifting economic power on society should be acknowledged. The challenges that exist in an increasingly linked globe must be comprehended, and intercultural understanding, an unwavering respect for freedom, and human rights must be promoted. In addition, the chapter also identified some issues, including inadequacies in ICT infrastructure and a start-up environment that is undercapitalised as other challenges that require mitigation.

10 OPPORTUNITIES FOR HIGHER EDUCATION IN THE FOURTH INDUSTRIAL REVOLUTION IN AFRICA

Several factors position Africa well to benefit from 4IR technologies. For example, the region has experienced a tremendous rise in mobile technology in recent years, with customers bypassing traditional development channels in favour of digital services, notably in the banking sector. Africa also has a disproportionately large population of young people, a demographic dividend that is already paying off in terms of the 4IR. More than 400 innovation clusters have sprouted up throughout the continent, due primarily to the efforts of young people, with three important cities gaining international recognition: Lagos, Nigeria; Nairobi, Kenya; and Cape Town, South Africa. The African Development Bank (AfDB) reported at the end of 2019 that the Internet of Things (IoT) has grown significantly in Africa, with high investment development in technology-driven fields. Indubitably, the revolutionary effects such technologies may have an effect on diverse aspects like agriculture, industry, healthcare, and government, the research found this unsurprising.

According to Kayembe and Nel (2019), the 4IR provides African educational institutions with the chance to foster a culture of creativity and innovation. New methods of science, technology, engineering, and mathematics (STEM) should be reconsidered in the 4IR STEM and STEAM curricula. In addition to 4IR literacy, new courses should emphasise 4IR collaboration abilities. Educational solutions to 4IR include reorganising institutions to enable them to deliver new interdisciplinary science programmes.

Technology may be used to address social exclusion concerns (Kayembe & Nel, 2019). To put it another way, new technological advances may be utilised to bridge the gap between the affluent and the poor, as well as between races. In this connection, the 4IR gives educational institutions the option to form collaborations with other stakeholders, particularly the government and commercial businesses. To allow students to gain competency in the fast-growing areas of genomics, data science, AI, robotics, and nanomaterials, significant reforms to the science and technology curriculum will be necessary. As a kind of 4IR literacy, a 4IR STEM curriculum would rethink the curriculum within the conventional "basic" sciences—biology, chemistry, and physics—and place a larger value on training in computer science and the arts disciplines (Berchin et al., 2021).

11 The Role of Higher Education in Harnessing the Fourth Industrial Revolution for Sustainable Development

World Economic Forum (2017) enunciates that industrialisation is to blame for a large number of the environmental issues the planet is facing today. Industrialisation is responsible for many environmental issues like deforestation, harmful air pollution levels, the depletion of fisheries, poisons in rivers and soils, an abundance of trash on land and in the ocean, loss of biodiversity, and climate change. The Internet of Things, blockchain, and other 4IR technologies are among those that are quickly entering the mainstream and enabling change of the entire networks and systems. Many changes can be observed in organisations, markets, nations, and society at large. These fundamental changes, which affect almost every industry and threaten both long-standing business models and completely new ones made possible by 4IR, are speeding up the pace and magnitude of market expansion.

At the global level, there has been increasing interest in the use of fourth industrial revolution technologies related to artificial intelligence to help achieve the Sustainable Development Goals (SDGs). Africa has a special opportunity to take advantage of the societal changes brought on by the 4IR to solve environmental problems and rethink how we manage our shared global environment. Yet, the 4IR may potentially increase current environmental security risks or generate brand-new hazards that must be considered and managed. In the view of Fataar (2020), a change in the "enabling environment," thus the governance frameworks and policy protocols, investment and financing models, the prevalent incentives for technology development, and the nature of societal engagement, will be necessary to take advantage of these opportunities and proactively manage these risks. This change won't take place without any human intervention. It will call for proactive cooperation between decision-makers in government, academia, business, civil society, and investors. New technologies have the potential to contribute significantly towards achieving Sustainable Development Goals (SDGs). However, this chapter ascertains that despite all the potential these technologies provide, they may also put more strain on the planet's resources and our society. Therefore, Africa needs to make sure that these technologies are used properly to realise their potential to revolutionise our world, transform people's lives, and open up new doors to prosperity, thereby accelerating the attainment of SDGs on a global scale.

Shenkoya and Kim (2023) declare that contextualising technology and digitalisation in a period of having to meet SDGs requires a paradigm shift, which calls for collaboration from multiple stakeholders. This chapter guarantees that HE is one of the requisite stakeholders that play a role of HE in fostering innovation within continental innovation systems and its significance for sustainable development. Therefore, the quality of HE must keep up with the necessary technological advances brought on by the digital transformation caused by these technologies as developed economies enter mature stages of the 4IR. By serving as innovative test sites and institutions of higher learning, universities highlight their role in influencing future technologies. The current rates of technological growth and industrial evolution are largely influenced by traditional education. But for HE to equip future generations with the appropriate knowledge and skills, it is essential to consider how the fourth industrial revolution will impact HEIs and how education will be delivered. For example, the opportunities that are accessible will determine the role that HE can play in the 4IR. Blending the advantages of traditional HE with the rising popularity of Massive Open Online Courses (MOOCs) is a potentially disruptive innovation, which is an essential step towards expanding high-quality education.

12 LESSONS FOR AFRICA

This chapter avers that the Internet of Things, robotics, virtual reality, and artificial intelligence are all part of the 4IR, and they are reshaping the way we live and work. Fields of science, technology, engineering, arts, and mathematics are seen as some of the most sought-after 4IR skills with high employment prospects. There is a growing sense of urgency to adopt the 4IR technologies like artificial intelligence (AI), robotics, the Internet of Things (IoT), and data analytics in HE in Africa. However, there appears to be no clear direction on what technologies should be adopted, how they should be integrated, or what effect their inclusion may have on the field of HE or learners (Ally, 2019). Therefore, Africa's HEIs must rethink their curricula and refocus their programmes on skills that will be useful in the 4IR. Through the functioning of 4IR emerging technologies, Nyagadza et al. (2022) claim that "Africa would be able to sustainably drive the digital human resources training transformations to be more understandable and tractable" (p. 6). They further posit that a lot of these technological changes have begun to take shape, particularly in emerging economies, in areas of agriculture, banking, education, energy and entertainment, and transport. While the 4IR technologies are spreading rapidly the onus are on developing countries to support HEIs to develop structure and systems to increase the uptake of new opportunities presented by the 4IR explosion.

The literature review has revealed that pockets of positive improvements have begun to emerge in HE, which has seen some success with regard to innovations, upgrades, and refocused ways of thinking about HE, resulting in greater outcomes for students and, as a result, the economy as a whole (Nyagadza et al., 2022; Eleyyan, 2021; Fataar, 2020). Therefore, HEIs need to reimagine curricula for the 4IR. New adaptable curriculum and teaching methodologies for varied contexts, as well as a shift away from a teleological conception of "skills," are essential if HE in Africa is to provide education that prepares students for the demands and challenges of the 4IR. Some of the possible solutions for HE in Africa in the context of industrial revolution 4.0 are as follows:

- *Raising awareness and renewing thinking about the HE development in the general development strategy of the continent:* HEIs in Africa must raise awareness of the importance of industrial revolution 4.0, changes in the job market, and the university's mission in preparing highly qualified human resources and participating in labour market restructuring to effectively take advantage of opportunities and overcome challenges posed by industrial revolution 4.0. The overall HE development strategy should emphasise the importance of teaching a workforce with professional qualifications, soft skills, creative thinking, and the ability to adapt to the global labour market's continual changes. The HE system must actively develop and produce with a long-term perspective, complete integration with the global HE system.

- *Innovate training models, programmes, and methods:* Training objectives need to change towards promoting creativity and developing personal capacity. Instead of teaching a standard curriculum, different programmes must be developed to customise training; each learner's strengths and weaknesses must be identified to construct an appropriate training programme. The curriculum system must be altered and updated regularly to support research and development of new fields (e.g., artificial intelligence, data analytics, intelligent ICT convergence).
- *Accelerate the digital transformation process and take the lead in applying new technologies:* Digital transformation must ensure four factors: empowering lecturers, exchange with students, organisational optimisation, and method innovation. University digital transformation takes place in all three phases: planning, independently formulating strategies and implementing innovations, and monitoring the impact of technology deployment.
- *Strengthen university-industry collaborations:* Establishing a high-level overall model based on a shared coherent pattern with multiple forms in a tight, interoperable, and supporting system is required. Training and technology transfer, or a combination of training, research, and implementation, are available at the university. Propagated for is a specific and separate model based on this overall model akin to linking university training by studying and working; theoretical training at university; skill practice at enterprises; training according to enterprise orders expanding training lecture halls from universities to industries.
- *Improve the quality of lecturers and HEI administrators:* Have policies and procedures in place for attracting and retaining highly qualified academics, and provide them with continuous professional development opportunities. Organise and prepare training

programmes and documentation for senior university administrators substitutable to the president of the university council, principal, vice-chancellor (and equivalent), and personnel affiliated with unit-level management of HEIs. Organise training sessions to help key managers and managers at linked HEIs improve their management skills.

- *Strengthen international, regional, and intercontinental cooperation and integration in training:* International, regional, and intercontinental cooperation and integration create opportunities for students to participate in exchange programmes or study abroad and have the freedom to develop personally; allow lecturers to learn management and educational methods from international universities and assist partners in better understanding HE in African states; create opportunities for transnational scientific research cooperation; create opportunities for internationalisation and regionalisation/Africanisation of HE.

13   Conclusion

HE in the 4IR is complex and offers new prospects that have the potential to improve society. Artificial intelligence will power the 4IR, which will shift the workplace from task-based to human-centred characteristics. Because of the convergence of man and machine, the gap between humanities and social science, as well as science and technology, will narrow. As the power of 4IR technologies for either positive societal impacts or devastating environmental damage approaches, the necessity for an HEI to respond is imperative. Therefore, Africa urgently needs to invest in HE infrastructure that will enable skill development, reskilling, and upskilling. Examination of the key skills and learning areas in the reformed curricula across all levels necessitates extensive study. Africa's future will be determined by the business/firm/dynamic industry's adaptable nature. Innovations will alter the future of work, exacerbating a long-standing skill bias. Short- and long-term public policy initiatives are needed. Even though there are shortcomings of 4IR, HE in Africa must adopt strategies to apply technologies and products to equip students with the necessary skills and competencies to navigate the future challenges in the digital economy. This process entails a transition and transformation from general knowledge and skills to liquid or soft skills.

REFERENCES

Africa Centre for Economic Transformation (ACET). (2018). *The future of work in Africa.*

Ally, M. (2019). Competency profile of the digital and online teacher in future education. *The International Review of Research in Open and Distance Learning, 20*(2), 302–318.

Berchin, I. I., Dutra, A. R., & Guerra, J. B. S. (2021). How do higher education institutions promote sustainable development? A literature review. *Sustainable Development, 29*(6), 1204–1222.

Eleyyan, S. (2021). The future of education according to the fourth industrial revolution. *Journal of Educational Technology & Online Learning, 4*(1), 23–30. Accessed March 01, 2022, from http://dergipark.org.tr/jetol

Fataar, A. (2020). The emergence of an education policy dispositif in South Africa: An analysis of educational discourses associated with the fourth industrial revolution. *Journal of Education,* (80), 5–24. https://doi.org/10.17159/2520-9868/i80a01

Jung, J. (2020). The fourth industrial revolution, knowledge production and higher education in South Korea. *Journal of Higher Education Policy and Management, 42*(2), 134–156.

Kayembe, C., & Nel, D. (2019). Challenges and opportunities for education in the fourth industrial revolution. *African Journal Public Affairs, 11*(3), 79–94.

Masindei, M., & Roux, P. A. (2020). Transforming South Africa's universities of technology: A roadmap through 4IR lenses. *Journal of Construction Project Management and Innovation, 10*(2), 30–50.

Matthews, A., McLinden, M., & Greenway, C. (2021). Rising to the pedagogical challenges of the fourth industrial age in the university of the future: An integrated model of scholarship. *Higher Education Pedagogies, 6*(1), 1–21.

Morsy, H. (2020). *How Africa can harness the fourth industrial revolution.* Accessed April 26, 2022, from https://www.project-syndicate.org/commentary/how-africa-can-close-education-skills-gap-by-hanan-morsy-2020-08

Nyagadza, B., Pashapa, R., Chare, A., Mazuruse, G., & Hove, P. K. (2022). Digital technologies, fourth industrial revolution (4IR) & global value chains (GVCs) nexus with emerging economies' future industrial innovation dynamics. *Cogent Economics & Finance, 10*(1), 2014654. https://doi.org/10.108 0/23322039.2021.2014654

Schwab, K. (2017). *The fourth industrial revolution.* World Economic Forum.

Shenkoya, T., & Kim, E. (2023). Sustainability in higher education: Digital transformation of the fourth industrial revolution and its impact on open knowledge. *Sustainability, 15*, 2473. https://doi.org/10.3390/su15032473

Ramos, R. P. (2019). *Fourth industrial revolution: Opportunities and challenges on higher education institutions (HEIs) towards 2030 sustainable development goals*

(SDGS) agenda. Accessed April 26, 2022, from http://www.brainitiativesph.com/uploads/7/7/6/4/77644974/ramos_rsu_pice2019.pdf

Wilkesmann, M., & Wilkesmann, U. (2018). Industry 4.0: Organizing routines or innovations? *VINE Journal of Information and Knowledge Management Systems, 48*(2), 238–254. https://doi.org/10.1108/VJIKMS-04-2017-0019

World Economic Forum. (2019). *The 4 biggest challenges to our higher education model- and what to do about them.* Accessed April 22, 2022, from https://www.weforum.org/agenda/2019/12/fourth-industrial-revolution-higher-education-challenges

World Economic Forum. (2020a). *How can Africa succeed in the Fourth Industrial Revolution?* Accessed April 22, 2022, from https://www.weforum.org/agenda/2020/08/africa-fourth-industrial-revolution-technology-digital-education/

World Economic Forum. (2020b). *Schools of the future: Defining new models of education of the fourth industrial revolution.* Accessed April 24, 2022, from https://www.weforum.org/reports/schools-of-the-future-defining-new-models-of-education-for-the-fourth-industrial-revolution

World Economic Forum. (2017). *Harnessing the fourth industrial revolution for sustainable emerging cities.* Accessed February 23, 2023, from https://www3.weforum.org/docs/WEF_Harnessing_the_4IR_for_Sustainable_Emerging_Cities.pdf

Xing, B. & Marwala, T. (2017). *Implications of the fourth industrial age on higher education.* Accessed May 19, 2022, from https://arxiv.org/ftp/arxiv/papers/1703/1703.09643.pdf

INDEX

A

Aamodt, P. O., 23
Abdous, M., 16–18, 36
Abiddin, N. Z., 226
Abu Daabes, A. S., 103
Acaali, Christopher Mukidi, 6
Academia, xi, 8, 123, 145, 252, 254, 268, 271, 308, 323, 339, 368
Academic, xiv, 2, 6, 18, 19, 29, 31–36, 38, 39, 44, 63–66, 74–76, 80–86, 97–101, 103, 104, 107, 108, 128, 129, 137, 145, 146, 148, 150, 153, 154, 166, 198, 205, 217, 233, 235, 252, 256–258, 260–262, 264, 266–268, 274, 288, 290, 295, 328, 333, 334, 339–343, 370
Academic dishonesty, 16, 33–35
Accra Declaration, 124
Acme, 1–10
Acquisition, 7, 103, 131, 165, 167, 178, 179, 184, 224, 226, 229, 231, 235, 236, 238, 241, 274, 320, 350, 361, 363

Adams, N. E., 163–166, 172, 183, 186
Adarkwah, M. A., 99
Adekunle, A., 315
Adeoji, J., 262
Adina-Petruța, P., 342
Adler, J., 169
Administrative, 3, 66, 119, 194, 270, 339, 343, 363
Affecting, 98, 124, 165, 167, 195
Africa, vii, xiii, xiv, 1, 5–9, 15–39, 43–68, 77, 92, 119, 122, 124, 128, 135–158, 187, 198, 212, 251–275, 281–302, 311, 315, 319, 320, 327–344, 349–371
Africa Centre for Economic Transformation (ACET), 353, 358
African Higher Education, 8, 37–38, 305–323, 363–365
African Higher Education Institutions (AHEIs), 125
African Union (AU), 86, 225, 257–259, 263, 275

© The Author(s), under exclusive license to Springer Nature Switzerland AG 2024
P. Neema-Abooki (ed.), *The Sustainability of Higher Education in Sub-Saharan Africa*, Sustainable Development Goals Series, https://doi.org/10.1007/978-3-031-46242-9

375

376 INDEX

Agenda 2030, 4, 7, 24, 79, 118, 241, 272, 298, 337
Agriculture, 126, 127, 316, 317, 366, 369
Agroecology, 125
Agyedu, G. O., 175
Ahfad, 119
Ahmad, N., 241
Airasian, P., 232
Airasian, P.W., 166, 172–173, 183
Akhyar, M., 229
Aktan, O., 164
Aktas, C., 307
Al Musadieq, M., 307
Alazzeh, D., 93, 94
Al-esmail, R., 229
Alexopoulos, E. C., 180
Alfrey, L., 339
Alhumaid, K., 92
Ali, A., 92
Ali, S., 259, 274
Alison, G., 253, 270, 272
Allahar, A. L., 284, 285, 299
Allais, S., 225
Ally, M., 365, 369
Al-Omairi, A. R. A., 91, 100
Altbach, P., 261, 262
Altbach, P. G., 289, 295
Altinyelken, H. K., 96
Altman, B. W., 21, 28
Alumni, 117, 121, 152, 153
Amin, M.E., 80
Amonya, D., 103
Anand, Sudhir, 117
Ananiadou, K., 229
Anderson, L. W., 166, 172, 173, 183
Andreoni, A., 226
Andrew, Griffiths, 117
Anindo, J., 232
Anke, M., 322, 323
Ann, L., 93
Ansell, C., 253, 270, 272

Apple, M., 294
Applied sciences, 201, 361
AP-TRIAS, 126
Arani, O., 307
Archer, M. S., 261
Architects, C., 318
Aremu, Adejare Yusuff, 122
Arndt, R. E. A., 314
Arnove, R. F., 333
Arts and Social Studies, 127
Asiyai, R. I., 76
Aslam, S., 100
Aspiunza, Alvaro Rica, 120, 122
Assié-Lumumba, N. T., 335
Association of African Universities (AAU), 124, 258, 259, 262, 263
Association of Universities and Colleges of Canada (AUCC), 262
Asumadu-Sarkodie, S, 308
Ataya, N., 145, 272, 274
Atistsogbe, K., 240
Atuahene, F., 261
AUC, 319
Audu, R., 232
Awuzie, B, 78, 79
Aynte, A., 74
Aziz, Z., 93

B
Baber, H., 93
Babury, M. O., 74
Bagale, S., 206
Bailey, M., 261
Bailey, Tracy, 124
Ball, S. J., 328, 329, 338
Bansel, P., 284, 291
Barakat, S., 74
Bari, A. N. A., 307
Barnett, S. T., 263
Barrie, S. C., 140
Bashir, S., 225

INDEX 377

Basnet, K., 242
Baumann, J., 86
Bawa, A. K., 181
Bawa'aneh, M. S., 108
Bayaga, Anass, 7
Becker, G., 227
Becker, S., 96
Begashaw, B., 337
Behrendt, M., 230
Belda-Miquel, S., 337
Bello, H., 67, 243
Benn, Suzzane, 116
Bennett, D., 140, 224
Berchin, I. I, 354, 358, 361, 367
Bercovitz, J., 261
Bertrand, W. E., 24, 27
Best practices, 5, 18, 26, 32, 67–68,
     76, 85, 95, 149, 153, 157, 158,
     186, 204, 274, 286, 333,
     334, 338
Biggs, J., 30
Billing, D., 86
Bin Saud, M., 243
Biomass energy, 310, 316–317
Blackmur, D., 86
Black Soldier, 125
Blackwell, A., 148, 149
Blended learning, 31, 38, 128, 129
Blessinger, P., 145, 146
Blewitt, J., 322, 323
Bloch, C., 86
Boateng, J. K., 343
Bodilly, S. J. B., 203
Boitshwarelo, B., 76
Boni, A., 337
Boone, D. A., 175
Boone, H. N., 175
Bornman, G. M., 77
Bossanyi, E., 313
Bowes, L., 148–149
Brazil, 120, 293
Brinkley, I., 229

Brodie, S., 149
Brown, J., 43
Brown, R., 311
Budiyomo, Fajaryati, N., 229
Bugema University, 6, 91–110
Buntat, Y., 243
Bunting, Ian, 124
Burton, K., 76
Burton, R., 261
Burton, T., 313
Business Administration, 127, 151
Buwama, 126
Bwalya, K. J., 93
Bwalya, T., 332

C
Calhoun, A, 309
Campbell, D. F., 308
Campbell, J. D., 101
Can, D., 164, 165, 167, 168, 173,
     186, 187
Canelon, J., 94, 105
Cantarello, E., 78
Caparida, L., 95
Capitalism, 284, 285, 297, 328,
     333, 344
Carayannis, E., 308
Catindig, M., 95
Cebrián, G., 92
Centre for Extra-Mural Studies
     (CEMS), 126
Chankseliani, M., 2, 118, 145
Chapman, E., 229
Chare, A., 369
Chen, T., 92, 98
Chen, Y.-Ch., 231
Chesbrough, H., 260
China, 107, 120, 286
Chinien, C., 242
Chomsky, N., 286
Chung, C. K., 74

378 INDEX

Cifuentes, L., 102
Civín, L., 309
Clark, H., 25
Clarke, M., 136, 139, 157
Claro, M., 229
Clayton, N., 229
Cloete, Nico, 124
Coetzee, M., 229
Cognition, 7, 163–188
Cohen, Nelson, 233, 260
Cole, D., 152
Cole, N. L., 22
College of Education, 215
Commonwealth of Learning (COL), 23, 136, 157
Communities, xiii, 7, 26, 28, 50, 51, 59, 62, 63, 79, 85, 96, 97, 102, 119, 121, 123–132, 139, 146, 186, 195, 196, 202, 228, 255, 261, 305, 307, 310, 312, 314–316, 321, 329, 336, 342, 353, 362, 366
Community engagements, 2, 49–51, 62, 78, 115, 119–121, 125–130, 132, 260
Community Lending and Outside Capital (CLOC), 127, 128
Competency-based curriculum (CBC), 217, 224, 245
Computer hardware, 3, 19, 61, 99, 106, 109
Computer software, 3, 19, 34, 61, 62, 95, 99, 106, 109, 232, 358, 362
Comyn, P., 241
Cong, G., 92, 98
Conn, K. M., 194
Constantine, J., 335
Contestation, 8, 281–302, 329
Contract cheating, 33, 35
Cooley, L., 196, 201
Cooley. J., 194–197, 201–205, 207, 209, 214–218

Cornford, R., 224, 228, 229, 241
Coronavirus disease 2019 (COVID-19), xiv, 33, 92, 94, 98, 99, 105, 107, 128, 131, 145, 193
pandemic, vii, 3, 16, 28, 31, 36, 91–93, 95, 97–98, 101, 107–109
Country, vii, xiii, 2, 7, 9, 20, 22, 26–29, 65, 73, 74, 76, 79, 86, 93, 94, 96, 98, 104, 107–109, 124, 128, 129, 145, 157, 158, 163–188, 197–201, 204, 208, 209, 212–218, 224–227, 242, 252, 255–263, 270, 274, 275, 282, 284–287, 294, 298, 299, 315, 328, 333–336, 339, 351–353, 357–359, 363–365, 369
Cranmer, S., 143
Creswell, J., 232
Crimmins, G., 76
Crouch, L., 205, 215
Cruikshank, K. A., 166, 172, 173, 183
Cullingford, C., 322, 323
Curricula, 7, 9, 55, 62, 75, 76, 78, 79, 87, 117, 121, 122, 129, 131, 136, 138, 141, 145–152, 158, 198, 217, 273, 359, 360, 363–365, 367, 369, 371
Curriculum, xiii, 1, 3, 6, 7, 23, 51, 66, 73–87, 98, 121, 129, 135–158, 194, 198, 203, 204, 209, 211, 214, 215, 217, 224, 225, 231, 238–240, 245, 260, 288, 290, 299, 300, 360, 361, 364, 365, 367, 369, 370
reform, 86
Curtis, D., 224, 228, 229, 241

D
Damayanti, R. C., 307
D'Andrea, V., 21

Daria, P., 255, 266, 268
Dasmani, A., 232, 242
Davies, B., 284, 291
Davis, D., 94
de Almeida, Mariana Rodrigues, 120
De Jong, W., 316
de Rezende, Julio Francisco
    Dantas, 120
Deane, E., 149
Dede, C., 194, 205
Degn, L., 23, 86
Denby, L., 78
Denoncourt, J., 95
Departmental curriculum review
    committee, 81
Department of Education (DoE), 166
Department of Education for
    Sustainable Development and
    Community Engagement, 126
DeStefano, J., 205, 215
Developing, 7, 8, 25, 27, 35, 49–55,
    63–65, 79, 80, 86, 97, 99, 102,
    107, 108, 124, 131, 132, 138,
    140, 147, 149, 151, 152, 158,
    163–188, 194, 196, 202, 215,
    218, 225, 230, 242, 253, 255,
    257, 261, 270, 275, 291–294,
    309, 335, 341–343, 355, 363,
    365, 369, 370
Development, vii, xiii, xiv, 2, 4–6, 8,
    16–18, 20–22, 24–26, 31, 32, 36,
    45, 48, 52, 61–66, 75–80, 91, 96,
    97, 107, 116–119, 122–126,
    128–132, 136–138, 142–144,
    153–155, 158, 166, 187, 198,
    200, 201, 203–206, 213, 215,
    218, 225–231, 236, 238, 240–242,
    244, 245, 252–256, 258–264, 267,
    268, 273–275, 281–302, 306,
    308–310, 313, 320, 328, 335–338,
    343, 350, 353, 355, 357–359,
    361–363, 365–371

Dhillon, S., 99
Didham, R. J., 78, 79
Diesendorf, M., 116
Digital content, 19
Digital divide, 27
Digitalisation of higher education, 16
Digital learning, 5, 15–39, 100
Digital learning environment,
    22, 35, 36
Digital skills, 26, 27, 365
Digital technologies, 16, 23, 34, 36,
    163, 195
Dill, D., 259
Directorate of Outreach, 126
Distance Learning Programmes, 126
Doevenspeck, M., 282, 289, 294
Donkor, F., 175
Dori, D., 94
Dori, Y. J., 94
Duraku, Z. H., 93, 94, 98, 102
Durnali, M., 164
Dutra, A. R., 354, 358, 361, 367
Dzisah, J., 253

E
Economic, xiv, 4, 5, 7, 24, 50, 62, 64,
    65, 76, 78, 79, 96, 97, 116–118,
    120, 121, 123–126, 128, 129,
    131, 132, 139, 146, 147, 149,
    151, 153, 158, 196, 201, 204,
    206, 225, 228, 241, 252, 253,
    255, 256, 258–260, 262,
    265–268, 275, 281, 283,
    288–290, 292, 296, 297, 299,
    301, 306, 317, 320, 323, 328,
    329, 335–338, 350, 351, 353,
    358, 366
Economic sustainability, 4, 6, 117, 308
Ecosystem, 118, 196, 314, 351, 352,
    359, 363, 365
Edith Cowan University, 148, 150

380 INDEX

Education, vii, xiii, 3, 15, 59, 74, 91, 116, 145, 166, 193, 225, 254, 283, 317, 329, 350
Educational, 5–7, 16, 18–20, 23, 25, 26, 28, 30, 36, 43, 66, 74, 78, 80, 85, 91–98, 100, 104, 108, 109, 117, 118, 163–188, 195, 204, 205, 225, 231, 237, 273, 285, 290, 302, 323, 332–336, 357, 361, 362, 364–367, 371
Educational institutions, 2, 74, 80, 91, 97, 98, 187, 228, 289, 366, 367
Education and outreach activities, 4, 321
Education For All (EFA), 22, 96, 118
Education for Sustainable Development (ESD), 4, 5, 78–80, 126, 198, 210, 307
Education for Sustainable Development and Community Engagement (ESD & CE), 126, 127
Education innovations, 7, 193–218
Edwards, J., 224, 228, 241
Egodawatte, G., 164, 165, 186, 187
Eicker, F., 226
Eizaguirre, Almudena, 120, 122
Ekaterina, A., 255, 266, 268
Ekene, Osuji Gregory, 119
Ekwue, C., 231
Elbarbary, S, 316
Eldabi, T., 229
ELearning, 103, 110
Eleyyan, S., 364, 369
El-Jardali, F, 145, 272, 274
Elliot, A. J., 165
Elliot, D., 231
Employability mechanisms, 223–245
Employability skills, xiii, 8, 77, 135, 139, 143, 144, 223–245
Emuze, F., 78, 79
Enders, J., 43, 261

Energy Petroleum and Regulatory Authority (EPRA), 321
Energy self-sufficiency, 308–323
Energy sustainability, 8, 305–323
Engineering, 102, 229, 263, 363, 364, 369
Eno, M., 74
Enright, E., 339
Entrepreneurship for Impact (E-4 Impact), 124, 129
Environmental, 4, 6, 24, 26, 78, 79, 116, 118, 145, 167, 201, 206, 306, 309, 314–316, 318, 319, 323, 337, 367, 368, 371
Environmental sustainability, 117, 118, 132
Equitable quality education, 4, 25, 78, 95, 193, 225, 337
Erdil, N., 307
Erjavec, Jure, 122
Estates, U. O. W., 320
Estevez, J., 224, 232, 243
Etzkowitz, H., 62, 253, 271, 272
Eun, T., 242
Eurocentric, 285, 286, 290, 291, 298, 299, 301
Eurocentric view of the Western world, 282
Ezati, B. A., 84

F
Factors, 3, 24, 78, 79, 83, 92, 93, 98, 101, 131, 136, 140, 145, 158, 165, 167, 168, 188, 194, 206, 216, 229, 230, 242, 256, 257, 261, 334, 366, 370
Faculty of Education, 29, 177
Fadlallah, R., 145, 272, 274
Fahimirad, M., 229
Fajaryati, N., 229
Fallows, S., 144

INDEX 381

Farisyi, S., 307
Fataar, A., 349, 350, 361, 364, 365, 368, 369
Faturoti, B., 91–93, 98, 99
Febro, J., 95
Feldman, M., 261
Feller, I., 261
Feng, J., 229
Fistula, 124
Fleisch, B., 204
Flexible learning, 24, 360
Fluency, 7, 165, 167, 170, 173, 177–179, 183, 184
Fly Larvae, 125
Förster, R., 79, 80
Foshay, R, 242
Fossland, T., 23
Fourth industrialized labour market, 74
Fourth Industrial Revolution (4IR), 9, 26, 292, 309, 349–371
Fowler, A.R., 316
Fowler, J., 77
France, D., 136
Francis (Pope), 117
Franco, I., 79
Franklin, T., 230
Fraser, M. W., 212
Fraser, S., 149
Freire, Paulo, 122, 296, 297, 344
French-Arnold, E., 229
Fuentes, M., 92
Fulgence, Katherine, 7, 8, 224, 241
Fullan, M., 204

G
Galinsky, M. J., 212
Gallagher, E., 194
Gallagher, M. J., 194
Gans, R., 289, 290
Garcia, A., 327, 329

Garcia-Feijoo, Maria, 120, 122
Gardner, P., 144
Gay, I. R., 232
Geary, D. C., 167, 173
Geiger, R., 260
Geleen, M., 204
Gender-sensitive approaches, 123, 131
Geoffrey, N. M., 26–28
George, S., 285
Geothermal energy, 310, 315–316
Gertz, G., 328–330
Gibson, E., 149
Girma, Z., 255, 268
Global hegemony and legitimise, 285
Global North, 322, 335, 336
Global South, 273, 322, 335
Gosling, D., 21
Goswami, Y., 317, 318
Götze, N., 86
Government, 8, 22, 23, 28, 51, 66–68, 79, 91, 97, 110, 117, 123, 129, 136, 145, 147, 158, 195–197, 200, 202, 205–208, 211–213, 216, 218, 226, 230, 232, 251–255, 257, 259, 260, 263, 266, 270–275, 286, 287, 289, 293, 308, 312, 314, 321–323, 329, 335, 336, 338, 339, 353, 366–368
Government of Mauritius, 312
Government of the Republic of Namibia (GRN), 294
Government Republic of Botswana (GRB), 294
Government Republic of Namibia (GRN), 294
Govindarajalu, C., 322, 323
Graham, M., 93
Grainger, P., 76
Greater Masaka, 126
Green, D., 19–21
Greenway, C., 362

382 INDEX

Grol, R., 332, 341
Grünberg, L., 22
Grund, L., 145
Guerra, J.B.S., 354, 358, 361, 367
Gulliver, J. S., 314

H
Haase, S., 86
Habes, M., 92
Hadidi, H. H. E., 256, 269
Haipinge, Erkkie, 5
Hall, S., 231
Handschuh, D., 319
Hartmann & Linn, 201
Harvey, D., 328–330, 338
Harvey, L., 19–21, 43
Haseloff, G., 226
Hassan, A., 93, 94
Hatakenaka, S., 321
Hawkes, M., 78
Hayward, F. M., 74
Health, 63, 78, 116, 127, 201, 206,
    210, 237, 284, 329, 350
Heaney, C., 284, 292, 293, 296
Heffernan, M., 337
Heimler, R., 140, 141
Henry, D., 315
Hermosura, J. B., 254, 256
Hester. V. K., 186
Hettinger, L. M. P., 253
Hew, S. H., 91, 100
Higher education (HE), vii, xiii, xiv,
    xvii–xx, 1–10, 15–39, 43–68,
    73–80, 84–86, 93, 97, 101,
    107–109, 115–120, 123, 124,
    132, 135–158, 165–169, 185,
    187, 217, 218, 227, 251–275,
    281–302, 306–308,
    327–344, 349–371
Higher education institutions (HEIs),
    vii, xi, xiv, 4–9, 15, 16, 19–22, 25,
26, 29, 34, 36, 38, 43, 44, 46–52,
    54–62, 64, 66–68, 73–87, 97,
    115, 116, 118–128, 130–132,
    136, 139, 140, 144–146, 149,
    152–154, 156–158, 194, 252,
    257, 259–263, 287, 289–293,
    305–323, 332, 336, 338, 339,
    341, 342, 350, 351, 360,
    362, 368–371
High school, 137, 164–169, 171,
    173–175, 178–180, 182–187
Hill, D., 284, 293, 297
Hill, J., 136
Hindi, N., 229
Hodgen, J., 169
Hombrados, J., 226
Honan, J., 205
Hopf, L., 21, 28, 40
Hove, P. K., 369
Hoxha, L., 93, 94, 98, 102
Hsia, T. L., 94
Huisman, J., 332, 341
Hulya, K. A., 186
Human, 2, 4, 6, 8, 55, 65, 67, 79,
    116–119, 125, 132, 177, 228,
    284, 287–289, 291–294, 301,
    306, 320, 323, 349, 353, 362,
    366, 368
Human capital, 7, 116–118, 124, 128,
    131, 137, 224, 227, 228, 253,
    254, 256, 265, 270, 287,
    291–294, 298, 336, 363
Human capital theory, 137, 227, 228,
    282, 283
Humanity, 116, 117, 335, 350,
    361, 371
Human resource, 47, 60, 96, 117,
    202, 203, 228, 266, 268, 343,
    369, 370
Hung, J., 228
Hurst, C., 77
Hydroelectric systems, 315

I

Iacob, E., 306
Ibrahim, A., 181
Idehai, O., 315
Idris, A., 241
IIEP, 74
Incubation centres, 130
Information and Computer
    Technologies (ICT), 93
Innovations, 1, 7, 8, 16, 24–26, 28,
    64, 65, 75, 84, 95, 96, 119, 130,
    132, 152, 153, 155, 193–218,
    229, 252, 254–256, 258, 260,
    262–264, 267, 268, 270–272,
    274, 275, 295, 298, 307, 308,
    317, 350, 351, 355, 360–363,
    365, 366, 368–371
Institutional expenditures, 309, 317–320
Instruction, 7, 18, 19, 21, 93, 98,
    104, 105, 165, 167, 168, 176,
    179, 184, 185, 204, 231, 265,
    357, 361
Internal quality curricular review
    mechanisms, 74, 75, 77, 81
Internal quality review mechanism, 81
International Labour Organisation
    (ILO), 224, 225, 229, 241
International Monetary Fund (IMF),
    284, 296
International Network of Quality
    Assurance Agencies, 85
Internet, 6, 18, 27, 28, 38, 92, 93, 95,
    96, 99, 100, 103, 104, 106, 107,
    109, 110, 128, 225, 242, 243,
    355, 366
Internet bandwidth, 100, 109
Internet connectivity, 19, 93, 95, 100,
    102, 106
Internship, 75, 127, 130, 144, 150,
    155, 231
Inter-University Council for East
    Africa (IUCEA), 46, 47, 49, 68

Irani, Z., 229
Ismail, A., 226
Ismail, K., 225
Ivory towers, 6, 119, 120, 123, 128,
    130, 132
Ivrine, U., 319

J

Jaapar, A., 307
Jackson, D., 148, 150
James, P. C., 102
Janos, A. C., 285
Jansen, J. D., 262
Jarmo, R., 259, 274
Jarvis, D. S. L., 333, 339
Jenkins, N., 313
Jingura, R., 332
Job creators, 122, 129
Job seekers, 122, 129
Joint Advisory Group (JAG), 123
Joke, M. V., 186
Jonck, P., 157
Jones, N., 261
Jung, I., 99
Jung, J., 350, 353, 358
Junjan, V., 306

K

Kadhila, Ngepathimo, 5, 7–9
Kaisara, G., 93
Kalman, Y. M., 94
Kalz, M., 94
Kamin, Y., 232
Kamusoko, R., 332
Kanie, N., 79
Kapoor, K., 229
Kark, A., 314
Kasozi, A.B.K., 75, 85, 123
Katende, David, 5
Kavanagh, M., 149

384 INDEX

Kavenuke, Patrick Severine, 9
Kawamoto, K., 311
Kayembe, C., 350, 351, 353, 355, 363–367
Kayombo, Joel Jonathan, 9, 338, 339
Kenayathulla, H., 241
Kentikelenis, A., 330, 331
Kestin, T., 78
Kharas, H., 328–330
Kharbat, F. F., 103
Kiambati, K., 125
Kibera, F., 261, 264
Kilburn, A., 94
Kilburn, B., 94
Kilo, O. M. M., 253
Kim, E., 368
Kim, J., 77
Kinyanjui, E. M., 205, 208
Kinyota, Mjege, 9
Kioupi, V., 78–80
Kirby, D. A., 256, 269
Kisige, Abdu, 6, 80, 84
Kitawi, Alfred Kirigha, 8, 306, 307
Kituntu, 126
Kizza, J., 103
Knight, P., 137, 140, 144, 146, 224, 228
Knight, P. T., 137
Knowledge, 2, 4, 20, 22, 24, 29, 36, 56, 58, 75, 77, 79, 83, 84, 86, 92, 93, 95, 96, 105, 116, 119–125, 130, 137–140, 143, 145, 147, 150, 151, 153, 154, 163–165, 168, 169, 171–173, 178, 184–187, 198, 200, 201, 205, 210–214, 216, 225, 226, 228–233, 240, 241, 252, 255–262, 264–269, 271–275, 287, 288, 292–296, 299–301, 307, 308, 321, 322, 335, 337, 339, 340, 350, 351, 358, 359, 361, 362, 368, 371

Koehn, P. H., 262
Kohl, R., 194, 195, 201, 205, 214, 215
Komugabe, A., 103
Koomey, J., 311
Kordrostami, M., 94, 105
Kotamjani, S., 229
Kovacs, P., 296, 297
Krathwohl, D. R., 166, 172, 173, 183
Kreith, F., 317, 318
Krejcie, R. V., 80
Kromydas, T., 281, 283, 288
Krücken, G., 86
Kucharcikova, A., 240
Kumar, N. P., 229
Kumar, P., 93

L
Laboratory, L. B. N., 318
Labs, K., 318, 319
Lang'at-Thoruwa, C., 255, 256
Larasati, R., 101
Laursen, K., 261
Lawrence, M., 282, 284, 293, 294, 297
Learners, xiii, 7, 15, 27, 94, 96, 100, 102, 103, 106, 108, 110, 123, 129, 131, 137, 163–188, 194, 205, 209, 211, 217, 242, 252, 365, 369, 370
Learner satisfaction, 94
Learning management system/s (LMS), 15, 34, 98, 100, 103, 104, 107
Legemaate, M., 332, 341
Legitimate employment, 74
Lennartz, B., 226
Leon, I., 224, 232, 243
Leong, R., 149
Lessons, 1, 10, 19, 23, 31, 35–38, 52, 60, 61, 65–67, 85–86, 101, 109,

110, 153, 157, 187, 195, 197, 200, 204, 209, 210, 218, 235, 238, 245, 274–275, 282, 297, 299–301, 335, 369–371
Leung, D., 93, 94
Level, 2, 3, 6, 19, 29, 31, 36, 45–52, 57, 59, 62, 65, 66, 74, 75, 79, 85, 97, 99, 101, 105, 107, 108, 110, 117, 119, 130, 137–139, 141, 143, 144, 146, 148, 156, 163–175, 177–188, 198–200, 202, 206, 208, 210, 211, 213–216, 218, 225, 227, 230–236, 240–244, 257, 258, 261, 262, 266, 268, 270, 271, 273, 282–284, 306, 308, 317, 330–332, 335, 342, 358, 366, 367, 371
Levin, J., 322, 323
Levy, F., 137
Lewin, J., 231
Li, W., 226
Liberalisation, 329, 339
Lice, 230
Likando, Gilbert, 9
Lindsay, C., 224
Linnenbrink, E. A., 165
Liu, Y., 94, 105
Lorenz, C., 339
Lotz-Sisitka, H., 225
Loukkola, T., 342
Lowden, K., 231
Lubbe, B. A., 259, 274
Luckett, K., 22
Lund, H., 310
Lynch, M., 231
Lyytikäinen, M., 261

M
Maassen, Peter, 124
Machalek, R., 253
MacKinnon, P., 224

Macleod, K., 94
Magni, D., 100
Mahdinezhad, M., 229
Mahmoud, W. H., 321
Makame, Mwaka Omar, 8
Maier, M. A., 165
Makgato, M., 166, 170, 174, 185
Makoni, M., 272, 273
Malaji, Kyashane Stephen, 8
Malatji, Makwalete Johanna, 8, 252
Malichova, E., 241
Malloy, J., 194
Mama, N., 240
Mamposa, C., 19
Managerialism, 331, 339
Manwell, J. F., 313
Maranga, Ignatius Waikwa, 8, 321
Marcel, B., 255, 266, 268
Marhani, M. A., 307
Marieta, C., 224, 232, 243
Martin, M., 253, 260, 262
Martín-Garin, A., 224, 232, 243
Martinez, E., 327, 329
Martinez-Vargas, C., 282, 284, 285, 289, 290, 295, 300, 301
Martino, P., 168, 173, 185
Marwala, T., 362
Masindei, M., 360
Mason, G., 143
Massaquoi, J. G. M., 262
Matengu, Kenneth Kamwi, 9
Mathematical, 7, 140, 141, 164–175, 177, 178, 182–188, 198
Matimolane, M., 289, 290
Matthews, A., 362
Mavis, A.G., 181
Mavuso, P. M., 252
Maximiliano, Ngabirano, 123
Mayer, R.E., 166, 172, 173, 183
Mazrui, A. A., 289, 290
Mazuruse, G., 369
Mbalassa, M. J., 125

386 INDEX

Mbembe, A. J., 291, 300
Mbiydzenyuy, N. E., 108
Mbugua, Z. K., 224, 225, 232, 243
McCarthy, I. P., 267
McCloskey, S., 337
McCowan, Tristan, 2, 118, 145
McEwan, P. J., 194
McGowan, J. G., 313
McGrath, S., 225
McKenzie, P., 224, 228, 229, 241
McLinden, M., 362
McMahon, E. A., 21, 28
McQuaid, R., 224
Means, B., 93, 95, 98, 106
Medina, P., 94, 105
Medina, R., 229
Mendoza Velazco, D. J., 94, 105
Menon, R., 204
Mensah, M. A., 77, 84, 85
Meusburger, P., 337
Mgaiwa, S. J., 339
Mhlanga, E., 31
Michael, Alexander, 7
Millán-García, J. A., 224, 232, 243
Millennium Development Goals
    (MDGs), 118, 132
Milton, S., 74
Ministry of Education and Sports,
    48, 66–67
Ministry of Education Science and
    Technology (MoEST), 199, 200,
    210, 211, 215, 227
Mintz, B., 329, 338
Misiaszek, L. I., 339
Mji, A., 166, 170, 174, 185
Mogas, J., 92
Mohd, R., 225
Monk, D., 225
Moorad F., 186
Morgan, D. W., 80
Morote, E.S., 140, 141
Morsy, H., 350, 361, 363
Mpigi, 126, 127

Mpungose, C. B., 99
Muftahu, M., 108
Muldoon, R., 144
Multiple Levels Analysis, 330–332
Mumbi, M., 306, 307
Mungoo, J., 186
Munishi, E., 225, 230, 242
Munyoki, J., 261, 264
Murnane, R. J., 137
Murray, N., 99
Musgrove, P., 313
Muthaa, G. M., 224, 225, 232, 243
Mweseli, W. N., 74
Mwiria, K., 261, 262
Mythili, G., 91, 92, 94

N

Nabaho, L., 125
Nabeshima, K., 262
Nabukenya, M., 125
Nabunya, Kulthum, 6
Nadarajah, Gunalan, 122
Nagarajan, S., 224, 228, 241
Naing, L., 102
Nakab, A., 228
Nakimuli, L., 103
Namboodiri, S., 98, 100
Namibia, 28, 146, 149, 152,
    158, 294
National Centre for Vocational
    Education Research (NCVER),
    228, 229
National Council for Higher
    Education (NCHE), 29, 49,
    66, 68, 122
National Council for Technical
    Education (NACTE), 227, 229
National Council for Technical
    Vocational Education and
    Training (NACTVET), 227
National Development Plan, 129,
    131, 294

INDEX

National Renewable Energy Laboratory, 310
Natural resources, 117, 118, 259, 263, 350
Neary, J., 272, 273
Neema-Abooki, Peter, vii, 80
Neisler, J., 92, 93, 95, 98, 106
Nel, D., 350, 351, 353, 355, 363–367
Neoliberal ideology, 8, 9, 283–287, 292, 293, 297
Neoliberalism, 284, 329
Neoliberal Western ideology, 282
Netshifhefhe, L., 19
News, A., 318
New Zealand, 120, 157
Ngugi Wa Thiongo, 291
Ngulube, B., 135
Nguyen, T. M., 101
Ni, A., 231
Nieuwenhuis, L., 332, 341
Nina, P.M., 125
Nindye Health Centre, 127
Nkirina, S. P., 224, 226
Nkozi, 126, 127
Nobongoza, V., 19
Non-government organizations (NGO), 123, 195, 196, 200, 205, 213, 215
Nordman, B., 311
Notre Dame, 126, 127
NRC, 185
NREL, 311, 312
Ntshoe, I. M., 289, 297, 300
Numerical, 7, 46, 163–188
Nutrition, 63, 116, 126
Nwazor, C., 231
Nyagadza, B., 369
Nyambe, John Kamwi, 8
Nyamnjoh, F. B., 291
Nyerere, J. K., 336
Nygaard, S., 86

O

Obasi, C., 93, 94
Obeng, S., 175
Obeta, C., 231
Ofei-Manu, P., 78, 79
Ofoefule, A., 316
Ogunwale, S., 257–259, 274, 275
Ogutu, M., 261, 264
Ojuro, I., 231
Okoroigwe, C., 316
Olssen, M., 339
Oluoch-Suleh, Everlyn, 119
Omar, Abukar Mukhtar, 6
Omar, M. Z., 252, 256
Ong'ele, S., 205, 208
Online assessment, 34, 100, 102
Online learning, 6, 16, 18, 21, 22, 27, 91–94, 97–102, 104–108, 110, 166
Onoh, C., 231
Onokpanu, O., 231
Oolbekkink–Marchand, H., 332, 341
Open, Distance and eLearning (ODeL), 31, 32
Openjuru, G., 225
Opie, C., 177
Orakci, S., 164
Oregi, X., 224, 232, 243
Organisation for Economic Co-operation and Development (OECD), 6, 48, 60, 194
Osborne, M., 272, 273
Osmani, M., 229
Otto, D., 96
Ouma, R., 97, 99, 102, 107
Owens, J., 151
Owusu, A. P., 308

P

Padayachee, K, 289, 290
Palau, R., 92

388 INDEX

Pallawi, Kumar M., 240
Pande, J., 91, 92, 94
Pandey, B., 314
Pangestu, F., 231, 242
Pari, P., 240
Pârlea, D., 22
Pashapa, R., 369
Paul, R., 94
Peace, T. A., 101
Pedagogy/pedagogies, xiv, 19, 36, 75, 77, 84, 124, 128, 129, 131, 198, 204, 216, 217, 224, 243, 291, 293, 297, 299–301, 361, 365
Pekrun, R., 165
Peng, L., 92, 98
Perez, M. L., 100
Performance, 2, 7, 20, 46, 49–51, 55, 56, 58, 61, 62, 66, 68, 92–94, 98, 101–103, 105, 141, 163–188, 198, 203, 211, 217, 228, 261, 265, 307, 317, 319, 334, 340, 358, 361
Perkmann, M., 255, 256, 268
Perrault, M. F., 253
Personalised learning, 23
Peters, L., 205
Peters, M. A., 339
Phillips, D., 333, 334
Pillars, vii, 24, 78, 115–118, 125, 128, 132, 335
Pillay, Pundy, 124
Pintrich, P. R., 166, 172, 173, 183
Piper, B., 205, 208
Pitt, L., 266, 267, 269
Plagiarism, 16, 34
Plano, V., 232
Policy transfer, 332–336
Pool, D. L., 138, 139, 141–143
Popoola, B. A., 20
Potgieter, I., 229
Pournara, C., 164–166, 168, 169, 173, 185, 187

Predictor, 7, 163–188
Pretoria, 119
Pretorius, R., 19
Price-Kelly, H., 194–196, 201, 202, 205–207, 215, 217
Prigge, G. W., 252, 264, 265, 267, 268, 270, 272
Primary, 7, 16, 22, 26, 110, 118, 127, 145, 163–165, 167–169, 171, 173, 177, 179, 183–188, 197, 199, 200, 209–211, 213–215, 218, 230, 242, 255, 266, 271, 272, 283, 286, 290, 350, 360
Prior, J., 95
Private higher educational institutions, 85, 97
Privatisation, 293, 329, 338, 339
Problem-solving skills, 123, 224, 229, 238, 241, 292, 360
Professional, vii, xiv, 5, 9, 25, 33, 36, 38, 39, 55, 63, 81, 84, 85, 95, 119, 124, 138, 144, 147, 148, 151–153, 155, 199, 200, 203–205, 218, 227, 231, 236, 240, 242, 244, 260, 269, 274, 362, 370
Puriwat, W., 92, 94, 106

Q

QA philosophies and practices, 9, 327, 344
Quality, vii, x, xi, xiii, xiv, 3, 5–7, 9, 16, 17, 19–22, 26–28, 30–36, 38, 39, 43, 44, 47, 48, 50, 52, 53, 58, 63–67, 73–87, 92, 94–102, 105, 109, 116, 125, 140, 153, 166, 176, 187, 188, 193, 197, 198, 203, 205, 206, 209, 211, 212, 218, 225–227, 241, 252, 257, 261, 270, 273, 293, 294, 298, 307, 314, 317, 318, 323, 332, 334, 337, 340–344, 361, 364, 368, 370

INDEX 389

Quality assurance (QA), vii, xiv, 1–10, 16–18, 21–24, 26–32, 34–39, 43, 44, 46–48, 53, 58, 67, 68, 76, 77, 80, 85, 86, 100, 116, 117, 176, 210, 327–344
Quality Assurance Agency for Higher Education, 144, 153
Quality culture, 2, 3, 18, 32, 327, 332, 333, 340–344
Quality Curriculum for higher education, 75
Quality enhancement, 30
Quality in higher education, 19–21, 43

R
Rabin, E., 94
Ramalu, Subramaniam Sri, 122
Ramos, R. P., 4, 349, 353, 355, 357, 360, 362
Ramsarup, P., 225
Ramsden, M., 228
RAND Corporation, 257, 261
Ranga, M., 253
Raphael, C., 229, 242, 243
Raths, J., 166, 172, 173, 183
Reed, A., 92
Regional Centre of Expertise, 126
Reichstein, T., 261
Reimers, F. M, 74
Religious, 119
Renewable energy sources, 64, 309–317, 322, 323
Research, 2–4, 7, 9, 28, 44, 49–51, 58–60, 62, 66, 79, 84, 101, 103, 104, 106, 115, 119–121, 123–125, 127–130, 132, 136, 139, 145, 165–170, 172–178, 183, 184, 186–188, 194, 197, 201, 206, 214, 216–218, 223, 232, 243–245, 251–253,

255–268, 270–275, 287, 294–296, 298, 305, 307, 308, 312, 321, 322, 340, 341, 343, 360–362, 366, 370, 371
Research and lifelong learning, 76
Research projects, 81, 119, 124, 125, 127, 258, 322
Revenue, 9, 309, 312, 313, 316–320
Richardson, S., 224
Rieser, R., 97
Roberts, G., 204
Robertson, J., 266, 267, 269
Rochford, C., 330, 331
Rodríguez de Céspedes, B., 76
Rofingatun, S., 101
Rogers, A. L., 313
Rojewski, J., 223
Rong, J., 92, 98
Rosenberg, S., 140, 141
Ross, K., 78
Rossier, J., 240
Roux, P. A., 360
Ruckman, K., 267
Rufai, A., 231
Russon, A., 225
Ruwoko, E., 257, 258, 274
Rwanda, U. O., 186, 315
Ryerson, R., 194
Rynne, S. B., 339

S
Sá, C., 260
Saad, M., 255, 268
Sahu, P., 102, 107
Saibi, H., 316
Saibi, K., 316
Saint, W., 75, 84
Saito, O., 79
Saleh, I., 223
Salleha, M.S., 252, 256
Samson, E. I., 316

INDEX

Sanchez-Martin, J., 308
Sanders, Y., 169
Sang, A. K., 224, 225, 232, 243
Sasman, M., 164–166, 168, 169, 173, 184, 187
Sattler, C., 341, 342
Saunders, V., 143
Savings and Internal Lending Communities (SILC), 127, 128
Saxena, C., 93
Scaling, 7, 8, 193–218
Scapens, G., 77
Schein, E. H., 342
Schein, P. A., 342
Schneijderberg, C., 86
Schöer, V., 204
School, 7, 18, 28, 63, 78, 95, 98, 100, 108, 110, 137, 153, 163–171, 173–175, 177–179, 183–188, 193–218, 234, 242, 290, 351, 352, 359
Schultz, T., 292
Schwab, K., 353
SDSN Australia/Pacific, 25, 120
Secretariat, C., 95
Sekaran, U., 177
Sen, Amartya, 117, 308
Serafini, Paula Goncalves, 120
Serdyukov, P., 194, 198, 214, 217
Sestino, A., 100
Setiadi, R., 231
Shalyefu, Rakel Kavena, 7
Shankland, A., 335
Sharma, S., 315
Sharpe, D., 313
Sharpe, S., 149
Shenkoya, T., 368
Sher, A., 101
Shettima, K., 337
Shiel, C., 78
Shuaib, B., 243
Sidhva, D., 93, 94

Shivoro, Romanus, 7
SIEBEN Foundation, 126
Silber, H., 242
Simunich, B., 21, 28
Singh, S., 240
Sites of Domination, 8, 281–302
Skills, xiii, 4, 6, 8, 20, 22, 25–27, 34–36, 53, 55, 63, 64, 74–77, 79, 83, 84, 87, 92, 95, 96, 107, 116, 120–124, 126–131, 135, 137–141, 143, 144, 147–153, 155, 157, 163–165, 167–170, 173, 178, 179, 183–185, 188, 198, 200, 205, 210, 212, 223–245, 256, 264, 268, 274, 275, 283, 288, 292, 293, 307, 308, 321, 322, 342, 350, 351, 358–365, 368–371
Sklair, L., 284, 285
Skolnik, M. L., 343
Smith, C., 148
Smith, N., 51
Snellman, Carita Lilian, 123
Snilstveit, B., 204
Social, xiii, xiv, 2–4, 6–8, 15, 20, 22, 24, 26, 32, 33, 62, 63, 65, 76, 78, 79, 97, 103, 116, 118–121, 124, 125, 128, 129, 131, 132, 136, 141, 146, 147, 157, 158, 166, 196, 201, 202, 204, 206, 237, 253, 272, 281–285, 288, 293, 294, 296, 297, 300, 301, 306–308, 320, 323, 329, 330, 332, 335–338, 344, 350, 353, 357, 358, 361, 362, 364–367, 371
Social Enterprise Project (SEP), 127
Social sustainability, 116
Societal development, xiii, 282, 283, 287, 289–291, 296, 299, 301
Society, 4, 8, 22, 29, 61, 74, 78, 93, 96, 116, 118–125, 128–132, 146, 150, 155, 194, 225, 226, 255,

272, 274, 282, 285–289, 294, 295, 302, 307, 308, 320, 323, 335, 336, 340, 350, 352, 355, 357, 360–362, 366–368, 371
Solar photovoltaic systems, 312
Somalia, xiii, 6, 73–87
Somali National University (SNU), 73
Sonkar, S. K., 100
Soroti, 124
South Africa, 7, 48, 77, 119, 163–166, 168–170, 185–188, 211, 257, 261, 273, 291, 294, 313, 314, 364, 366
South African, 164–171, 184, 186, 187
Sovet, L., 240
Sparks, D. L., 263
Spencer, D., 101
Stabilization, 329
Staff development, 131
Start-ups, 130, 366
Steiner-Khamsi, G., 333, 334
Steinhardt, I., 86
Stensaker, B., 43
Stephen, J. K., 229, 242, 243
Steven, C., 144
Stevenson, J., 204
Strategic Partnerships for Higher Education Innovation and Reform (SPHEIR), 122
Strathmore University, 312, 321, 322
Strielkowski, W., 309
Struggle, 198, 281–302, 336
Student-centred, 18, 102, 103, 122, 231
Suarsana, L., 337
Subekti, A. S., 101
Sudan, 119, 364
Sukardi, 231, 242
Sumbodo, W., 231
Suskie, L., 27
Sustainability, vii, x, xiii, xvii, xix, 1–10, 52, 77–80, 96–98, 115–132, 146, 152, 166–167,

197, 206, 212, 213, 216, 218, 272, 305–323, 337, 343
Sustainability in higher education, 306–308
Sustainability plan, 309–310, 323
The Sustainability Tracking, 319
Sustainable Development Goals (SDGs), xiii, xiv, 2, 4, 6, 7, 10, 24, 25, 61–65, 77–80, 95–96, 115, 116, 118–121, 128, 129, 131, 132, 145–146, 163–188, 193, 199, 225, 241, 253, 272–274, 282, 298–299, 306, 307, 337–338, 367, 368
Sustainable transport, 320
Swart, W., 94
Swell, P., 138, 139, 141, 143

T
Tabula rasa, 122
Tabulawa, R. T., 292
Takemoto, K., 79
Tambwe, M. A, 230
Tan, H., 225
Tan, L. C., 229
Tanaka, N., 225
Tanuraharjo, H. H., 92
Tarkhanova, E., 309
Tavares, O., 19, 21
Taylor, R., 322, 323
Teacher-centred, 122, 170, 211, 225
Teaching, 1, 2, 6, 8, 24, 28, 30–32, 36, 38, 44, 49, 50, 52, 59, 75, 78, 79, 81, 83, 86, 92–94, 99, 101–107, 109, 119–125, 128, 131, 143, 145, 194, 197–200, 209–212, 216, 217, 224, 225, 231, 232, 236, 240, 243, 244, 255, 264, 265, 267, 271, 272, 290, 305, 307, 308, 312, 340, 341, 343, 360, 369, 370

392 INDEX

Teaching and learning, 3, 6, 18, 21, 23, 29–33, 38, 43, 44, 51, 52, 73–87, 97, 99, 100, 108, 109, 115, 116, 120–123, 128, 129, 131, 132, 150, 163, 170, 174, 185, 188, 200, 201, 203, 205, 209, 218, 225, 231, 232, 239, 240, 243, 301, 306, 308, 360, 361
Team, T., 253
Technological, 3, 9, 23, 92, 93, 101, 118, 141, 145, 208, 245, 256, 258, 260, 265, 272, 274, 275, 283, 295, 323, 349, 350, 353, 355, 358, 359, 361, 365–369
Teichler, U., 261
Temmerman, N., 18, 19, 22–24
Temple, T., 101
Tennyson, R. D., 94
Tertiary education, 5, 15, 78, 80, 118, 298
Thapa, B., 320
Third mission, 62, 125, 267, 272, 307
Third stream funding, 317, 321–323
Thompson, K. M., 92
Thornton, A., 204
Thwaites, J., 78
Tibby, M., 151, 152
Times Higher Education (THE), 120
Tokarcikova, E., 241
Tømte, C.E., 23
Tongji University, 120
Toor, A., 101
Torraco, R. J., 252, 264, 265, 267, 268, 270
Torres, C. A., 338
Toyibah, D., 338, 340
Tram, N., 101
Tran, Q. H., 101
Tran, T., 136
Transforming Employability for Social Change in East Africa (TESCEA), 122, 123, 128, 131

Triple mission, 6, 115–132
Tripney, S., 226
Tripopsakul, S., 92, 94, 106
Tuah, N. A. A., 102
Tuan, L., 101
Tuma, F., 101
Tumuti, D. W., 255, 256
Turner, T., 145
Turyasingura, W., 125
Tvaronaviciene, M., 309
Twinomuhwezi, I., 125

U
Udemba, F., 231
Uganda, 48, 49, 65, 66, 68, 94–97, 99, 102, 103, 108, 109, 122, 124, 129, 315, 316
Uganda Martyrs University (UMU), 6, 115–132
Umar, I. N., 101
Underperformance, 7, 164–169, 171–173, 177–179, 183–185, 187, 340
Uniqueness, 1, 9–10, 217
United Nations (UN), 4, 24, 61, 63, 115, 118–120, 132, 145, 166, 298
United Nations Children's Fund (UNICEF), 226, 231
United Nations Educational Scientific and Cultural Organization (UNESCO), 2, 4, 46, 47, 49, 51, 62–65, 96, 124, 166, 198, 199, 223, 227, 298
United Republic of Tanzania (URT), 225–227
University, 2, 18, 44, 73, 91, 115–132, 136, 166, 205, 252, 283, 307, 333, 360
University management, 84, 156, 269, 340

INDEX   393

University of Auckland, 120, 307
University of California, 318–320
University of Dar es Salaam, 232, 336
University of Lagos (UNILAG), 312
University of Namibia (UNAM),
    28–35, 146–152
University of Pretoria, 119
University of Sao Paulo, 120
University of South Africa, 119, 314
University Partnership for Research and
    Development (UNFORD), 127
Unwin, T., 53
URT, see United Republic of Tanzania
Utami, N. H., 307
Uzodinma, E., 316

V
van den Belt, M., 78
Van Der Bank, C. M., 20
Van Ommen, R. J., 316
Van Wart, M., 94, 105
Vaughter, P., 79
Vedovello, C., 261
Vemuri, S., 76
Verma, A. K., 24
VEROZON, 127
Vidart, D., 228
Videira, C. S. P., 19, 21
Video conferencing, 103
Vilcea, M. A., 342
Virtual learning environment (VLE),
    6, 92–94, 98–101
Vispute, S., 320
Vlăsceanu, L., 22
Vocational education, 8, 223–245
Vocational Education and Training
    Authority (VETA), 225, 227,
    229, 232, 236, 241
Volmink, J., 261
Voulvoulis, N., 78–80
Vught, F., 44, 259

W
Waheed, A., 92
Walkington, H., 136
Walsh, K., 255, 256, 268
Walzberg, J., 306
Wanderi, P. M., 255, 256
Wang, T., 74, 98, 100–102, 107
Watson, D., 318, 319
Weber, H., 337, 338
Wedekind, V., 225
Weerakkody, V., 229
Welch, A., 260, 261
Wengrowicz, N., 94
Western neoliberal hegemonic
    interests, 282
Whereat, J., 79
Wilhelm, S., 79, 80
Wilkesmann, M., 352
Wilkesmann, U., 352
William, W., 322, 323
Williams, G., 143
Wilton, N., 136, 148
Wind energy, 310, 313–314
Wiranto, T., 229
Wittrock, M. C., 166, 172,
    173, 183
Woldegiorgis, E. T., 85, 86, 282,
    289, 294
Wong, D. K. Y., 149
Woodward, T., 95
World Bank, 39, 51, 66, 124,
    166, 251, 257, 258, 262,
    265, 268, 269, 275,
    284, 296
World Economic Forum
    (WEF), 350–352, 358,
    363, 367
Wotto, M., 107
Woyo, E., 223
Wrenn, M. V., 329
Wright, J., 229
Wu, J. H., 94

Y

Yang, J., 92, 98
Yazan, M. D., 306
Yetkin Özdemir, I. E., 164, 165, 167, 168, 173, 186, 187
Yin, X., 48, 74
Yorke, M., 136, 137, 139, 140, 144, 146, 224, 228
Young, J., 229
Yu, Y., 306
Yudiono, H., 231
Yunusa, A. A., 101
Yusuf, S., 262

Z

Zahid, E., 92
Zamora-Polo, F., 308
Zamzami, I., 102
Zan, R., 168, 173, 185
Zawawi, M., 307
Zeelen, J., 225
Zeng, X., 98, 100–102, 107
Zhang, J., 94, 105
Zhang, T., 342
Zimmerman, W. A., 21, 28
Zimmermann, A. B., 79, 80
Zizka, L., 78
Zuzel, K., 143